MULTIPARAMETRIC STATISTICS

MULTIPARAMETRIC STATISTICS

BY
VADIM I. SERDOBOLSKII

Moscow State Institute of Electronics and Mathematics
Moscow, Russia

ELSEVIER

AMSTERDAM • BOSTON • HEIDELBERG • LONDON
NEW YORK • OXFORD • PARIS • SAN DIEGO
SAN FRANCISCO • SINGAPORE • SYDNEY • TOKYO

Elsevier
Radarweg 29, PO Box 211, 1000 AE Amsterdam, The Netherlands
Linacre House, Jordan Hill, Oxford OX2 8DP, UK

First edition 2008

Library of Congress Cataloging-in-Publication Data
A catalog record for this book is available from the Library of Congress

British Library Cataloguing in Publication Data
A catalogue record for this book is available from the British Library

ISBN: 978-0-444-53049-3

For information on all Elsevier publications
visit our website at *www.books.elsevier.com*

08 09 10 11 12   10 9 8 7 6 5 4 3 2 1

ON THE AUTHOR

Vadim Ivanovich Serdobolskii is a professor at Applied Mathematics Faculty, Moscow State Institute of Electronics and Mathematics.

He graduated from Moscow State University as physicist theoretician.

In 1952, received his first doctorate from Nuclear Research Institute of Moscow State University in investigations in resonance theory of nuclear reactions.

In 2001, received the second (advanced) doctoral degree from the Faculty of Calculation Mathematics and Cybernetics of Moscow State University in the development of asymptotical theory of statistical analysis of high-dimensional observations.

He is the author of the monograph "Multivariate Statistical Analysis. A High-Dimensional Approach," Kluwer Academic Publishers, Dordrecht, 2000.

Internet page: serd.miem.edu.ru

E-mail: vvserd@mail.ru

CONTENTS

Chapter 5. Multiparametric Discriminant Analysis   193

Chapter 6. Theory of Solution to High-Order Systems of Empirical Linear Algebraic Equations   239

FOREWORD

This monograph presents the mathematical theory of statistical methods different by treating models defined by a number of parameters comparable in magnitude with sample size and exceeding it. This branch of statistical science was developed in later decades and until lately was presented only in a series of papers in mathematical journals, remaining practically unknown to the majority of statisticians.

The first attempt to sum up these investigations has been made in the author's previous monograph "Multivariate Statistical Analysis. A High-Dimensional Approach," Kluwer Academic Publishers, 2000.

The proposed new monograph is different in that it pays more attention to fundamentals of statistics, refinines theorems proved in the previous book, solves a number of new multiparametric problems, and provides the solution of infinitely dimensional problems.

This book is written, first, for specialists in mathematical statistics who will find here new settings, new mathematical methods, and results of a new kind urgently required in applications. Actually, a new branch of statistics is originated with new mathematical problems.

Specialists in applied statistics creating statistical packages for statistical software will be interested in implementing new more efficient methods proposed in the book. Advantages of these methods are obvious: the user is liberated from the permanent uncertainty of possible degeneration of linear methods and gets approximately unimprovable algorithms whose quality does not depend on distributions.

Specialists applying statistical methods to their concrete problems will find in the book a number of always stable, approximately unimprovable algorithms that will help them solve better their scientific or economic problems.

Students and postgraduates may be interested in this book in order to get at the foremost frontier of modern statistical science that would guarantee them the success in their future carrier.

Fortunately, as is shown in this book, this difficulty of principle may be overcome in the asymptotics of the increasing number of parameters. This asymptotical approach was proposed by A. N. Kolmogorov in 1968–1970. He suggested to consider a sequence of statistical problems in which sample size N increases along with the number of parameters n so that the ratio $n/N \rightarrow c > 0$. This setting is different in that a concrete statistical problem is replaced by a sequence of problems so that the theory may be considered as a tool for isolating leading terms that describe approximately the concrete problem. The ratio n/N shows the boundary of the applicability of traditional methods and the origin of new multiparametric features of statistical phenomena. Their essence may be understood if we prescribe the magnitude $1/n$ to contribution of a separate parameter and compare it with the variance of standard estimators that is of the order of magnitude of $1/N$. Large ratios n/N show that the contribution of separate parameters is comparable with "noise level" produced by the uncertainty of sample data. In this situation, the statistician has no opportunity to seek more precise values of unknown parameters and is obliged to take practical decisions over the accessible data.

The proposed theory of multiparametric statistics is advantageous for problems where n and N are large and variables are boundedly dependent. In this case, the inaccuracies of estimation of a large number of parameters become approximately independent and their summation produces an additional mechanism of averaging (mixing) that stabilizes functions uniformly dependent on arguments. It is of importance that this class of functions includes standard quality functions. As a consequence, their variance proves to be small, and in the multiparametric approach, we may say not on estimation of quality functionals but on their *evaluation*. This property changes radically the whole problem of estimator dominance and methods of search for admissible estimators. We may say that a new class of mathematical investigations in statistics appears whose purpose is a systematical construction of *asymptotically better* and *asymptotically unimprovable solutions*.

Another specifically multiparametric phenomenon is the appearance of stable limit relations between sets of observable quantities and sets of parameters. These relations may have the form of the Fredholm integral equations of the first kind with respect to unknown functions of parameters. Since they are produced by random distortions in the process of observation, these relations may be called *dispersion equations*. These equations present functional relations between sets of first and second moments of variables and sets of their estimators. Higher order deviations are averaged out, and this leads to a remarkable essentially multiparametric effect: the leading terms of standard quality functions depend on only two first moments of variables. Practically, it means that for a majority of multivariate (regularized) statistical procedures, for a wide class of distributions, standard quality functions do not differ much from those calculated under the normality assumption, and the quality of multiparametric procedures is approximately population free. Thus, three drawbacks of the existing methods of multivariate statistics prove to be overcome at once: the instability is removed, approximately unimprovable solutions are found, and the population–free quality is obtained.

Given approximately nonrandom quality functions, we may solve appropriate extremum problems and obtain methods of constructing statistical solutions that are *asymptotically unimprovable independently of distributions.*

The main result of investigations presented in the book is a regular multiparametric technology of improving statistical solutions. Briefly, this technology is as follows. First, a generalized family of always stable statistical solutions is chosen depending on an a priori vector of parameters or an a priori function that fixes the chosen algorithm. Then, dispersion equations are derived and applied to calculate the limit risk as a function of only population parameters. An extremal problem is solved, and the extremum a priori vector or function is calculated that define the asymptotically ideal solution. Then, by using dispersion equations once more, this ideal solution is approximated by statistics and a practical procedure is constructed providing approximately maximum quality. Another way is, first, to isolate the leading term of the risk function and then to minimize some statistics approximating the risk. It remains to estimate the influence of small inaccuracies produced by asymptotic approach.

These technologies are applied in this book to obtain asymptotically improved and unimprovable solutions for a series of most usable statistical problems.

They include the problem of estimation of expectation vectors, matrix shrinkage, the estimation of inverse covariance matrices, sample regression, and discriminant analysis. The same technology is applied for the minimum square solutions to large systems of empirical linear algebraic equations (over a single realization of random coefficient matrix and random right-hand side vector).

For practical use, the implementation of simplest two-parametric shrinkage-ridge versions of existing procedures may be especially interesting. They save the user from the danger of a degeneration, and provide solutions certainly improved for large n and N. Asymptotically unimprovable values of shrinkage and ridge parameters for problems mentioned above are written out in the book. These two-parametric solutions improve over conventional ones and also over one-parametric shrinkage and ridge regularization algorithms, only insignificantly increase the calculation job, and are easy for programming.

Investigations of specific phenomena produced by estimation of a large number of parameters were initiated by A. N. Kolmogorov. Under his guidance in 1970–1972, Yu. N. Blagovechshenskii, A. D. Deev, L. D. Meshalkin, and Yu. V. Arkharov carried out the first but basic investigations in the increasing-dimension asymptotics. In later years, A. N. Kolmogorov was also interested and supported earlier investigations of the author of this book. The main results exposed in this book are obtained in the Kolmogorov asymptotics.

The second constituent of the mathematical theory presented in this book is the spectral theory of increasing random matrices created by V. A. Marchenko, L. A. Pastur (1967), and V. L. Girko (1975–1995) that was later applied to sample covariance matrices by the author. This theory was developed independently under the same asymptotical approach as the Kolmogorov asymptotics. Its main achievements are based on the method of spectral functions that the author learned from reports and publications by V. L. Girko in 1983. The spectral function method is used in this book for developing a general technology of construction of improved and asymptotically unimprovable statistical procedures distribution free for a wide class of distributions.

I would like to express my sincere gratitude to our prominent scientists Yu. V. Prokhorov, V. M. Buchstaber, and S. A. Aivasian, who appreciated the fruitfulness of multiparametric approach in the statistical analysis from the very beginning, supported my investigations, and made possible the publication of my books. I appreciate highly the significant contribution of my pupils and successors: my son Andrei Serdobolskii, who helped me much in the development of the theory of optimum solutions to large systems of empirical linear equations, and also V. S. Stepanov and V. A. Glusker who performed tiresome numeric investigations that presented convincing confirmation of practical applicability of the theory developed in this book.

Moscow   Vadim I. Serdobolskii

January, 2007

CHAPTER 1

INTRODUCTION

THE DEVELOPMENT OF MULTIPARAMETRIC STATISTICS

Today, we have the facilities to create and analyze informational models of high complexity including models of biosystems, visual patterns, and natural language. Modern computers easily treat information arrays that are comparable with the total life experience (near 10^{10} bits). Now, objects of statistical investigation are often characterized by very large number of parameters, whereas, in practice, sample data are rather restricted. For such statistical problems, values of separate parameters are usually of a small interest, and the purpose of investigation is displaced to finding optimal statistical decisions.

Some examples.

1. Statistical analysis of biological and economic objects

These objects are characterized by a great complexity and a considerable nuisance along with bounded samples. Their models depend on a great number of parameters, and the standard approach of mathematical statistics based on expansion in the inverse powers of sample size does not account for the problem specificity. In this situation, another approach proposed by A. N. Kolmogorov seems to be more appropriate. He introduced an asymptotics in which the dimension n tends to infinity along with sample size N, allowing to analyze the effects of inaccuracies accumulation in estimating a great number of parameters.

2. Pattern recognition

Today, we must acknowledge that the recognition of biological and economic objects requires not so much data accumulation as the extraction of regularities and elements of structure against the noise background. These structure elements are then used as features for recognition. But the variety of possible structure elements is measured by combinatorial large numbers, and the new mathematical problem arises of efficient discriminant analysis in space of high dimension.

3. Interface with computer using natural language

This problem seems to become the central problem of our age. It is well known that the printed matter containing the main part of classical literature requires rather moderate computer resources. For example, a full collection of A. S. Pushkin's compositions occupies only 2–5 megabytes, while the main corpus of Russian literature can be written on 1-gigabyte disk. The principal problem to be solved is how to extract the meaning from the text. Identifying the meaning of texts with new information and measuring it with the Shannon measure, we can associate the sense of a phrase with the statistics of repeating words and phrases in the language experience of a human. This sets a problem of developing a technology of search for repeating fragments in texts of a large volume. A specific difficulty is that the number of repetitions may be far from numerous; indeed, the human mind would not miss even a single coincidence of phrases.

Traditionally, the statistical investigation is related to a cognition process, and according to the R. Fisher conception, the purpose of statistical analysis is to determine parameters of an object in the process of analyzing more and more data. This conception is formalized in the form of an asymptotics of sample size increasing indefinitely, which lays in the foundation of well-developed theory of asymptotic methods of statistics. The most part of investigation in mathematical statistics deals with one-dimensional observations and fixed number of parameters under arbitrarily large sample sizes. The usual extension to many-dimensional case is reduced to the replacement of scalars by vectors and matrices and to studying formal relations with no insight into underlying phenomena.

The main problem of mathematical statistics today remains the study of the consistency of estimators and their asymptotic properties under increasing sample size. Until recently, there is no fruitful approach to the problem of quality estimation of the statistical procedures under fixed samples. It was only established that nearly all popular statistical methods allow improvement and must be classified as inadmissible. In many-dimensional statistics, this conclusion is much more severe: nearly all consistent multivariate linear procedures may have infinitely large values of risk function. These estimators should be called "essentially inadmissible."

Meanwhile, we must acknowledge that today the state of methods of multivariate statistical analysis is far from satisfactory. Most popular linear procedures require the inversion of covariance matrix. True inverse covariance matrices are replaced by consistent estimators. But sample covariance matrices (dependently on data) may be degenerate and their inversion can be impossible (even for the dimension 2). For large dimension, the inversion of sample covariance becomes unstable, and that leads to statistical inferences of no significance. If the dimension is larger than sample size, sample covariance matrices are surely degenerate and their inversion is impossible. As a consequence, standard consistent procedures of multivariate statistical analysis included in most packages of statistical software do not guarantee neither stable nor statistically significant results, and often prove to be inapplicable. Common researchers applying methods of multivariate statistical analysis to their concrete problems are left without theoretical support confronted by their difficulties. The existing theory cannot recommend them nothing better than ignoring a part of data artificially reducing the dimension in hope that this would provide a plausible solution (see [3]).

This book presents the development of a new special branch of mathematical statistics applicable to the case when the number of unknown parameters is large. Fortunately, in case of a large number of boundedly dependent variables, it proves to be possible to use specifically many-parametric regularities for the construction of improved procedures. These regularities include small variance of standard quality functions, the possibility to estimate them reliably from sample data, to compare statistical procedures by their efficiency and choose better ones. Mathematical theory developed in this book offers a number of more powerful versions of most usable statistical procedures providing solutions that are both reliable and approximately unimprovable in the situation when the dimension of data is comparable in magnitude with sample. The statistical analysis appropriate for this situation may be qualified as *the essentially multivariate analysis* [69]. The theory that takes into account the effects produced by the estimation of a large number of unknown parameters may be called *the multiparametric statistics*.

The first discovery of the existence of specific phenomena arising in multiparametric statistical problems was the fact that standard sample mean estimator proves to be inadmissible, that is, its square risk can be diminished.

The Stein Effect

In 1956, C. Stein noticed that sample mean is not a minimum square risk estimator, and it can be improved by multiplying by a scalar decreasing the length of the estimation vector. This procedure was called "shrinkage," and such estimators were called "shrinkage estimators." The effect of improving estimators by shrinkage was called the "Stein effect." This effect was fruitfully exploited in applications (see [29], [33], [34]). Let us cite the well-known theorem by James and Stein.

Let $\mathbf{x}$ be an n-dimensional observation vector, and let $\bar{\mathbf{x}}$ denote sample mean calculated over a sample of size N. Denote (here and below) by I the identity matrix.

PROPOSITION 1. *For $n > 2$ and $\mathbf{x} \sim \mathbf{N}(\vec{\mu}, I)$, the estimator*

$$\widehat{\mu}^{\mathrm{JS}} = \left(1 - \frac{n-2}{N\bar{\mathbf{x}}^2}\right)\bar{\mathbf{x}} \tag{1}$$

has the quadratic risk

$$\mathbf{E}(\vec{\mu} - \widehat{\mu}^{\mathrm{JS}})^2 = \mathbf{E}(\vec{\mu} - \bar{\mathbf{x}})^2 - \left(\frac{n-2}{N}\right)^2 \mathbf{E}\frac{1}{\bar{\mathbf{x}}^2} \tag{2}$$

(here and in the following, squares of vectors denote squares of their length).

Proof. Indeed,

$$R^{\mathrm{JS}} = \mathbf{E}(\vec{\mu} - \widehat{\mu}^{\mathrm{JS}})^2 = y + 2y_2\,\mathbf{E}\frac{(\vec{\mu} - \bar{\mathbf{x}})^T\bar{\mathbf{x}}}{\bar{\mathbf{x}}^2} + y_2^2\,\mathbf{E}\frac{1}{\bar{\mathbf{x}}^2}, \quad (3)$$

where $y = n/N$ and $y_2 = (n-2)/N$. Let f be the normal distribution density for $\mathbf{N}(\vec{\mu}, I/N)$. Then, $\vec{\mu} - \bar{\mathbf{x}} = (Nf)^{-1}\nabla f$, where ∇ is the differentiation operator in components of $\bar{\mathbf{x}}$. Substitute this expression in the second addend of (3), and note that the expectation can be calculated by the integration in $f\,d\bar{\mathbf{x}}$. Integrating by parts we obtain (2). $\square$

The James–Stein estimator is known as a "remarkable example of estimator inadmissibility."

This discovery led to the development of a new direction of investigation and a new trend in theoretical and applied statistics, with hundreds of publications and effective applications [33].

In following years, other versions of estimators were offered that improved as standard sample mean estimator as the James–Stein estimator. The first improvement was offered in 1963 by Baranchik [10]. He proved that the quadratic risk of the James–Stein estimator can be decreased by excluding negative values of the shrinkage estimator (positive-part shrinkage). A number of other shrinkage estimators were proposed subsequently, decreasing the quadratic risk one after the other (see [31]).

However, the James–Stein estimator is singular for small $\bar{\mathbf{x}}^2$. In 1999, Das Gupta and Sinha [30] offered a robust estimator

$$\widehat{\mu}^G = \left(1 - \frac{n}{n + \bar{\mathbf{x}}^2}\right)\bar{\mathbf{x}},$$

that for $n \geq 4$ dominates $\widehat{\mu} = \bar{\mathbf{x}}$ with respect to the quadratic risk and dominates $\widehat{\mu}^{\mathrm{JS}}$ with respect to the absolute risk $\mathbf{E}|\vec{\mu} - \widehat{\mu}^G|$ (here and in the following, the absolute value of a vector denotes its length). In 1964, C. Stein suggested an improved estimator for $x \sim \mathbf{N}(\vec{\mu}, dI)$, with unknown $\vec{\mu}$ and d. Later, a series of estimators were proposed, subsequently improving his estimator (see [42]). A number of shrinkage estimators were proposed for the case $\mathbf{N}(\vec{\mu}, \Sigma)$ (see [49]).

The shrinkage was also applied in the interval estimation. Using shrinkage estimator, Cohen [15] has constructed confidence intervals that have the same length but are different by a uniformly greater probability of covering. Goutis and Casella [27] proposed other confidence intervals that were improved with respect to both the interval length and the covering probability. These results were extended to many-dimensional normal distributions with unknown expectations and unknown covariance matrix.

The Stein effect was discovered also for distributions different from normal. In 1979, Brandwein has shown that for spherically symmetrical distributions with the density $f(|\mathbf{x} - \theta|)$, where θ is a vector parameter, for $n > 3$, the estimator $\widehat{\mu} = (1 - a/\bar{\mathbf{x}}^2)\bar{\mathbf{x}}$ dominates $\widehat{\mu} = \bar{\mathbf{x}}$ for a such that $0 < a < a_{\max}$. For these distributions, other improved shrinkage estimators were found (see the review by Brandwein and Strawderman [13]).

For the Poisson distribution of independent integer-valued variables $k_i, i = 1, 2, \ldots, n$ with the vector parameter $\lambda = (\lambda_1, \lambda_2, \ldots, \lambda_n)$, the standard unbiased estimator is the vector of rates $(f_1, f_2, \ldots, f_n)$ of events number $i = 1, 2, \ldots, n$ in the sample. Clevenson and Zidek [14] showed that the estimator $\widehat{\lambda}$ of the form

$$\widehat{\lambda}_i = \left(1 - \frac{a}{a+s}\right)f_i, \quad i = 1, 2, \ldots, n \quad \text{where} \quad s = \sum_{i=1}^{n} f_i \quad (4)$$

for $n > 2$ has the quadratic risk less than the vector $(f_1, f_2, \ldots, f_n)$, if $n-1 \le a \le 2\,(n-1)$. A series of estimators were found improving the estimator (4).

Statistical meaning of shrinkage

For understanding the mechanism of risk reduction, it is useful to consider the expectation of the sample average square. For distributions with the variance d of all variables, we have $\mathbf{E}\bar{\mathbf{x}}^2 = \bar{\mu}^2 + yd > \bar{\mu}^2$, where $y = n/N$, and, naturally, one can expect that shrinkage of sample average vectors may be useful. The shrinkage effect may be characterized by the magnitude of the ratio $\bar{\mu}^2/yd$. This ratio may be interpreted as the "signal-to-noise" ratio. The shrinkage is purposeful for sufficiently small $\bar{\mu}^2/yd$. For $d = 1$ and restricted dimension, $y \approx 1/N$, and shrinkage is useful only for the vector length less than $1/\sqrt{N}$. For essentially many-dimensional statistical problems with $y \approx 1$, the shrinkage can be useful only for bounded vector length when $\vec{\mu}^2 \approx 1$, and its components have the order of magnitude $1/\sqrt{N}$. This situation is characteristic of a number of statistical problems, in which vectors $\vec{\mu}$ are located in a bounded region. The important example of these is high-dimensional discriminant analysis in the case when the success can be achieved only by taking into account of a large number of weakly discriminating variables.

Let it be known a priori that the vector $\vec{\mu}$ is such that $\vec{\mu}^2 \leq c$. In this case, it is plausible to use the shrinkage estimator $\widehat{\mu} = \alpha\bar{\mathbf{x}}$, with the shrinkage coefficient $\alpha = c/(c + y)$. The quadratic risk

$$R(a) = \mathbf{E}(\vec{\mu} - a\bar{\mathbf{x}})^2 = y\,\frac{\vec{\mu}^2 y + c^2}{(c + y)^2} \leq y\,\frac{c}{c + y} < R(1).$$

It is instructive to consider the shrinkage effect for simplest shrinkage with nonrandom shrinkage coefficients. Let $\widehat{\mu} = \alpha\bar{\mathbf{x}}$, where nonrandom positive $\alpha < 1$. For $\mathbf{x} \sim \mathbf{N}(\vec{\mu}, I)$, the quadratic risk of this a priori estimator is

$$R = R(\alpha) = (1 - \alpha)^2 \vec{\mu}^2 + \alpha^2 y, \tag{5}$$

$y = n/N$. The minimum of $R(\alpha)$ is achieved for $\alpha = \alpha^0 = \vec{\mu}^2/(\vec{\mu}^2 + y)$ and is equal to

$$R^0 = R(\alpha^0) = y\vec{\mu}^2/(\vec{\mu}^2 + y). \tag{6}$$

Thus, the standard quadratic risk y is multiplied by the factor $\vec{\mu}^2/(\vec{\mu}^2 + y) < 1$. In traditional applications, if the dimension is not high, the ratio y is of order of magnitude $1/N$, and the shrinkage effect proves to be insignificant even for a priori bounded localization of parameters. However, if the accuracy of measurements is low and the variance of variables is so large that it is comparable with N, the shrinkage can considerably reduce the quadratic risk.

The shrinkage "pulls" estimators down to the coordinate origin; this means that the shrinkage estimators are not translation invariant. The question arises of their sensitivity to the choice of coordinate system and of the origin. In an abstract setting, it is quite not clear how to choose the coordinate center for "pulling" of estimators. The center of many-dimensional population may be located, generally speaking, at any faraway point of space and the shrinkage may be quite not efficient. However, in practical problems, some restrictions always exist on the region of parameter localization, and there is some information on the central point. As a rule, the practical investigator knows in advance the region of the parameter localization. In view of this, it is quite obvious that the standard sample mean estimator must be improvable, and it may be improved, in particular, by shrinkage. Note that this reasoning has not attracted a worthy attention of researches as yet, and this fact leads to the mass usage of the standard estimator in problems, where the quality of estimation could be obviously improved.

It is natural to expect that as much as the shrinkage coefficient in the James–Stein estimator is random, it can decrease the quadratic risk less efficiently than the best nonrandom shrinkage. Compare the quadratic risk R^{JS} of the James–Stein estimator (2) with the quadratic risk R^0 of the best a priori estimator (6).

PROPOSITION 2. *For n-dimensional observations* $\mathbf{x} \sim \mathbf{N}(\vec{\mu}, I)$ *with $n > 2$, we have*

$$R^{JS} \leq R^0 + 4\,\frac{n-1}{N^2}\,\frac{1}{\vec{\mu}^2 + y} \leq R^0 + 4/N.$$

Proof. We start from Proposition 1. Denote $y_2 = (n-1)/N$. Using the properties of moments of inverse random values, we find that

$$R^{JS} = y - y_2^2 \mathbf{E}(\bar{\mathbf{x}}^2)^{-1} \leq y - y_2^2\,(\mathbf{E}\bar{\mathbf{x}}^2)^{-1} = R^0 + 4(n-1)N^{-2}/(\vec{\mu}^2 + y).$$

$\square$

Thus, for large N, the James–Stein estimator practically is not worse than the unknown best a priori shrinkage estimator that may be constructed if the length of the vector is known.

Application in the regression analysis

Consider the regression model

$$Y = X\beta + \varepsilon, \quad \varepsilon \sim \mathbf{N}(0, I),$$

where X is a nonrandom rectangular matrix of size $N \times n$ of a full rank, $\beta \in \mathbb{R}^n$, and I is the $n \times n$ identity matrix. The standard minimum square solution leads to the estimator $\widehat{\beta}_0 = (X^T X)^{-1} X^T Y$, which is used in applied problems and included in most applied statistical software. The effect of application of the James–Stein estimator for shrinking of vectors $\widehat{\beta}_0$ was studied in [29]. Let the (known) plan matrix is such that $X^T X$ is the identity matrix. Then, the problem of construction of regression model of best quality (in the meaning of minimum sum of residual squares) is reduced to estimation of the vector $\beta = \mathbf{E} X^T Y$ with the minimum square risk. The application of the Stein-type estimators allows to choose better versions of linear regression (see [33], [84]).

This short review shows that the fundamental problem of many-dimensional statistics, estimation of the position of the center of population, is far from being ultimately solved. The possibility of improving estimators by shrinking attracts our attention to the improvement of solutions to other statistical problems.

Chapter 2 of this book presents an attempt of systematical advance in the theory of improving estimators of expectation vectors of large dimension.

In Section 2.1, the generalized Stein-type estimators are studied, in which shrinkage coefficients are arbitrary functions of sample mean vector length. The boundaries of the quadratic risk decrease are found. In Section 2.2, it is established that in case when the dimension is large and comparable with sample size, shrinking of a wide class of unbiased estimators reduces the quadratic risk independently of distributions. In Section 2.3, the Stein effect is investigated for infinite-dimensional estimators. In Section 2.4, "component-wise" estimators are considered that are defined by arbitrary "estimation functions" presenting some functional transformation of each component of sample mean vector. The quadratic risk of this estimator is minimized with the accuracy to terms small for large dimension and sample size.

The Kolmogorov Asymptotics

In 1967, Andrei Nikolaevich Kolmogorov was interested in the dependence of errors of discrimination on sample size. He solved the following problem. Let $\mathbf{x}$ be a normal observation vector, and $\bar{\mathbf{x}}_\nu$ be sample averages calculated over samples from population number $\nu = 1, 2$. Suppose that the covariance matrix is the identity matrix. Consider a simplified discriminant function

$$g(\mathbf{x}) = (\bar{\mathbf{x}}_1 - \bar{\mathbf{x}}_2)^T \left(\mathbf{x} - (\bar{\mathbf{x}}_1 + \bar{\mathbf{x}}_2)/2\right)$$

and the classification rule $w(\mathbf{x}) > 0$ against $w(\mathbf{x}) \leq 0$. This function leads to the probability of errors $\alpha_n = \Phi(-G/\sqrt{D})$, where G and D are quadratic functions of sample averages having a noncentral χ^2 distribution. To isolate principal parts of G and D, Kolmogorov proposed to consider not one statistical problem but a sequence of n-dimensional discriminant problems in which the dimension n increases along with sample sizes N_ν, so that $N_\nu \to \infty$ and $n/N_\nu \to \lambda_\nu > 0$, $\nu = 1, 2$. Under these assumptions, he proved that the probability of error α_n converges in probability

$$\operatorname*{plim}_{n \to \infty} \alpha_n = \Phi\left(-\frac{J - \lambda_1 + \lambda_2}{2\sqrt{J + \lambda_1 + \lambda_2}}\right), \tag{7}$$

where J is the square of the Euclidean limit "Mahalanobis distance" between centers of populations. This expression is remarkable by that it explicitly shows the dependence of error probability on the dimension and sample sizes. This new asymptotic approach was called the "Kolmogorov asymptotics."

Later, L. D. Meshalkin and the author of this book deduced formula (7) for a wide class of populations under the assumption that the variables are independent and populations approach each other in the parameter space (are contiguous) [45], [46].

In 1970, Yu. N. Blagoveshchenskii and A. D. Deev studied the probability of errors for the standard sample Fisher–Andersen–Wald discriminant function for two populations with *unknown common* covariance matrix. A. D. Deev used the fact that the probability of error coincides with the distribution function $g(\mathbf{x})$. He obtained an exact asymptotic expansion for the limit of the error probability α. The leading term of this expansion proved to be especially interesting. The limit probability of an error (of the first kind) proved to be

$$\alpha = \Phi\left(-\Theta\frac{J - \lambda_1 + \lambda_2}{2\sqrt{J + \lambda_1 + \lambda_2}}\right),$$

where the factor $\Theta = \sqrt{1 - \lambda}$, with $\lambda = \lambda_1\lambda_2/(\lambda_1 + \lambda_2)$, accounts for the accumulation of estimation inaccuracies in the process of the covariance matrix inversion. It was called "the Deev formula." This formula was thoroughly investigated numerically, and a good coincidence was demonstrated even for not great n, N.

Note that starting from Deev's formulas, the discrimination errors can be reduced if the rule $g(\mathbf{x}) > \theta$ against $g(\mathbf{x}) \leq \theta$ with $\theta = (\lambda_1 - \lambda_2)/2 \neq 0$ is used. A. D. Deev also noticed [18] that the half-sum of discrimination errors can be further decreased by weighting summands in the discriminant function.

After these investigations, it became obvious that by keeping terms of the order of n/N, one obtains a possibility of using specifically multidimensional effects for the construction of improved discriminant and other procedures of multivariate analysis. The most important conclusion was that traditional consistent methods of multivariate statistical analysis should be improvable, and a new progress in theoretical statistics is possible, aiming at obtaining nearly optimal solutions for fixed samples.

The Kolmogorov asymptotics (increasing dimension asymptotics [3]) may be considered as a calculation tool for isolating leading terms in case of large dimension. But the principal role of the Kolmogorov asymptotics is that it reveals specific regularities produced by estimation of a large number of parameters. In a series of further publications, this asymptotics was used as a main tool for investigation of *essentially many-dimensional* phenomena characteristic of *high-dimensional statistical analysis*. The constant n/N became an acknowledged characteristics in many-dimensional statistics.

In Section 5.1, the Kolmogorov asymptotics is applied for the development of theory allowing to improve the discriminant analysis of vectors of large dimension with independent components.

The improvement is achieved by introducing appropriate weights of contributions of independent variables in the discriminant function. These weights are used for the construction of asymptotically unimprovable discriminant procedure. Then, the problem of selection of variables for discrimination is solved, and the optimum selection threshold is found.

But the main success in the development of multiparametric solutions was achieved by combining the Kolmogorov asymptotics with the spectral theory of random matrices developed independently at the end of 20th century in another region.

Spectral Theory of Increasing Random Matrices

In 1955, the well-known physicist theoretician E. Wigner studied energy spectra of heavy nuclei and noticed that these spectra have a characteristic semicircle form with vertical derivatives at the edges. To explain this phenomenon, he assumed that very complicated hamiltonians of these nuclei can be represented by random matrices of high dimension. He found the limit spectrum of symmetric random matrices of increasing dimension $n \to \infty$ with independent (over-diagonal) entries W_{ij}, zero expectation, and the variance $\mathbf{E}W_{ii}^2 = 2v^2$, $\mathbf{E}W_{ij} = v^2$ for $i \neq j$ [88]. The empirical distribution function (counting function) for eigenvalues λ_i of these matrices

$$F_n(u) = n^{-1} \sum_{i=1}^{n} \mathrm{ind}(\lambda_i \leq u)$$

proved to converge almost surely to the distribution function $F(u)$ with the density

$$F'(u) = (2\pi v^2)^{-1}\sqrt{4v^2 - u^2}, \qquad |u| \leq 2|v|$$

(limit spectral density). This distribution was called Wigner's distribution.

In 1967, V. A. Marchenko and L. A. Pastur published the well-known paper [43] on the convergence of spectral functions of random symmetric Gram matrices of increasing dimension $n \to \infty$. They considered matrices of the form

$$B = A + N^{-1} \sum_{m=1}^{N} \mathbf{x}_m \mathbf{x}_m^T,$$

where A are nonrandom symmetric matrices with converging counting functions $F_{An}(u) \to F_A(u)$, and $\mathbf{x}_m$ are independent random vectors with independent components x_{mi} such that $\mathbf{E}x_{mi} = 0$ and $\mathbf{E}x_{mi}^2 = 1$. They assumed that the ratio $n/N \to y > 0$, distribution is centrally symmetric and invariant with respect to components numeration, and the first four moments of x_{mi} satisfy some tensor relations. They established the convergence $F_{Bn}(u) \to F_B(u)$, where $F_{Bn}(u)$ are counting functions for eigenvalues of B, and derived a specific nonlinear relation between limit spectral functions of matrices A and B. In the simplest case when $A = I$ it reads

$$h(t) = \int (1 + ut)^{-1} dF_B(u) = (1 + ts(t))^{-1},$$

where $s(t) = 1 - t + yh(t)$. By the inverse Stilties transformation, they obtained the limit spectral density

$$F_B'(u) = \frac{1}{\sqrt{2\pi yu}} \sqrt{(u_2 - u)(u - u_1)}, \qquad u_1 \leq u \leq u_2,$$

where $u_2, u_1 = (1 \pm \sqrt{y})^2$. If $n > N$, the limit spectrum has a discrete component at $u = 0$ that equals $1 - N/n$.

From 1975 to 2001, V. L. Girko created an extended limit spectral theory of random matrices of increasing dimension that was published in a series of monographs (see [22]–[26]). Let us describe in general some of his results. V. L. Girko studied various matrices formed by linear and quadratic transformations from initial random matrices $X = \{x_{mi}\}$ of increasing dimensions $N \times n$ with independent entries. The aim of his investigation is to establish the convergence of spectral functions of random matrices to some

limit nonrandom functions $F(u)$ and then to establish the relation
between $F(u)$ and limit spectral functions of nonrandom matrices.
For example, for $B = A^T X X^T A$, the direct functional relation is
established between limit spectra of nonrandom matrices A and
random B. V. L. Girko calls such relations "stochastic canonical
equations" (we prefer to call them "dispersion equations").

In the first (Russian) monograph "Random Matrices" published in 1975, V. L. Girko assumes that all variables are independent, spectral functions of nonrandom matrices converge, and
the generalized Lindeberg condition holds: for any $\tau > 0$,

$$\lim_{n \to \infty} N^{-1} \sum_{m=1}^{N} n^{-1} \sum_{i=1}^{n} x_{mi}^2 \, \text{ind}(x_{mi}^2 \geq \tau) \xrightarrow{\text{P}} 0.$$

The main result of his investigations in this monograph was a
number of limit equations connecting spectral functions of different random matrices and underlying nonrandom matrices.

In monograph [25] (1995), V. L. Girko applied his theory specifically to sample covariance matrices. He refines his theory by with-
drawing the assumption on the convergence of spectral functions of
true covariance matrices. He postulates a priori some "canonical"
equations, proves their solvability, and only then reveals their con-
nection with limit spectra of random matrices. Then, V. L. Girko
imposes more restrictive requirements to moments (he assumes
the existence of four uniformly bounded moments) and finds limit
values of separate (ordered) eigenvalues.

Investigations of other authors into the theory of random Gram
matrices of increasing dimension differ by more special settings
and by less systematic results. However, it is necessary to cite
paper by Q. Yin, Z. Bai, and P. Krishnaia (1984), who were first
to establish the existence of limits for the least and the largest
eigenvalues of Wishart matrices. In 1998, Bai and Silverstein [9]
discovered that eigenvalues of increasing random matrices stay
within the boundaries of the limit spectrum with probability 1.

Spectral Functions of Sample Covariance Matrices

Chapter 3 of this book presents the latest development in the
spectral theory of sample covariance matrices of large dimension.
Methods of spectral theory of random matrices were first applied

to sample covariance matrices in paper [63] of the author of this monograph (1983). The straightforward functional relation was found between limit spectral functions of sample covariance matrices and limit spectral functions of unknown true covariance matrices. Let us cite this result since it is of a special importance to the multiparametric statistics. Spectra of true covariance matrices Σ of size $n \times n$ are characterized by the "counting" function

$$F_{0n}(u) = n^{-1} \sum_{m=1}^{n} \text{ind}(\lambda_i \leq u), \quad u \geq 0$$

of eigenvalues λ_i, $i = 1, 2, \ldots, n$. Sample covariance matrices calculated over samples $\mathfrak{X} = \{\mathbf{x}_m\}$ of size N have the form

$$C = N^{-1} \sum_{m=1}^{N} (\mathbf{x}_m - \bar{\mathbf{x}})(\mathbf{x}_m - \bar{\mathbf{x}})^T,$$

where $\bar{\mathbf{x}}$ are sample average vectors.

THEOREM 1.   *If n-dimensional populations are normal* $\mathbf{N}(0, \Sigma)$, $n \to \infty$, $n/N \to \lambda > 0$, *and functions* $F_{0n}(u) \to F_0(u)$, *then for each $t \geq 0$, the limit exist*

$$h(t) = \lim_{n \to \infty} \mathbf{E} n^{-1} \text{tr}(I + tC)^{-1} = \int (1 + ts(t)u)^{-1} dF_0(u),$$

$$\text{and} \quad \mathbf{E}(I + tC)^{-1} = (I + ts(t)\Sigma)^{-1} + \Omega_n, \tag{8}$$

where $s(t) = 1 - \lambda + \lambda h(t)$ and $\|\Omega_n\| \to 0$ (here the spectral norms of matrices are used).

In 1995, the author of this book proved that these relations remain valid for a wide class of populations restricted by the values of two specific parameters: the maximum fourth central moment of a projection of $\mathbf{x}$ onto nonrandom axes (defined by vectors $\mathbf{e}$ of unit length)

$$M = \sup_{|\mathbf{e}|=1} \mathbf{E}\left(\mathbf{e}^T \mathbf{x}\right)^4 > 0$$

and measures of dependence of variables

$$\nu = \sup_{\|\Omega\|=1} \mathrm{var}(\mathbf{x}^T \Omega \mathbf{x}/n), \quad \text{and} \quad \gamma = \nu/M,$$

where Ω are nonrandom, symmetric, positive semidefinite matrices of unit spectral norm. Note that for independent components of $\mathbf{x}$, the parameter $\nu \leq M/n$. For normal distribution, $\gamma \leq 2/3n$. The situation when the dimension n is large, sample size N is large, the ratio n/N is bounded, the maximum fourth moment M is bounded, and γ is small may be called the situation of the *multiparametric statistics applicability.*

Section 3.1, the latest achievements of spectral theory of large Gram matrices and sample covariance matrices are presented. Theorem 1 is proved under weakest assumptions for wide class of distributions. Analytical properties of $h(z)$ are investigated, and finite location of limit spectra is established. In Section 3.2, the dispersion equations similar to (8) are derived for infinite-dimensional variables.

Note that the regularization of the inverse sample covariance matrix C^{-1} by an addition of a positive "ridge" parameter $\alpha > 0$ to the diagonal of C before inversion produces the resolvent of C involved in Theorem 1. Therefore, the ridge regularization of linear statistical procedures leads to functions admitting the application of our dispersion equations with remainder terms small in the situation of multiparametric statistics applicability. Theorems proved in Sections 3.1 allow to formulate the *Normal Evaluation Principle,* presented in Section 3.3. It states that limiting expressions of standard quality functions for regularized multivariate statistical procedures are determined by only two moments of variables and may be approximately evaluated under the assumption of populations normality.

We say that function $f(\mathbf{x})$ of variable $\mathbf{x}$ from population $\mathfrak{S}$ allows ε-*normal evaluation* in the square mean, if for $\mathfrak{S}$ some normal distribution exists $\mathbf{y} \sim \mathbf{N}(\vec{\mu}, \Sigma)$ with $\vec{\mu} = \mathbf{E}\mathbf{x}$ and $\Sigma = \mathrm{cov}(\mathbf{x}, \mathbf{x})$ such that

$$\mathbf{E}(f(\mathbf{x}) - f(\mathbf{y}))^2 \leq \varepsilon.$$

PROPOSITION 3. *Under conditions of multiparametric statistics applicability (large N, bounded n/N and M, and small γ), the principal parts of a number of standard quality functions of regularized linear procedures allow ε-normal evaluation with small $\varepsilon > 0$.*

This means that, in the multiparametric case, it is possible to develop (regularized) statistical procedures such that

1. their standard quality functions have a small variance and allow reliable estimation from sample data;

2. the quality of these procedures is only weakly depending on distributions.

Constructing Multiparametric Procedures

In Chapter 4, the spectral theory of sample covariance matrices is used for systematical construction of practical approximately unimprovable procedures. Using dispersion equations, one can calculate leading terms of quality functions in terms of parameters excluding dependence on random values, or on the opposite, to express quality functions only in terms of observable data excluding unknown parameters. To choose an essentially multivariate statistical procedure of best quality, one may solve two alternative extremum problems:

(a) find an a priori best solution of statistical problem using the expression of quality function as a function only on parameters;

(b) find the best statistical rule using the quality function presented as a function of only observable data.

For $n \ll N$, all thus improved multiparametric solutions pass to standard consistent ones.

In case of large n and N, the following practical recommendation may be offered.

1. For multivariate data of any dimension, it is desirable to apply always stable and not degenerate, approximately optimal, multiparametric solutions instead of traditional methods consistent only for fixed dimension.

2. It is plausible to compare different multivariate procedures theoretically for large dimension and for large sample size by quality function expressed in terms of first two moments of variables.

3. Using the multiparametric technique, it is possible to calculate principal parts of quality functions from sample data, compare different versions of procedures, and choose better ones for treating concrete samples.

Let us describe the technology of construction of unimprovable multiparametric procedures.

1. Standard multivariate procedure is regularized, and a wide class of regularized solutions is chosen.

2. The quality function is studied, and its leading term is isolated. Then one of two tactics, (a) or (b), is followed.

Tactic "a"

1. Using dispersion equation, the observable variables are excluded and the principal part of quality function is presented as a function only on parameters.

2. The extremum problem is solved, and an a priori best solution is found.

3. The parameters in this solution are replaced by statistics (having small variance), and a consistent estimator of the best solution is constructed.

4. It is remains to prove that this estimator leads to a solution whose quality function approximates well the best quality function.

Tactic "b"

1. Using dispersion equations, the unknown parameters are excluded and the principal part of quality function is expressed as a function only on statistics.

2. An extremum problem is solved, and the approximately best solution is obtained depending only on observable data.

3. It is proved that this extremum solution provides the best quality with accuracy to remainder terms of the asymptotics.

In Chapter 4, this multiparametric technique is applied to construction of a number of multivariate statistical procedures that are surely not degenerate and are approximately optimal independently of distributions. Among these are problems of optimal estimation of the inverse covariance matrices, optimal matrix shrinkage for sample mean vector, and minimizing quadratic risk of sample linear regression.

In 1983, the author of this book found [63] conditions providing the minimum of limit error probability in the discriminant analysis of large-dimensional normal vectors $\mathbf{x} \sim \mathbf{N}(\vec{\mu}_\nu, \Sigma)$, $\nu = 1, 2$, within a generalized class of linear discriminant function. The inverted sample covariance matrix C in the standard discriminant "plug-in" linear discriminant function is replaced by the matrix

$$\Gamma(C) = \int_{t>0} (I + tC)^{-1} d\eta(t),$$

where $\eta(t)$ are arbitrary functions of finite variation. In [63] the extremum problem is solved, and the Stilties equation is derived for the unimprovable function $\eta(t) = \eta_0(t)$. This equation was used in [82] by V. S. Stepanov for treating some real (medical and economical) data. He found that it provides remarkably better results even for not great n, $N \approx 5 - 10$. In Section 5.2, this method is extended to a wide class of distributions.

Optimal Solution to Empirical Linear Equations

Chapter 6 presents the development of statistical approach for finding minimum square pseudosolutions to large systems of linear empiric equations whose coefficients are random values with known distribution function. The standard solution to system of linear algebraic equations (SLAE) $A\mathbf{x} = \mathbf{b}$ using known empiric random matrix of coefficients R and empiric right-hand side vector $\mathbf{y}$ can be unstable or nonexisting if the variance of coefficients is sufficiently large. The minimum square solution $\widehat{\mathbf{x}} = (R^T R)^{-1} R^T \mathbf{y}$ with empiric matrix R and empiric right-hand sides also can be unstable or nonexisting. These difficulties are produced by incorrect solution and the inconsistency of random system. The well-known Tikhonov regularization methods [83] are based on a rather artificial requirement of minimum complexity; they guarantee the existence of a pseudosolution but minimize neither the quadratic risk nor the residuals. Methods of the well-known confluent analysis [44] lead to the estimator $\widehat{\mathbf{x}} = (R^T R - \lambda I)^{-1} R^T \mathbf{y}$, where $\lambda \geq 0$ and I is the identity matrix. These estimators are even more unstable and surely do not exist when the standard minimum square solution does not exist (due to additional estimating of the coefficients matrices).

In Chapter 6, the extremum problem is solved. The quadratic risk of pseudosolutions is minimized within a class of arbitrary linear combinations of regularized pseudosolutions with different regularization parameters. First, in Section 6.1, an a priori best solution is obtained by averaging over all matrices A with fixed spectral norm and all vectors $\mathbf{b}$ of fixed length. Section 6.2 presents the theoretical development, providing methods of the construction of asymptotically unimprovable solutions of unknown SLAE $A\mathbf{x} = \mathbf{b}$ from empiric coefficient matrix R and the right-hand side vector $\mathbf{y}$ under the assumption that all entries of the matrix R and components of the vector $\mathbf{b}$ are independent and normally distributed.

CHAPTER 2

FUNDAMENTAL PROBLEM OF STATISTICS
MULTIPARAMETRIC CASE

Analyzing basic concepts of statistical science, we must acknowledge that there is an essential difference between goals and settings of a single-parameter statistical investigations and statistical problems involving a number of unknown parameters. In case of one parameter, only one problem arises, the problem of achieving more and more accurate approximation to its unknown magnitude: and the most important property of estimators becomes their consistency.

In case of many-parameter models, the investigator starts with the elucidation of his interest in the statistical problem and formulation of criteria of quality and choosing quality function. Separate parameters are no more of interest, and the problem of the achievement of maximum gain in quality arises. The possibility of infinite increase of samples is replaced by the necessity of optimal statistical decisions under fixed samples available for the investigator. Thus, many-parameter problems require a more appropriate statistical setting that may be expressed in terms of the gain theory and of statistical decision functions. Such approach was proposed by Neumann and Morgenstern (1947) and by A. Wald (1970) and led to the concept of estimators dominance and search for uniformly unimprovable (admissible) estimators minimizing risk function (see [91]).

Suppose $\mathfrak{D}$ is a class of estimators and $R(\theta, d)$ is the risk function in estimating the parameter θ from the region Θ with the estimator d. The estimator d_1 is said to dominate the estimator d_2 if $R(\theta, d_1) \leq R(\theta, d_2)$ for all $\theta \in \Theta$. The conventional approach to theoretical developments is the construction of estimators dominating some classes of estimators. The subclass $\mathfrak{K} \in \mathfrak{D}$ of estimators

21

is called complete if for each estimator $d \in \mathfrak{D} - \mathfrak{K}$, there exists an estimator $d^* \in \mathfrak{K}$ such that d^* dominates d. The main progress achieved in the theory of estimator dominance is connected with the Cramer–Rao theorem and minimizing the quadratic risk. The quadratic risk function proved to be most convenient and most acceptable, especially for many-dimensional problems.

In 1956, C. Stein found that for the dimension greater 2, a simple transformation of standard sample average estimator of expectation vector by introducing a common scalar multiple (shrinking) may reduce the quadratic risk uniformly in parameters, presenting an improved estimator. This discovery showed that well-developed technique of consistent estimation is not best for many-dimensional problems, and a new statistical approach is necessary to treat many-parameter case. Practically, for example, in order to locate better the point in three-dimensional space by a single observation, the observer should refuse from straightforward interpretation of the observed coordinates and multiply them by a shrinkage factor less 1.

A series of investigations followed (see Introduction), in which it was established that nearly all most popular estimators allow an improvement, and this effect appears for a wide class of distributions. The gain could be measured by the ratio of the number of parameters to sample size (the Kolmogorov ratio). However, the authors of most publications analyze only simple modifications of estimators like shrinking, and till now, an infinite variety of other possible improvements remains not explored.

In this chapter, we present a more detailed investigation of shrinkage effect and its application in the fundamental problem of unimprovable estimation of expectation vectors.

2.1. SHRINKAGE OF SAMPLE MEAN VECTORS

Assume that n-dimensional observation vector $\mathbf{x}$ has independent components, the expectation vector $\vec{\mu} = \mathbf{Ex}$ exists, and variance of all components is 1. Let $\bar{\mathbf{x}}$ be a sample average vector calculated over a sample of size N. Denote $y = n/N$. Let the absolute value of a vector denote its length, and the square of a vector be the square of its length. We have the expectation $\mathbf{E}\bar{\mathbf{x}}^2 = \vec{\mu}^2 + y$.

Remark 1. In the family $\mathfrak{K}^{(0)}$ of estimators with a priori shrinkage of the form $\hat{\mu} = \alpha\bar{\mathbf{x}}$, where $\bar{\mathbf{x}}$ is sample average vector and α is a nonrandom scalar, the minimum is achieved for $\alpha = \alpha_0 = 1 - y/(\vec{\mu}^2 + y)$ and the quadratic risk is

$$\mathbf{E}(\vec{\mu} - \alpha_0\bar{\mathbf{x}})^2 = R_0 \stackrel{def}{=} \mathbf{E}(\vec{\mu} - \bar{\mathbf{x}})^2 = y\,\frac{\vec{\mu}^2}{\vec{\mu}^2 + y}. \qquad (1)$$

Definition 1. Let a class of estimators $\mathfrak{K}^1$ be chosen for vectors $\vec{\mu}$. We call an estimator of $\vec{\mu}$ *composite a priori estimator* if it is a collection of estimators from $\mathfrak{K}^0$ with nonintersecting subsets of components of $\vec{\mu}$ that includes all components.

THEOREM 2.1. *For composite a priori estimators from $\mathfrak{K}^{(0)}$, the sum of quadratic risks is not larger than the quadratic risk of estimating all components with a single a priori shrinkage coefficient.*

Proof. We write R_0 in the form $y - y^2/(\vec{\mu}^2 + y)$. Let angular brackets denote averaging over subsets $\vec{\mu}(j)$, $j = 1, \ldots, k$ of the set of all components $(\mu_1, \ldots, \mu_n)$ with weights $n(j)/n$, $j = 1, \ldots, k$. Denote $y(j) = n(j)/N$ for all j. Using the concavity property of the inverse function, we find that

$$y\left\langle \frac{y(j)}{\vec{\mu}^2(j) + y(j)} \right\rangle = y\left\langle \frac{1}{\vec{\mu}^2(j)/y(j) + 1} \right\rangle \geq \frac{y}{\langle \vec{\mu}^2(j)/y(j) + 1 \rangle} =$$

$$= \frac{y^2}{\vec{\mu}^2 + y}.$$

Q.E.D. $\square$

Now, let a half of the vector $\vec{\mu}$ components be equal to zero. Then in the family $\mathfrak{K}^{(0)}$, a composite estimator can have the quadratic risk nearly twice as small.

Shrinkage for Normal Distributions

Consider an n-dimensional population $\mathbf{N}(\vec{\mu}, \Sigma)$ with the identity covariance matrix $\Sigma = I$. Consider a class of estimators $\widehat{\mu}$ of vectors $\vec{\mu} = (\mu_1, \ldots, \mu_n)$ over samples $\mathfrak{X}$ of size N with $\mathbf{E}\widehat{\mu}^2 < \infty$ having the form $\widehat{\mu} = \alpha\bar{\mathbf{x}}$, where $\alpha = \alpha(\mathfrak{X})$ is a scalar function. Let us restrict ourselves with the case when the function $\alpha(\cdot)$ is differentiable with respect to all arguments. Since $\bar{\mathbf{x}}$ is a sufficient statistics, it suffices to consider a complete subclass of estimators of the form $\widehat{\mu} = \varphi(|\bar{\mathbf{x}}|)\bar{\mathbf{x}}$. Denote $y_\nu = (n - \nu)/N$, $\nu = 1, 2, 3, 4$.

Let $\mathfrak{K}$ be a set of estimators of the form $\widehat{\mu} = \varphi(|\bar{\mathbf{x}}|)\bar{\mathbf{x}}$, where $\varphi(r) < 1$ is a differentiable function of $r = |\bar{\mathbf{x}}| > 0$ such that the product $r^2\varphi(r)$ is uniformly bounded on any segment $[0, b]$. This class includes the "plug-in" estimator $\widehat{\mu}_1 = \varphi_1(|\bar{\mathbf{x}}|)\bar{\mathbf{x}}$, where $\varphi_1(r) = 1 - y/r^2$ for $r > 0$, and the James–Stein estimator $\widehat{\mu}_S = \varphi_S(|\bar{\mathbf{x}}|)\bar{\mathbf{x}}$, where $\varphi_S(r) = 1 - y_2/r^2$, and the majority of shrinkage estimators considered in the reviews [29], [30], and [33].

Let us compare the quadratic risk $R_1 = R(\varphi_1) = \mathbf{E}[\vec{\mu} - \varphi_1(|\bar{\mathbf{x}}|)\bar{\mathbf{x}}]^2$ of the estimator $\widehat{\mu}_1$ with the quadratic risk

$$R_S = R(\varphi_S) = \mathbf{E}[\vec{\mu} - \varphi_S(|\bar{\mathbf{x}}|)\bar{\mathbf{x}}]^2 \text{ of the estimator } \widehat{\mu}_S.$$

Remark 2. For $n \geq 3$, the moment $\mathbf{E}|\bar{\mathbf{x}}|^{-2}$ exists, $R_1 = y - yy_4\mathbf{E}|\bar{\mathbf{x}}|^{-2}$ and $R_S = y - y_2^2\,\mathbf{E}|\bar{\mathbf{x}}|^{-2}$.

Indeed, rewrite the quadratic risk of $\widehat{\mu}_1$ as follows:

$$R_1 = \mathbf{E}[\vec{\mu} - \varphi_1(|\bar{\mathbf{x}}|)\bar{\mathbf{x}}]^2 = \mathbf{E}[(\vec{\mu} - \bar{\mathbf{x}})^2 + 2y(\vec{\mu} - \bar{\mathbf{x}})^T\bar{\mathbf{x}}/\bar{\mathbf{x}}^2 + y^2/\bar{\mathbf{x}}^2].$$

Here in the second term under the sign of $\mathbf{E}$, we have $(\vec{\mu} - \bar{\mathbf{x}}) = N^{-1}(\nabla f)/f$, where f is the normal distribution density for $\bar{\mathbf{x}}$. Integrating by parts, we find that this term equals

$$\frac{2y}{N}\int_0^\infty \frac{\bar{\mathbf{x}}^T\nabla f}{\bar{\mathbf{x}}^2}\, d\bar{\mathbf{x}} = -2\frac{n-2}{N}\mathbf{E}\frac{1}{\bar{\mathbf{x}}^2}.$$

Thus, $R_1 = \mathbf{E}[y - 2yy_2/|\bar{\mathbf{x}}|^2 + y^2/\bar{\mathbf{x}}^2] = y - yy_4\mathbf{E}|\bar{\mathbf{x}}|^{-2}$. The calculation of R_S is similar. $\square$

Remark 3. For $n \geq 4$, we have

$$\mathbf{E}\frac{1}{\bar{\mathbf{x}}^2} = \int_0^\infty \exp(-u|\bar{\mathbf{x}}|^2)\,du =$$

$$= \frac{N}{2}\int_0^\infty (1+t)^{-n/2}\exp\left(-\bar{\mu}^2\frac{N}{2}\frac{t}{1+t}\right)dt =$$

$$= \frac{N}{2}\int_0^1 (1-z)^{n/2-2}\exp\left(-\bar{\mu}^2\frac{N}{2}z\right)dz, \tag{2}$$

and the relations hold true

$$(\bar{\mu}^2 + y)^{-1} \leq \mathbf{E}|\bar{\mathbf{x}}|^{-2} \leq \min[y_2^{-1}, (\bar{\mu}^2 + y_4)^{-1}]; \tag{3}$$
$$\text{if } \bar{\mu} = 0, \quad \text{then } \mathbf{E}|\bar{\mathbf{x}}|^{-2} = y_2^{-1}.$$

Indeed, the first inequality in (3) follows from the inequality $\mathbf{E}Z^{-1} \geq (\mathbf{E}Z)^{-1}$ for positive Z, and the second inequality in (3) is a consequence of a monotone dependence on $\bar{\mu}^2$. $\square$

Remark 4. For $n \geq 4$, we have $R_0 < R_S < R_1$.

Indeed, let us substitute $z = 2x/N$ in (2) and write the difference $R_S - R_0$ in the form

$$y^2 \int_0^{N/2} h(x) \exp(-\bar{\mu}^2 x - yx)\,dx + y^2 \int_{N/2}^\infty \exp(-\bar{\mu}^2 x - yx)\,dx,$$

where

$$h(x) = 1 - \left(1 - \frac{2}{n}\right)^2\left(1 - \frac{2x}{N}\right)^{n/2-2}\exp(xy).$$

Here $h(0) = 1 - (1 - 2/n)^2 > 0$, and $h(N/2) = 1$. The derivative $h'(x)$ vanishes only if $x = y/2$. For this x, the function

$$h(x) = h(y/2) = 1 - (1 - 2/n)^2(1 - 4/n)^{n/2-2}\exp(2).$$

This expression decreases monotonously from 1 at $n = 4$ to 0 as $n \to \infty$. It follows $h(y/2) > 0$. We conclude that $h(x) > 0$ for any $0 \le x \le N/2$. Consequently, we have $R_S > R_0$. The difference $R_1 - R_S = 4N^{-2} \mathbf{E}|\bar{\mathbf{x}}|^{-2}$. The remark is justified. $\square$

Remark 5. For $n \ge 4$, the risk functions R_1 and R_S of the estimators $\hat{\mu}_1$ and $\hat{\mu}_S$ increase monotonously as $|\vec{\mu}|$ increases;
if $\vec{\mu} = 0$, then $R_1 = 2[1 + 2/(n-2)]/N$, $R_S = 2/N$;
if $\vec{\mu} \ne 0$, the inequalities hold

$$y - yy_4/(\vec{\mu}^2 + y_4) \le R_1 \le y - yy_4/(\vec{\mu}^2 + y),$$
$$y - y_2^2/(\vec{\mu}^2 + y_4) \le R_S \le y - y_2^2/(\vec{\mu}^2 + y).$$

Note that composite James–Stein estimators can have a greater summary quadratic risk than the quadratic risk of all-component estimator: for $\vec{\mu} = 0$, we have the quadratic risk $R_S = 2/N$ independently of $n \ge 3$, while the quadratic risk of the composite estimator can be some times greater.

Now we consider arbitrary estimators of the form $\hat{\mu} = \varphi(|\bar{\mathbf{x}}|) \, \bar{\mathbf{x}}$.

LEMMA 2.1. *For $n \ge 3$ and $\hat{\mu} \in \mathfrak{R}$, the quadratic risk*

$$R(\varphi) = \mathbf{E}(\vec{\mu} - \hat{\mu})^2 = \mathbf{E}(\vec{\mu} - \varphi(r) \, \bar{\mathbf{x}})^2 =$$
$$= \mathbf{E}[\vec{\mu}^2 + 2y\varphi(r) - 2r^2\varphi(r) + 2r\varphi'(r)/N + r^2\varphi^2(r)], \quad (4)$$

where $r = |\bar{\mathbf{x}}|$.

Proof. We transform the addend of $R(\varphi)$ including $\bar{\mathbf{x}}^T \vec{\mu}$ by integrating by parts. Replace

$$\bar{\mathbf{x}}^T \vec{\mu} = \bar{\mathbf{x}}^2 + \frac{1}{N} \, \bar{\mathbf{x}}^T \nabla \ln \, f(\bar{\mathbf{x}}),$$

where $f(\bar{\mathbf{x}})$ is the normal density for the vector $\bar{\mathbf{x}}$, and ∇ is a vector of derivatives with respect to components of $\bar{\mathbf{x}}$. We find that

$$\mathbf{E}\varphi(r)\bar{\mathbf{x}}^T \nabla \ln \, f(\bar{\mathbf{x}}) = -\mathbf{E}\nabla^T(\varphi(r) \, \bar{\mathbf{x}}) = -\mathbf{E}[r\varphi'(r) + n\varphi(r)]$$

since the product $\varphi(|\bar{\mathbf{x}}|) \, f(\bar{\mathbf{x}}) \bar{\mathbf{x}} \to 0$ as $|\bar{\mathbf{x}}| \to \infty$. The lemma statement follows. $\square$

Denote by $g(r)$ the distribution density of random value $r = |\bar{\mathbf{x}}|$ for $\bar{\mathbf{x}} \sim \mathbf{N}(\vec{\mu}, I/N)$. We calculate the logarithmic derivative

$$\frac{g'(r)}{g(r)} = N\left(\frac{y_2}{r} - r + \vec{\mu}\operatorname{cth}(|\mu|Nr)\right). \qquad (5)$$

Denote

$$\varphi_E(r) = 1 - \frac{y_1}{r^2} + \frac{1}{rN}\frac{g'(r)}{g(r)} = \frac{|\vec{\mu}|}{r}\operatorname{cth}(|\vec{\mu}|Nr) - \frac{1}{Nr^2}. \qquad (6)$$

For $r \to 0$, we have $\varphi_E(r) \to 2/3\ \vec{\mu}^2 N$, and there exists the derivative $\varphi'_E(+0)$. Let us prove that the function $\varphi_E(r)$ provides the minimum of $R(\varphi)$. Denote $y_\nu = (n - \nu)/N, \quad \nu > 0$.

THEOREM 2.2. *For $n \geq 3$, the estimator $\widehat{\mu}_E = \varphi_E(r)\bar{\mathbf{x}}$, with $\varphi_E(r)$ of the form (6) and $r = |\bar{\mathbf{x}}|$ belonging to the class $\mathfrak{K}$, and its quadratic risk*

$$R_E = R(\varphi_E) = \mathbf{E}(\vec{\mu} - \widehat{\mu}_E)^2 = \vec{\mu}^2 - \mathbf{E}r^2\varphi_E^2(r) =$$

$$= y_1\left(1 - y_2\ \mathbf{E}\frac{1}{r^2}\right) + \mathbf{E}\left(\frac{1}{r^2N^2} - \frac{\vec{\mu}^2}{\operatorname{sh}^2(|\mu|Nr)}\right).$$

The quadratic risk of any estimator $\widehat{\mu} \in \mathfrak{K}$ equals

$$R(\varphi) = R(\varphi_E) + \mathbf{E}r^2(\varphi(r) - \varphi_E(r))^2.$$

Proof. We start from (4). First, integrating by parts, we calculate the contribution of the summand preceding to the last. We find that

$$\mathbf{E}r\varphi'(r) = \int r\varphi(r)g(r)\,dr = -\int r\varphi(r)g'(r)\,dr - \mathbf{E}\varphi(r) =$$

$$-\mathbf{E}r\varphi(r)\frac{g'(r)}{g(r)} - \mathbf{E}\varphi(r).$$

The substitution in (4) transforms this expression to

$$R(\varphi) = \vec{\mu}^2 + \mathbf{E}\left(r^2\varphi^2(r) - 2r\varphi(r)\varphi_E(r)\right) =$$

$$= \vec{\mu}^2 - \mathbf{E}r^2\varphi_E^2(r) + \mathbf{E}(\varphi(r) - \varphi_E(r))^2.$$

We obtain the first expression for R_E and the expression for $R(\varphi)$ in the formulation of this theorem. It remains to calculate R_E. Substituting (6) we find that

$$R_E = \bar{\mu}^2 - \mathbf{E}\left(r - \frac{y_2}{r} + \frac{1}{N}\frac{g'(r)}{g(r)}\right)^2 =$$
$$= \bar{\mu}^2 - \mathbf{E}\left(r - \frac{y_2}{r}\right)^2 - \frac{2}{N}\int\left(r - \frac{y_2}{r}\right)g'(r)\, dr - \mathbf{E}\left(\frac{1}{N}\frac{g'(r)}{g(r)}\right)^2.$$

We calculate the integral integrating by parts. Obviously, it equals $-1 + y_2\mathbf{E}|\bar{\mathbf{x}}|^{-2}$. We obtain

$$R_E = y - y_1 y_3 \mathbf{E}\frac{1}{r^2} - \mathbf{E}\left(\frac{1}{N}\frac{g'(r)}{g(r)}\right)^2. \tag{7}$$

Also integrating by parts, we calculate the last term

$$\mathbf{E}\left(\frac{1}{N}\frac{g'(r)}{g(r)}\right)^2 = \frac{1}{N}\int_{+0}^{\infty}\left[\frac{y_2}{r} - r + \mu\,\mathrm{cth}(\mu N r)\right]g'(r)\, dr =$$
$$= \frac{1}{N}\left(1 + y_2\mathbf{E}\frac{1}{r^2}\right) + \mathbf{E}\frac{\mu^2}{\mathrm{sh}^2(|\mu|Nr)},$$

where $\mu = |\bar{\mu}|$, in view of the fact that contributions of small $r \to 0$ and $r \to \infty$ vanish. The second expression for R_E follows. Theorem 2.1 is proved. $\square$

Remark 6. For $n \geq 3$, the following inequalities are valid:

$$\frac{1}{N}\left(1 + y_2\,\mathbf{E}\frac{1}{r^2}\right) \leq \mathbf{E}\left(\frac{1}{N}\frac{g'(r)}{g(r)}\right) \leq \frac{1}{N}\left(1 + y_1\mathbf{E}\frac{1}{r^2}\right),$$
$$y_1 - y_1 y_2\mathbf{E}\frac{1}{r^2} \leq R_E \leq y_1 - y_1 y_2\mathbf{E}\frac{1}{r^2} + N^{-2}\mathbf{E}\frac{1}{r^2},$$

where $r = |\bar{\mathbf{x}}|$.

Remark 7. For $n \geq 4$, we have

$$y_1 \left(1 - \frac{y_2}{\mu^2 + y_4} \right) \leq R_E \leq y_1 \left(1 - \frac{y_2}{\mu^2 + y} \right) + \frac{1}{2N},$$
$$\frac{1}{N} \leq R_1 - R_E \leq \frac{4}{N}, \quad \frac{1}{N} \leq R_S - R_E \leq \frac{2}{N},$$

where $\mu = |\vec{\mu}|$.

These inequalities follow from (7), Remarks 5 and 6, and inequalities (3). $\square$

Since the estimator $\hat{\mu}_E$ has the minimum risk in $\mathfrak{K}$, one can conclude that, in this class, the estimator $\hat{\mu}_1$ provides the minimum quadratic risk with the accuracy up to $4/N$, and the quadratic risk of the James–Stein estimator is minimum with the accuracy up to $2/N$.

Shrinkage for a Wide Class of Distributions

From [72], it is known that in case of a large number of boundedly dependent variables, smooth functionals of sample moments approach functions of normal variables. We use the techniques developed in [71] for studying the effect of shrinkage of sample averages in this general case.

Let us restrict distributions only by the following two requirements.

A. There exist expectation vectors and the fourth moments of all components of the observation vector $\mathbf{x}$.

B. The covariance matrix $\text{cov}(\mathbf{x}, \mathbf{x}) = I$, where I is the identity matrix.

The latter condition is introduced for the convenience.

Denote $\vec{\mu} = \mathbf{Ex}$, $\overset{\circ}{\mathbf{x}} = \mathbf{x} - \vec{\mu}$. Let us introduce two parameters (see Introduction):

$$M = \sup_{\mathbf{e}} \mathbf{E}(\mathbf{e}^T \overset{\circ}{\mathbf{x}})^4 > 0 \quad \text{and} \quad \gamma = \sup_{\|\Omega\|} \text{var}(\overset{\circ}{\mathbf{x}}^T \Omega \overset{\circ}{\mathbf{x}}/n)/M, \quad (8)$$

where the supremes are calculated for all nonrandom unit vectors $\mathbf{e}$ and for all nonrandom, symmetric, positively semidefinite matrices of unit spectral norm. Note that in the case of normal observations, under condition B, we have $M = 3$, $\gamma = 2/3n$, and for any distribution with $M < \infty$ and independent components of $\mathbf{x}$, we have the inequality $\gamma \leq 2/n + 1/n^2$.

Denote $y = n/N$, $r_0^2 = \bar{\mu}^2 + y$.

Remark 8. Under Assumptions A and B,

$$\sigma_0^2 \overset{def}{=} \text{var}(\bar{\mathbf{x}} - \bar{\mu})^2 \leq y[2 + y(1 + M\gamma)]/N,$$

$$\sigma^2 \overset{def}{=} \text{var } \bar{\mathbf{x}}^2 \leq r_0^2[10 + y(1 + M\gamma)]/N.$$

Indeed, estimating the variance of $\overset{\circ}{\bar{\mathbf{x}}}{}^2$ from the above we have

$$\mathbf{E}(\overset{\circ}{\bar{\mathbf{x}}}{}^2)^2 = N^{-4} \sum \overset{\circ}{\mathbf{x}}{}_{m1}^T \overset{\circ}{\mathbf{x}}_{m2} \overset{\circ}{\mathbf{x}}{}_{m3}^T \overset{\circ}{\mathbf{x}}_{m4},$$

where the centered sample vectors $\overset{\circ}{\mathbf{x}}$ have the indexes m_1, m_2, m_3, m_4 running from 1 to N. Nonzero contribution is provided by the summands with $m_1 = m_2 = m_3 = m_4$ and other summands that have pairwise coinciding indexes. Terms with $m_1 = m_2$ and $m_3 = m_4$ do not contribute to the variance. We find that

$$\sigma_0^2 = \text{var}(\overset{\circ}{\bar{\mathbf{x}}}{}^2) < N^{-3}\mathbf{E}(\overset{\circ}{\mathbf{x}}{}_1^2)^2 + 2N^{-2}\text{tr}\Sigma^2 \leq N^{-1}y[2 + y\,(1 + M\gamma)],$$

$$\sigma^2 = \text{var}(\bar{\mathbf{x}}^2) \leq 8\,\text{var}(\bar{\mu}^T \overset{\circ}{\bar{\mathbf{x}}}) + 2\text{var}(\overset{\circ}{\bar{\mathbf{x}}}{}^2) \leq [10 + y(1 + M\gamma)]\,r_0^2/N.$$

The remark is justified. $\square$

We may interpret the number $b = n\gamma$ as a measure of dependence. We say that the variables are *boundedly dependent* if the product $y\gamma < b/N < 1$.

In the general case without restrictions on distribution, it is convenient to replace the traditional Stein estimator by a regularized Stein estimator. Denote

$$\hat{\mu}_R = \varphi_R(|\bar{\mathbf{x}}|)\bar{\mathbf{x}}, \qquad \varphi_R(r) \overset{def}{=} 1 - yr^{-2}\,\text{ind}(r^2 > \tau), \qquad \tau \geq 0,$$

$$1/2 \leq \varphi_R(r) \leq 1.$$

THEOREM 2.3. *Under Assumptions A and B for* $\tau = y/2$, *the quadratic risk of the estimator* $\widehat{\mu}_R$ *is*

$$R(\varphi_R) \leq R_0 + 72[1 + y(1 + M\gamma)/6]/N. \tag{9}$$

Proof. Let us write $\mathbf{E}(\vec{\mu} - \widehat{\mu}_R)^2$ in the form

$$R(\varphi_R) = y - y^2 \mathbf{E}i|\bar{\mathbf{x}}|^{-2} + 2y\mathbf{E}i(y - \overset{\circ}{\mathbf{x}}{}^2 - \vec{\mu}^T \overset{\circ}{\mathbf{x}})/|\bar{\mathbf{x}}|^2, \tag{10}$$

where $i = \operatorname{ind}(|\bar{\mathbf{x}}|^2 > y/2)$. Denote $p = \mathbf{E}(1 - i) = \mathbf{P}(|\bar{\mathbf{x}}|^2 \leq y/2)$, where the probability

$$p = \mathbf{P}(\bar{\mathbf{x}}^2 \leq y/2) \leq \mathbf{P}(|\xi| > y/2) \leq$$
$$\leq 4\mathbf{E}\operatorname{ind}(|\xi| > y/2)\, \xi^2/y^2 \leq 4\sigma^2/y^2,$$

and $\xi = \vec{\mu}^2 + y - \bar{\mathbf{x}}^2$. We find that

$$\mathbf{E}i|\bar{\mathbf{x}}|^{-2} = \mathbf{E}i(|\bar{\mathbf{x}}|^{-2} - r_0^{-2}) + r_0^{-2} - p/r_0^2.$$

Here in the first term of the right-hand side, we may set $\bar{\mathbf{x}}^2 > y/2$ and therefore it is greater than p/r_0^2. We find that $\mathbf{E}i|\bar{\mathbf{x}}|^{-2} > r_0^{-2}$, and it follows that the sum of the first two terms in the right-hand side of (10) is not greater $R_0 = y - y^2/r_0^2$.

The third term in the right-hand side of (10) presents a correction to R_0. Since $\mathbf{E}(y - \overset{\circ}{\mathbf{x}}{}^2 - \vec{\mu}^T \overset{\circ}{\mathbf{x}}) = 0$, this term equals

$$2y\mathbf{E}i(y - \overset{\circ}{\mathbf{x}}{}^2 - \vec{\mu}^T \overset{\circ}{\mathbf{x}})\, r_0^{-2} + 2y\mathbf{E}i(y - \overset{\circ}{\mathbf{x}}{}^2 - \vec{\mu}^T \overset{\circ}{\mathbf{x}})\, (|\bar{\mathbf{x}}|^{-2} - r_0^{-2}) =$$
$$= 2y\mathbf{E}j\frac{y - \overset{\circ}{\mathbf{x}}{}^2 - \vec{\mu}^T \overset{\circ}{\mathbf{x}}}{r_0^2} + 2y\mathbf{E}i(y - \overset{\circ}{\mathbf{x}}{}^2 - \vec{\mu}^T \overset{\circ}{\mathbf{x}})\frac{r_0^2 - |\bar{\mathbf{x}}|^2}{|\bar{\mathbf{x}}|^2 r_0^2}, \tag{11}$$

where $j = i - 1$. Let us apply the Schwarz inequality. The first summand of the right-hand side of (11) in absolute value is not greater than

$$2yr_0^{-2}\sqrt{p}\left[\sqrt{\operatorname{var}(\overset{\circ}{\mathbf{x}}{}^2)} + \sqrt{\operatorname{var}(\vec{\mu}^T \overset{\circ}{\mathbf{x}})}\,\right], \tag{12}$$

where $\mathrm{var}(\overset{\circ}{\mathbf{x}}{}^2) = \sigma_0^2$ and $\mathrm{var}(\vec{\mu}^T \overset{\circ}{\mathbf{x}}) = \mu^2/N$. We find that (12) is not greater than

$$4(\sigma_0 + |\vec{\mu}|/\sqrt{N})\sigma/r_0^2 \leq 4\sqrt{40 + 24x + 2x^2}/N \leq 6(6 + x)/N,$$

where $x = y(1 + M\gamma)$. In the second term of the right-hand side of (11), we have $iy|\overline{\mathbf{x}}|^{-2} \leq 2$. Similarly, this term does not exceed the quantity $4(\sigma_0 + |\vec{\mu}|/\sqrt{N})\,\sigma/r_0^2$. Thus, the expression (11) is no more than twice as large. Theorem 2.3 is proved. $\square$

For symmetric distributions, the upper estimate of the remainder terms in (9) is approximately twice as small. The numeric coefficient in (9) can be decreased by increasing the threshold τ.

Conclusions

Thus, in all considered cases, the shrinkage of expectation vector estimators reduces the quadratic risk approximately by the factor $\vec{\mu}^2/(\vec{\mu}^2 + y)$, where $\vec{\mu}^2$ is the square of the expectation vector length, the parameter $y = n/N$, n is the observation vector dimension, and N is the sample size.

For normal distributions with unit covariance matrix, Theorem 2.2 presents a general formula for the quadratic risk of shrinkage estimators from the class $\mathfrak{K}$, with the shrinkage coefficient depending only on sample average vector length. This theorem also states the lower boundary of quadratic risks. The majority of improved estimators suggested until now belong to the class $\mathfrak{K}$. All these estimators reduce the quadratic risk as compared with James–Stein estimator by no more than $2/N$ (Remark 7). The shrinkage effect retains for a wide class of distributions with all four moments of all variables. In Theorem 2.3, it is proved that for these distributions with unit variance of all variables, the quadratic risk is the same as for normal distributions with accuracy up to c/N. For a wide class of distributions, the coefficient c depends neither on n nor on distributions as long as the ratio n/N is bounded.

2.2. SHRINKAGE OF UNBIASED ESTIMATORS

The effect of shrinking sample mean vectors was well studied in a number of investigations published after the famous paper [80] by Stein in 1964. Most of them present an exact evaluation of the quadratic risk for the Stein-type estimators for normal and some centrally symmetric distributions. In this section, we set the problem of studying shrinkage effect for wide classes of statistical estimators.

In the previous section, we showed that the reduction of the quadratic risk by shrinking is essentially multiparametric phenomenon and can be measured by the ratio of parameter space dimension n to sample size N. For small n, the gain is small and its order of magnitude is $1/N$. Therefore, it is of a special interest to study the quadratic risk reduction for n comparable to N in more detail. For this purpose, the Kolmogorov asymptotics ($N \to \infty$, $n/N \to y > 0$) is appropriate. In this situation, the shrinkage effects are at their most. Fortunately, as is shown in [71], under these conditions, standard risk functions including the square risk function only weakly depend on distributions for a large class of populations. In this section, we study the shrinkage effect for the dimension of observations comparable to the sample size.

First, we study shrinkage of unbiased estimators for normal distributions.

Special Shrinkage of Normal Estimators

In this section, we assume that n-dimensional vector of an original estimator $\widehat{\theta}$ is distributed as $\mathbf{N}(\theta, I/N)$, where I is the identity matrix. We consider a parametric family $\mathfrak{K}^{(0)}$ of shrinkage estimators of the form $\alpha\widehat{\theta}$, where α is a nonrandom scalar. For these, the quadratic risk is

$$R = R(\alpha) = \mathbf{E}(\theta - \alpha\widehat{\theta})^2 \qquad (1)$$

(here and in the following, squares of vectors denote squares of their lengths). Denote $y = n/N$. For the original estimator, $R = R(1) = y$. Minimum of $R(\alpha)$ is reached at $\alpha = \alpha^{(0)} =$

$\theta^2/(\theta^2 + y)$, and it equals $R(\alpha^{(0)}) = \alpha^{(0)} R(1)$. As an estimator of $\alpha^{(0)}$, we choose the statistics $\widehat{\alpha}^{(0)} = 1 - y\widehat{\theta}^2$. Let us find the quadratic risk principal part for the estimators $\widehat{\theta}^{(0)} = \widehat{\alpha}^{(0)}\widehat{\theta}$ and compare it with the quadratic risk of the James–Stein estimator $\widehat{\theta}^S = \widehat{\alpha}^S\widehat{\theta}$, for which $\alpha^S = 1 - (n-2)/N\widehat{\theta}^2$.

THEOREM 2.4. *For $\widehat{\theta} \sim \mathbf{N}(\theta, I/N)$ with $\theta \neq 0$ and $n \geq 4$, the estimator $\widehat{\theta}^{(0)}$ has the quadratic risk* (1) *satisfying the inequalities*

$$\frac{\theta^2}{\theta^2 + y - 4/N}\, y \leq R(\widehat{\alpha}^{(0)}) \leq \frac{\theta^2 + 4/N}{\theta^2 + y + 4/N}\, y; \qquad (2)$$

for the estimator $\widehat{\theta}^S = \widehat{\alpha}^{(0)}\widehat{\theta}$ we have

$$\frac{\theta^2 - 4/nN}{\theta^2 + y - 4/N}\, y \leq R(\widehat{\alpha}^S) \leq \frac{\theta^2 + 4(1 - 1/n)/N}{\theta^2 + y}\, y. \qquad (3)$$

Proof. The quadratic risk is

$$R(\widehat{\alpha}^{(0)}) = \mathbf{E}(\vec{\mu} - \widehat{\theta} + y\,\widehat{\theta}/\widehat{\theta}^2)^2 = y\mathbf{E}(-1 + 2(\theta, \widehat{\theta})/\widehat{\theta}^2 + y/\widehat{\theta}^2), \quad (4)$$

where the expectation is calculated with respect to the distribution $\mathbf{N}(\theta, I)$, $\theta \in \mathbb{R}^n$. Integrating by parts, we obtain the equality $\mathbf{E}(\theta, \widehat{\theta})/\widehat{\theta}^2 = 1 - (n-2)N^{-1}\mathbf{E}(\widehat{\theta}^2)^{-1}$. We find that

$$R(\widehat{\alpha}^{(0)}) = y\left(1 - \frac{n-4}{N}\, \mathbf{E}\frac{1}{\widehat{\theta}^2}\right).$$

The inequality $\mathbf{E}(\widehat{\theta}^2)^{-1} \geq 1/\mathbf{E}(\widehat{\theta}^2)$ presents the first upper estimate in the theorem formulation. Similarly, for the James–Stein estimator with $\widehat{\alpha}^S = 1 - (n-2)/N\widehat{\theta}^2$, we get

$$R(\widehat{\alpha}^S) = y - \frac{(n-2)^2}{N^2}\, \mathbf{E}\frac{1}{\widehat{\theta}^2}.$$

The inequality $\mathbf{E}(\widehat{\theta}^2)^{-1} \geq 1/(\theta^2 + y)$ provides the second upper estimate.

Further, for $n > 4$ we find

$$\mathbf{E}\frac{1}{\widehat{\theta^2}} = \int_0^\infty \mathbf{E}\exp(-u\widehat{\theta}^2)\,du = \frac{N}{2}\int_0^1 (1-t)^{n/2-2}\exp(-\theta^2 Nt/2)\,dt$$

$$\leq \frac{N}{2}\int \exp[-(n/2-2)\,t - \theta^2 Nt/2]\,dt = \frac{1}{\theta^2 + y - 4/N}.$$

This provides both lower estimates in the theorem formulation. $\square$

Theorem 2.4 shows that the estimators $\widehat{\theta}^{(0)}$ and $\widehat{\theta}^S$ dominate the class of estimators $\widehat{\mathfrak{K}}^{(0)}$ with the accuracy up to $4/N$ and $4(n-1)/nN$, respectively.

One can see that the upper and lower estimates in (3) are narrowed down as compared with (2) by a quantity of the order of $1/n^2$.

Theorem 2.4 can be applied in a special case when the observation vector $\mathbf{x} \sim \mathbf{N}(\bar{\mu}, I)$, and the original estimator is the sample mean vector $\bar{\mathbf{x}}$ calculated over a sample $\mathfrak{X} = \{\mathbf{x}_m\}$ of size N.

For the distribution of the form $\mathbf{N}(\bar{\mu}, dI)$, where the parameter $d > 0$ is unknown, to estimate $\alpha^{(0)}$, one can propose the statistics $\widehat{\alpha}^{(1)} = 1 - n\widehat{d}/N\bar{\mathbf{x}}^2$, where

$$\widehat{d} = n^{-1}(N-1)^{-1}\sum_{m=1}^N (\mathbf{x}_m - \bar{\mathbf{x}})^2.$$

If $n \geq 4$, the quadratic risk of this statistics is

$$R = R(\widehat{\alpha}^{(1)}) \leq \frac{\bar{\mu}^2 + 4/N + 2/nN}{\bar{\mu}^2 + y}\,y.$$

Shrinkage of Arbitrary Unbiased Estimators

We consider the original class of unbiased estimators $\widehat{\theta}$ of n-dimen-sional vectors θ restricted by the following requirements.

1. For all components of $\widehat{\theta}$, there exists the expectation and all moments of the fourth order.

2. The expectation exists $\mathbf{E}(\widehat{\theta}^2)^{-2}$.

Let the estimator $\widehat{\theta}$ be calculated over a sample $\mathfrak{X}$ of size N. We introduce the parameters

$$Q = N^2 \sup_{|e|=1} \mathbf{E}[(e, \widehat{\theta} - \theta)^2]^2, \quad b^2 = \mathbf{E}\,(1/\widehat{\theta}^2)^2,$$

$$\text{and} \quad M = \max((\theta^2)^2, Q), \tag{5}$$

where the supremum is calculated with respect to all nonrandom unit vectors e. The quantity Q presents the maximum fourth moment of the projection of the centered estimation vector $\widehat{\theta}$ onto nonrandom axes. Denote $y = n/N$.

Let us characterize variance of the estimator $\widehat{\theta}$ by the matrix $D = N \operatorname{cov}(\widehat{\theta}, \widehat{\theta})$. The spectral norm $\|D\|$ presents the maximal second moment of projections of the centered vector $\widehat{\theta}$ on to nonrandom axes, $\|D\|^2 \leq Q$. Denote $d = n^{-1}\operatorname{tr} D \leq \|D\|$ and (for $yd > 0$)

$$\gamma = \sup_{\|\Omega\| \leq 1} \frac{\operatorname{var}[(\widehat{\theta} - \theta)^T \Omega (\widehat{\theta} - \theta)]}{Qy^2} \leq 1,$$

where the supremum is calculated for all nonrandom, symmetric, positively semidefinite matrices Ω with the spectral norm not greater 1. The quantity $\gamma \leq 1$ restricts the variance of the estimator vector components. In case of independent components of $\widehat{\theta}$, it is easy to see that the quantity $\gamma \leq 1/n$.

We will isolate the principal part of the quadratic risk and search for upper estimates of the remainder terms that are small for bounded M, b, and y, for large N and small γ. For the original estimators, the quadratic risk (1) equals $R(1) = yd$. Let $yd > 0$. Minimum of $R(\alpha)$ is reached for $\alpha^{(0)} = \theta^2/(\theta^2 + y)$ and equals $\alpha^{(0)}R(1)$.

Let us find estimators of θ^2 and $\mathbf{E}\widehat{\theta}^2$ and prove that they have small variance.

First, let d be known. As an estimator of $\alpha^{(0)}$, we choose the statistics $\widehat{\alpha}^{(1)} = 1 - d/\widehat{\theta}^2$. Denote by $\mathfrak{R}^{(1)}$ a class of estimators θ of the form $\widehat{\alpha}^{(1)}\widehat{\theta}$.

Denote $d = n^{-1}\operatorname{tr} D$, $s = \theta^2 + yd$, $\delta = \widehat{\theta} - \theta$, $\varepsilon = y^2\gamma + 1/N$.

We have $d^2 \leq \|D\|^2 \leq Q \leq M$, $\mathbf{E}(1/\widehat{\theta}^2) \leq b$, $\mathbf{E}\delta^2 = yd$, $\mathbf{E}(\delta^2)^2 \leq y^2 Q$, $\mathbf{E}\delta\delta^T = D/N$, $\operatorname{var}\delta^2 \leq Q\,y^2\gamma$.

Denote all (possibly different) numeric coefficients by a single letter a. We find that

$$\mathrm{var}(\widehat{\theta}^2) = \mathrm{var}[2(\theta,\delta) + \delta^2] \leq a \ (\theta^2\|D\|/N + Q \ y^2\gamma) \leq aM\varepsilon^2. \quad (7)$$

LEMMA 2.2. *Under Assumptions 1 and 2, we have*

$$\mathbf{E}\,\frac{1}{\widehat{\theta}^2} = \frac{1}{s} + \omega_1, \quad \mathbf{E}\,\frac{(\theta,\widehat{\theta})}{\widehat{\theta}^2} = \frac{\theta^2}{s} + \omega_2, \quad (8)$$

where $|\omega_1| \leq abs^{-1}\sqrt{M}\varepsilon$, $|\omega_2| \leq ab\sqrt{M}\varepsilon$.

Proof. We find that

$$|\omega_1| \leq s^{-1}|\mathbf{E}(s - \widehat{\theta}^2)/\widehat{\theta}^2| \leq s^{-1}\sqrt{\mathbf{E}(\widehat{\theta}^2)^{-2}}\,\sqrt{\mathrm{var}\widehat{\theta}^2} \leq ab/s\sqrt{M}\varepsilon.$$

For the scalar product $(\theta,\widehat{\theta})$, we have $\mathrm{var}(\theta,\widehat{\theta}) \leq \theta^T D\theta/N$ and

$$|\omega_2| \leq s^{-1}|\mathbf{E}(s(\theta,\widehat{\theta}) - \theta^2\widehat{\theta}^2)/\widehat{\theta}^2| \leq$$
$$\leq a\sqrt{\mathbf{E}(\widehat{\theta}^2)^{-2}}\,\sqrt{\mathrm{var}(\theta,\widehat{\theta}) + \mathrm{var}(\widehat{\theta}^2)} \leq ab\sqrt{M}\varepsilon.$$

The lemma is proved. $\square$

THEOREM 2.5. *Under Assumptions 1 and 2 for known $d > 0$ and $y > 0$, we have*

$$R(\widehat{\alpha}^{(1)}) = \frac{\theta^2}{\theta^2 + yd}\,yd + \omega, \quad (9)$$

where $|\omega| \leq abyd\sqrt{M(\gamma y^2 + 1/N)}$, *and a is a number.*

Proof. The quadratic risk

$$R(\widehat{\alpha}^{(1)}) = \mathbf{E}(\theta - \widehat{\theta} + d\widehat{\theta}/\widehat{\theta}^2)^2 = yd\mathbf{E}(-1 + 2(\theta,\widehat{\theta})/\widehat{\theta}^2 + yd/\widehat{\theta}^2)^2.$$

By Lemma 2.2, the second term in parentheses equals $2\theta^2/s$ with the accuracy up to $ab\sqrt{M}\varepsilon$; the third term is equal to yd/s with the same accuracy. It follows that $R(\widehat{\alpha}^{(1)})$ equals $yd\,\theta^2/s$ with the accuracy up to $abyd\sqrt{M}\varepsilon$. The theorem is proved. $\square$

We conclude that under Assumptions 1 and 2, the estimators $\widehat{\theta}^{(1)}$ are dominating the class of estimators $\mathfrak{K}^{(0)}$ with the accuracy up to $abyd\sqrt{M(\gamma y^2 + 1/N)}$.

Now, suppose that the variance d of the original estimators is unknown. Instead of estimating d, we estimate θ^2. For the parameter θ^2, we propose the unbiased estimator of the form $\widehat{\theta}_1^T\widehat{\theta}_2$, where $\widehat{\theta}_1$ and $\widehat{\theta}_2$ are original estimators of θ over two half-samples $\mathfrak{X}_1$ and $\mathfrak{X}_2$ of equal sizes $N/2$ (for even N).

Consider the statistics $\widehat{\alpha}^{(2)} = (\widehat{\theta}_1, \widehat{\theta}_2)/\widehat{\theta}^2$. Denote by $\mathfrak{K}^{(2)}$, the class of estimators for θ of the form $\widehat{\alpha}^{(2)}\widehat{\theta}$. To estimate the quadratic risk $R(\widehat{\alpha}^{(2)})$ of the estimator $\widehat{\alpha}^{(2)}$, we prove the following two lemmas.

Assume that the estimators $\widehat{\theta}$ are calculated over half-samples. Then, the new values of L, M, $\|D\|$, d, and γ differ from the original ones by no more than a numeric factor.

LEMMA 2.3. *Under Assumptions 1 and 2, we have*

1. $\mathrm{var}(\widehat{\theta}_1, \widehat{\theta}_2) \leq as\|D\|/N$
2. $\mathbf{E}(\widehat{\theta}_1^2)^2 \leq aM(1 + y^2)$
3. $\left|\mathbf{E}[(\widehat{\theta}_1, \widehat{\theta}_2)^2] - (\theta^2)^2\right| \leq a(1 + y)M/\sqrt{N}$
4. $\mathrm{var}[(\widehat{\theta}_1, \widehat{\theta}_2)^2] \leq aM^2(1 + y^2)\varepsilon^2$,

where a are numeric coefficients.

Proof. Denote $\delta_1 = \widehat{\theta}_1 - \theta$, $\quad \delta_2 = \widehat{\theta}_2 - \theta$. We have

$$\mathrm{var}(\widehat{\theta}_1, \widehat{\theta}_2) \leq a\,\mathrm{var}[(\theta, \delta_1) + (\theta, \delta_2) + (\delta_1, \delta_2)]$$
$$\leq a(\theta^T D\,\theta/N + \mathrm{tr}\,D^2/N^2) \leq as\|D\|/N.$$

The quantity $\mathbf{E}(\widehat{\theta}_1^2)^2$ is not greater than

$$a\mathbf{E}[\theta^2 + (\theta, \delta_1) + (\delta_1)^2]^2 \leq$$
$$\leq a[(\theta^2)^2 + \theta^T D\theta + y^2 d^2 + Qy^2] \leq aM(1 + y^2).$$

The second statement is proved.

Now, let us substitute $(\widehat{\theta}_1, \widehat{\theta}_2) = \theta^2 + \Delta$ to the left-hand side of the third inequality. We obtain

$$|\mathbf{E}(\widehat{\theta}_1, \widehat{\theta}_2)^2 - \theta^2| = \mathbf{E}|(2\theta^2 + \Delta)\Delta| \leq \sqrt{\mathbf{E}(2\theta^2 + \Delta)^2}\sqrt{\mathbf{E}\Delta^2}.$$

Here

$$\mathbf{E}\Delta^2 = \mathbf{E}[(\theta, \delta_1) + (\theta, \delta_2) + (\delta_1, \delta_2)]^2 \leq$$

$$\leq a(\theta^T D\theta/N + \mathrm{tr}D^2/N^2) \leq as \,\|D\|/N.$$

We obtain the third statement.

Let us prove the fourth statement. Denote $\varphi = (\widehat{\theta}_1, \widehat{\theta}_2)^2$. Using the independence of $\widehat{\theta}_1$ and $\widehat{\theta}_2$, we obtain

$$\mathrm{var}\, \varphi \leq 2[\mathbf{E}\, \mathrm{var}(\varphi|\theta_1) + \mathrm{var}\, \mathbf{E}(\varphi|\theta_1)]. \tag{10}$$

For $\theta_1 = \mathbf{b} = \mathrm{const}$, we have

$$\mathrm{var}[(\mathbf{b}, \theta_2)^2] = \mathrm{var}[2\,(\mathbf{b}, \theta)(\mathbf{b}, \delta_2) + (\mathbf{b}, \delta_2)^2] \leq$$

$$\leq a[\mathbf{b}^2\theta^2(\mathbf{b}^T D\mathbf{b})/N + Qy^2(\mathbf{b}^2)^2\gamma] \leq a(\mathbf{b}^2)^2 M\varepsilon^2.$$

By Statement 2, the first addend of the left-hand side of (10) is not greater than $aM^2(1+y^2)\varepsilon^2$. In the second addend of (10), we have $\mathbf{E}(\varphi|\theta_1) = (\theta, \widehat{\theta}_1)^2 + (\widehat{\theta}_1^T D\widehat{\theta}_1)/N$. Here the variance of the first term of the right-hand side is not greater than

$$\mathrm{var}[2\theta^2(\theta, \delta_1) + (\theta, \delta_1)^2] \leq a(\theta^2)(\theta^T D\theta/N + My^2\gamma) \leq a(\theta^2)^2 M\varepsilon^2.$$

The variance of the second term is (let $\Omega = D/\|D\|$)

$$N^{-2}\mathrm{var}(\widehat{\theta}_1^T D\widehat{\theta}_1)^2 \leq N^{-2}\|D\|\mathrm{var}(2\theta^T\Omega\,\delta_1 + \delta_1^T\Omega\,\delta_1) \leq$$

$$\leq N^{-2}\|D\|^2(\theta^T\Omega D\Omega\theta/N + My^2\gamma) \leq M\|D\|^2 N^{-2}\varepsilon^2.$$

It follows that the variance $\mathbf{E}(\varphi|\widehat{\theta}_1)$ is not greater than $M^2\varepsilon^2$. Consequently, the left-hand side of (10) is not greater than $aM^2(1+y^2)\,\varepsilon^2$. Thus, we obtained the fourth lemma statement. $\square$

LEMMA 2.4. *Under Assumptions 1 and 2,*

$$1. \ \mathbf{E} \, \frac{(\widehat{\theta}_1, \widehat{\theta}_2)}{\widehat{\theta^2}} = \frac{\theta^2}{\theta^2 + yd} + \omega_1, \tag{11}$$

$$2. \ \mathbf{E} \, \frac{(\widehat{\theta}_1, \widehat{\theta}_2)}{\widehat{\theta^2}} \, (\theta, \widehat{\theta}) = \frac{(\theta^2)^2}{\theta^2 + yd} + \omega_2, \tag{12}$$

$$3. \ \mathbf{E} \, \frac{(\widehat{\theta}_1, \widehat{\theta}_2)^2}{\widehat{\theta^2}} = \frac{(\theta^2)^2}{\theta^2 + yd} + \omega_3, \tag{13}$$

where $|\omega_1| \leq ab\sqrt{M}\varepsilon, \ |\omega_2| \leq ab(1 + \sqrt{y})\sqrt{M}\varepsilon, \ |\omega_3| \leq ab$
$(1 + y)\sqrt{M}\varepsilon.$

Proof. By virtue of Lemma 2.2, we have

$$|\omega_1| \leq s^{-1} \left| \mathbf{E}[s(\widehat{\theta}_1, \widehat{\theta}_2) - \theta^2 \widehat{\theta}^2]/\widehat{\theta}^2] \right|$$
$$\leq a \sqrt{\mathbf{E}(\widehat{\theta^2})^{-2}} \, \sqrt{\mathrm{var}(\widehat{\theta}_1, \widehat{\theta}_2) + \mathrm{var} \, \widehat{\theta^2}}.$$

Using Statements 1 and 2 of Lemma 2.3, we obtain the upper estimate of ω_1 in Lemma 2.4. We substitute $(\widehat{\theta}_1, \widehat{\theta}_2) = \theta^2 + \Delta$ to the left-hand side of (12). The principal term equals $(\theta^2)^2/s + \theta^2 \omega_1$, and the term with Δ is not greater than $\sqrt{\mathbf{E}\theta^2/\widehat{\theta}_2^2} \, \sqrt{\mathbf{E}\Delta^2}$. Here the first multiple does not exceed $\theta^2 b$, and by Lemma 2.3, and the second multiple is not greater than $a \, \sqrt{as \, \|D\| \, M/N}$. Consequently, $|\omega_2| \leq a(1 + \sqrt{y})bM/\varepsilon$.

Let us prove Statement 3 of Lemma 2.4. Denote $\varphi = (\widehat{\theta}_1, \widehat{\theta}_2)^2$, and let $\varphi = \mathbf{E}\varphi + \Delta$. We have

$$\mathbf{E} \, \frac{\varphi}{\widehat{\theta^2}} = \mathbf{E} \, \varphi \mathbf{E} \, \frac{1}{\widehat{\theta^2}} + \mathbf{E} \, \frac{\Delta}{\widehat{\theta^2}},$$

where by Statement 3 of Lemma 2.3 in view of Lemma 2.2, we obtain $\mathbf{E}\varphi = (\theta^2)^2 + r$, where $|r| \leq a(1 + y)M/\sqrt{N}$, and $\mathbf{E}(1/\widehat{\theta^2}) = 1/s + \omega_1$, where $|\omega_1| \leq ab/s \, \sqrt{M}\varepsilon$. We find that

$$|\omega_3| \le \mathbf{E}(1/\widehat{\theta}^2) + (\theta^2)^2\omega_1 \le ab(1+y)M/\sqrt{N} + ab\theta^2\sqrt{M}\varepsilon$$
$$\le ab(1+y)M\varepsilon.$$

The proof of Lemma 2.4 is complete. $\square$

THEOREM 2.6. *Under Assumptions 1 and 2 for yd > 0, the quadratic risk is*

$$R(\widehat{\alpha}^{(2)}) = \frac{\theta^2}{\theta^2 + yd}\, yd + \omega, \tag{14}$$

where $|\omega| \le abM(1 + y)\sqrt{\gamma y^2 + 1/N}$, *and a is a numeric coefficient.*

The statement of Theorem 2.6 immediately follows from (12) and (13).

One can draw the conclusion that under conditions 1 and 2, the estimators $\widehat{\theta}^{(2)} = \alpha^{(2)}\widehat{\theta}$ are dominating the class of estimators $\mathfrak{K}^{(0)}$ with the accuracy up to $ab\,(1+y)M\sqrt{\gamma y^2 + 1/N}$.

Example. Let $\widehat{\theta} \sim \mathbf{N}(\theta, I/N)$. We have $Q = 3$, $D = I$, $d = 1$, $\gamma = 2/(3n)$ independently on N. For $n \ge 6$, we have

$$b^2 = \frac{N^2}{4}\int_0^1 t(1-t)^{n/2-3}\exp(-\theta^2 Nt/2)\ dt \le \frac{1}{(\theta^2 + y - 6/N)^2}. \tag{15}$$

In this case, Theorem 2.6 provides the principal part of the quadratic risk of estimators from $\mathfrak{K}^{(1)}$ and $\mathfrak{K}^{(2)}$ being equal to $y\widehat{\theta}^2/(\widehat{\theta}^2 + y)$; this quantity fits the boundaries of the quadratic risk established in Theorem 2.4 with the accuracy up to $4/N$.

Limit Quadratic Risk of Shrinkage Estimators

Now let us formulate our assertions in the form of limit theorems. We consider a sequence of problems of estimating vector θ

$$\mathfrak{P}_N = (n,\ \theta,\ \widehat{\theta},\ D,\ M,\ b,\ \gamma, \widehat{\alpha}^{(1)},\ \widehat{\alpha}^{(2)},\ R)_N,\quad N = 1, 2, \ldots, \tag{16}$$

in which (we do not write indexes N for arguments of (16)) the populations are determined by n-dimensional parameters θ. Unbiased estimators $\widehat{\theta}$ having the properties 1 and 2 are calculated over samples of size N. These estimators are characterized by the parameters M, b, D, and γ. The quadratic risk R of the form (1) is calculated for two shrinkage estimators: $\widehat{\alpha}^{(1)}\widehat{\theta}$ and $\widehat{\alpha}^{(2)}\widehat{\theta}$.

THEOREM 2.7. *Assume that for each N in (16), the conditions 1 and 2 hold and, moreover, $b \le c_1$, $M \le c_2$, where the constants c_1 and c_2 do not depend on N, and suppose that the limits exist*

$$\lim_{N\to\infty} n/N = y_0 > 0; \quad \lim_{N\to\infty} d = d_0 > 0; \quad \lim_{N\to\infty} \theta^2 = \theta_0{}^2; \quad \lim_{N\to\infty} \gamma = 0.$$

Then,

$$\lim_{N\to\infty} R(\alpha^{(1)}) = \lim_{N\to\infty} R(\alpha^{(2)}) = \frac{\theta_0^2}{\theta_0^2 + y_0 d_0} \, y_0 d_0. \qquad (17)$$

We conclude that the estimators $\widehat{\theta}^{(1)} = \widehat{\alpha}^{(1)}\widehat{\theta}$ and $\widehat{\theta}^{(2)} = \widehat{\alpha}^{(2)}\widehat{\theta}$ are asymptotically dominating the class of estimators $\mathfrak{K}^{(0)}$ in $\mathfrak{P}$ as $N \to \infty$.

Note that each partition of the set of components θ in (16) to nonintersecting subsets allows to introduce a set of shrinkage coefficients $\widehat{\alpha}^{(1)}$ and $\widehat{\alpha}^{(2)}$, which can be chosen optimally by Theorems 2.5 and 2.6.

THEOREM 2.8. *Let the sequence of problems (16) be partitioned to k subsequences*

$$\mathfrak{P}_j = (n_i, \; \theta_j, \; \widehat{\theta}_j, \; D_j, \; M_j, \; b_j, \; \gamma_j, \; \widehat{\alpha}_j^{(1)}, \; \widehat{\alpha}_j^{(2)}, \; R_j)_N,$$

$j = 1,\ldots,k$, such that for each N, we have $n = n^1 + \ldots + n^k$, $\theta = (\theta^1,\ldots,\theta^k)$, $\widehat{\theta} = (\widehat{\theta}_1,\ldots,\widehat{\theta}_k)$. If each of the subsequences $\mathfrak{P}_j$ satisfies conditions of Theorem 2.7, then

$$R(\widehat{\alpha}^{(\nu)}) \ge \sum_{j=1}^{k} R_j(\widehat{\alpha}_j^{(\nu)}), \quad \nu = 1, 2.$$

Proof. To prove this theorem, it is sufficient to estimate the sum of limit risks from above. Let a_j denote the limit value of θ_j^2, and $y_j > 0$ be a limit ratio n_j/N, $j = 1, \ldots, k$. Then, $a = a_1 + \ldots + a_k = \theta_0^2$. Denote $\rho_j = y_j/y$, $j = 1, \ldots, k$. We have

$$\sum_{j=1}^{k} \frac{a_j y_j}{a_j + y_j} = y - y \sum_{j=1}^{k} \frac{\rho_j}{1 + r_j}, \tag{18}$$

where $r_j = a_j/y_j$. Using the concavity of the function $f(r) = 1/(1 + r)$, we find that the right-hand side of (17) is not greater than $y - y/(1 + \bar{r})$, where the mean $\bar{r} = a/y$. This proves the theorem. $\square$

We can draw a general conclusion that in the sequence (16) for each (sufficiently large) partition of the set of parameters, the optimal shrinkage of subvectors can decrease the limit quadratic risk. This allows to suggest a procedure of the sequential improvement of estimators by multiple partition to subsets and using different shrinkage coefficients.

The problem of the purposefulness of partitions when n and N are fixed requires further investigations.

Conclusions

It is well known that superefficient estimators decrease the quadratic risk nonuniformly with respect to parameters, and this decrease can be appreciable only in a bounded region (at "superefficiency points"). In multidimensional problems, the domain of parameters is often bounded a priori, whereas their number may be considered as infinite. For example, in the discriminant analysis, the distance between centers of populations is bounded by the origin of the discrimination problem.

For a fixed number of parameters and standard risk functions, the consistency of estimators guarantees the zero-limit risk. As compared with the James–Stein estimator, most of shrinkage estimators provide only an infinitesimal gain as $N \to \infty$. Theorems 2.6 and 2.7 provide the substantial decrease of the quadratic risk if the number of parameters is large and comparable with N.

Equations (6) and (7) determine values of N and γ that guarantee a finite-risk decrease or the risk minimization with a given accuracy. Theorem 2.7 states the property of the substantial asymptotic dominance for the estimators $\widehat{\alpha}^{(1)}\widehat{\theta}$ and $\widehat{\alpha}^{(2)}\widehat{\theta}$ when n is comparable with N. Theorem 2.8 guarantees the dominating property of estimators constructed by partitions.

2.3. SHRINKAGE OF INFINITE-DIMENSIONAL VECTORS

In using statistical models of great complexity, a class of problems arise in which the number of unknown parameters is large and essentially exceeds sample sizes. The actuality of these problems was discussed by Kolmogorov in his program note [41].

In [69] and [71], a number of specific effects are described characteristic of high-dimensional statistics, and these effects were used for the improvement of multivariate statistical procedures. One of these effects is the possibility to decrease the quadratic risk by shrinking estimators. In a series of papers (see review [33]), this effect was investigated thoroughly and applied for the improvement of some practical solutions. In most of these investigations, only normal distribution is considered, with the identical variance known a priori.

As an extension of these investigations, we assume here that the observation vectors are infinite dimensional and have different variances of components.

Let $\vec{\mu} = (\mu_1, \mu_2, \ldots)$ be vector of unknown parameters that is estimated over a sample $\mathfrak{X} = \{\mathbf{x}_m, \ m = 1, \ldots, N\}$ of infinite-dimensional observations $\mathbf{x} = (x_1, x_2, \ldots)$ with expectation $\vec{\mu} = \mathbf{E}\mathbf{x}$. Suppose there exist all four moments of all components of the vector $\mathbf{x}$. Denote $d_i = \mathrm{var}(x_i)$, $i = 1, 2, \ldots$. We rank these quantities so that $d_1 \geq d_2 \geq d_3 \geq \ldots$. Suppose the vector $\vec{\mu}$ has a finite length and the convergence holds $d_1 + d_2 + \ldots = d$. Let $d > 0$. Let $\bar{\mathbf{x}}$ denote a sample average vector calculated over $\mathfrak{X}$.

Let us study the quadratic risk of infinite-dimensional shrinkage estimator

$$\widehat{\mu} = \widehat{\alpha}^0 \bar{\mathbf{x}}, \quad \text{with} \quad \widehat{\alpha}^0 = 1 - \frac{\widehat{d}}{N\bar{\mathbf{x}}^2}, \tag{1}$$

$$\text{where} \quad \widehat{d} = (N-1)^{-1} \sum_{m=1}^{N} (\mathbf{x}_m - \bar{\mathbf{x}})^2 \tag{2}$$

is sample variance (here and in the following, squares of vectors mean squares of their length). The estimator (1) is similar to the James–Stein estimator and differs by the "plug-in" variance $\widehat{d}$. Obviously, $\mathbf{E}\bar{\mathbf{x}}^2 = \vec{\mu}^2 + d/N$. For the sake of convenience, denote

$\beta_i = d_i/N$, $i = 1, 2, \ldots$, $\beta = d/N$, and introduce the quantities $\rho_i = d_i/d$ that characterize the relative contribution of components of $\mathbf{x}$ to the sum of variances, and let $\rho_1 \geq \rho_2 \geq \rho_3 \geq \ldots$.

Normal Distributions

Independent components

Suppose that random vectors $\mathbf{x}$ have independent components $x_i \sim \mathbf{N}(\bar{\mu}_i, d_i)$, $i = 1, 2, \ldots$. Denote $\overset{\circ}{\mathbf{x}} = \mathbf{x} - \bar{\mu}$, $\overset{\circ}{\bar{\mathbf{x}}} = \bar{\mathbf{x}} - \bar{\mu}$. Obviously,

$$\mathrm{var}(\overset{\circ}{\mathbf{x}}{}^2) = 2\sum_i d_i^2, \quad \mathrm{var}(\overset{\circ}{\bar{\mathbf{x}}}{}^2) = 2\sum_i \beta_i^2$$

(here and in the following, the indexes i in sums are assumed to run from 1 to infinity). In the case when $d_1 = 1$, and $d_i = 0$ for $i > 1$, the multiparametric problem degenerates. If the quantity $\rho = \rho_1$ is small, the variance of $\mathbf{x}^2$ is defined by contributions of many variables, and this produces effects we study in this paper.

First, consider a class of a priori estimators of $\bar{\mu}$ of the form $\hat{\mu} = \alpha\bar{\mathbf{x}}$, where α is an a priori chosen nonrandom scalar. The quadratic risk of the estimator $\hat{\mu}$ is

$$R(\alpha) = \mathbf{E}(\bar{\mu} - \alpha\bar{\mathbf{x}})^2 = (1 - 2\alpha)\bar{\mu}^2 + \alpha^2(\bar{\mu}^2 + \beta).$$

Obviously, $R(1) = \beta$. The minimum of $R(\alpha)$ is achieved for $\alpha = \alpha_0 = \bar{\mu}^2/(\bar{\mu}^2 + \beta)$ and is equal to

$$R^0 \overset{def}{=} R(\alpha_0) = \frac{\bar{\mu}^2}{\bar{\mu}^2 + \beta}\,\beta. \tag{3}$$

The problem is reduced to estimation of the parameter α_0.

Special case 1

Consider a special case where the variance $d_i = 0$ for all $i > n$ so that the problem becomes finite dimensional. Let all $d_i = 1$ for $i \leq n$. Then, $\beta = n/N$, and to estimate α_0 for $n > 2$, we may suggest the James–Stein estimator $\widehat{\mu}^{\mathrm{JS}} = (1 - (n-2)/N\bar{\mathbf{x}}^2)\bar{\mathbf{x}}$. Its quadratic risk is as follows (see Introduction)

$$R^{\mathrm{JS}} = y - (n-2)^2/N^2\mathbf{E}|\bar{\mathbf{x}}^2|^{-1} \le R^0 + 4(n-1)/nN.$$

If $\vec{\mu} = 0$, we have $R^{\mathrm{JS}} = N/(n-2)$.

Now we prove that the estimator (1) of α_0 produces the quadratic risk approaching R^0 in the infinite-dimensional case. Let us write the quadratic risk of $\widehat{\mu} = \widehat{\alpha}^0\bar{\mathbf{x}}$ in the form

$$R(\widehat{\alpha}^0) = \mathbf{E}(\vec{\mu} - \widehat{\alpha}^0\bar{\mathbf{x}})^2 = -\beta + 2\mathbf{E}\,\frac{\widehat{\beta}}{\bar{\mathbf{x}}^2}\,\vec{\mu}^T\bar{\mathbf{x}} + \mathbf{E}\,\frac{\widehat{\beta}^2}{\bar{\mathbf{x}}^2},$$

where $\widehat{\beta} = \widehat{d}$. Denote $\rho = \rho_1$ and

$$\sigma^2 = \sum_{i=1}^{\infty} \rho_i^2 \le 1. \tag{4}$$

Lemma 2.5. *For $\beta > 0$, $n \ge 3$, and $\rho < 1/3$, we have $\sigma^2 \le \rho$ and*

$$\frac{1}{\vec{\mu}^2 + \beta} \le \mathbf{E}\,\frac{1}{\bar{\mathbf{x}}^2} \le \frac{1}{\beta} + \frac{1}{\beta}\,\frac{1}{\sqrt{1/3 - \sigma^2}}. \tag{5}$$

Proof. The first lemma statement is an obvious property of inverse moments. Now we integrate over the distribution $\bar{x}_i \sim \mathbf{N}(\mu_i, \beta_i)$, $i = 1, 2, \ldots$, and obtain that

$$\mathbf{E}\,\frac{1}{\bar{\mathbf{x}}^2} = \int_0^{\infty} \mathbf{E}\exp(-t\bar{\mathbf{x}}^2)\,dt =$$

$$= \int_0^{\infty} \prod_{i=1}^{\infty} \frac{1}{\sqrt{1 + 2\beta_i t}} \cdot \exp\left(-\frac{t\mu_i^2}{1 + 2\beta_i t}\right)\,dt. \tag{6}$$

If $\rho_2 = 0$, then $\rho_1 = 1$ in controversy to the lemma conditions. If $\rho_3 = 0$, then $\rho_1 + \rho_2 = 1$ with $\rho_1 < 1/3$, which is impossible. Therefore, under our lemma conditions, $\rho_1 \ge \rho_2 \ge \rho_3 > 0$. Let us majorize (6), leaving in the common square root only the sum of products of three multiples. We find that

$$\mathbf{E}\frac{1}{\bar{\mathbf{x}}^2} \le \int_0^\infty [1 + 8\,t^3 \sum_{i<j<k} \beta_i\beta_j\beta_k]^{-1/2}dt,$$

where (and below) the indexes i, j, k run over all natural numbers under the restrictions $1 \le i < j < k$. Let us isolate the integration region $[0, \beta^{-1})$. We obtain

$$\mathbf{E}\frac{1}{\bar{\mathbf{x}}^2} \le \beta^{-1} + \int_{1/\beta}^\infty (8t^3\beta^3 \sum_{i<j<k} \rho_i\rho_j\rho_k)^{-1/2}dt \le$$

$$\le \beta^{-1}\left[1 + (2\sum_{i<j<k} \rho_i\rho_j\rho_3)^{-1/2}\right]. \tag{7}$$

Note that

$$6\sum_{i<j<k} \rho_i\rho_j\rho_k = \sum_{i,j,k=1}^\infty \rho_i\rho_j\rho_k - 3\sum_{i=1}^\infty \rho_i^2 + 2\sum_{i=1}^\infty \rho_i^3 \ge 1 - 3\sigma^2.$$

This proves the lemma. □

For special case 1, the inequality (5) can be sharpened.

LEMMA 2.6. *For $\beta > 0$ and $n \ge 3$,*

$$\mathbf{E}\frac{\vec{\mu}^T\bar{\mathbf{x}}}{\bar{\mathbf{x}}^2} = 1 - \beta\mathbf{E}\frac{1}{\bar{\mathbf{x}}^2} + 2\sum_i \beta_i\mathbf{E}\frac{\bar{x}_i^2}{(\bar{\mathbf{x}}^2)^2}. \tag{8}$$

Proof. Denote by

$$f_i = \frac{1}{(2\pi\beta_i)^{1/2}}\exp[-\frac{(\bar{x}_i - \mu_i)^2}{2\beta_i}], \quad f_i' = \frac{\partial \ln f_i}{\partial \bar{x}_i}$$

the probability density for $\bar{x}_i$ and its derivative, $i = 1, 2, \ldots$. We find that

$$\mathbf{E}\frac{\vec{\mu}^T\bar{\mathbf{x}}}{\bar{\mathbf{x}}^2} = \mathbf{E}\frac{1}{\bar{\mathbf{x}}^2}\sum_i \bar{x}_i(\bar{x}_i + \beta_i\frac{f_i'}{f_i}) = 1 + \sum_i \beta_i\int \frac{\bar{x}_i}{\bar{\mathbf{x}}^2}f_i'd\bar{x}_i.$$

Integrating by parts, we obtain (for $n \geq 3$) that this expression equals to

$$1 - \sum_i \beta_i \mathbf{E}\left(\frac{1}{\bar{\mathbf{x}}^2} - 2\, \frac{\bar{x}_i^2}{(\bar{\mathbf{x}}^2)^2}\right).$$

This coincides with the right-hand side of (8). The lemma is proved. $\square$

Denote $\widehat{\beta} = \widehat{d}/N$, where $\widehat{d}$ is defined by (2).

LEMMA 2.7. *If* $N > 1$ *we have*

$$\operatorname{var} \widehat{\beta} = \frac{2}{N-1}\, \beta^2 \sigma^2.$$

Proof. For normal variables $\mathbf{x}_m$, one can use the linear transformation (the Helmert transformation) such that

$$\sum_{m=1}^{N} (\mathbf{x}_m - \bar{\mathbf{x}})^2 = \sum_{m=1}^{N-1} \widetilde{\mathbf{x}}_m^2,$$

where $\widetilde{\mathbf{x}}_m$ are independent normal vectors distributed identically as $\mathbf{x}_m$. From (2), we obtain

$$\operatorname{var} \widehat{\beta} = \frac{1}{N^2(N-1)^2} \sum_{m=1}^{N-1} \operatorname{var}(\widetilde{\mathbf{x}}^2) = \frac{2}{N-1} \sum_i \beta_i^2.$$

This is the lemma statement. $\square$

Normal distribution: general case

Let $\mathbf{x} = (x_1, x_2, \ldots) \sim \mathbf{N}(\vec{\mu}, \Sigma)$, where the infinite-dimensional matrix Σ can be diagonalized by orthogonal transformations. One can diagonalize the matrix Σ and denote the eigenvalues by d_i, $i = 1, 2, \ldots$. Suppose that their sum $d = \operatorname{tr}\Sigma$ is finite. Then, Lemmas 2.6 and 2.7 remain valid with $\beta = d/N$.

THEOREM 2.9. *Let an infinite-dimensional vector* $\mathbf{x} \sim \mathbf{N}(\vec{\mu}, \Sigma)$ *and* $0 < \operatorname{tr}\Sigma < \infty$. *If* $d_1 < d/3$, $n > 2$ *and* $N > 1$.

Then, $\sigma^2 < 1/3$, and for estimator (1) we have

$$\mathbf{E}\,(\vec{\mu} - \widehat{\mu})^2 = R(\widehat{\alpha}^0) \leq R^0 + 6\beta\sigma[1 + (1/3 - \sigma^2)^{-1/2}]. \quad (9)$$

Proof. Let us start from (4). For normal distributions, random values $\bar{\mathbf{x}}$ and $\widehat{\beta}$ are independent. We substitute $\mathbf{E}\widehat{\beta} = \beta$ and $\mathbf{E}(\vec{\mu}^T\mathbf{x})/\bar{\mathbf{x}}^2$ from (8). It follows

$$R(\widehat{\alpha}^0) = \beta - \beta^2\mathbf{E}\frac{1}{\bar{\mathbf{x}}^2} + 4\beta\sum_i \beta_i\mathbf{E}\frac{\bar{x}_i^2}{(\bar{\mathbf{x}}^2)^2} + \mathbf{E}\frac{1}{\bar{\mathbf{x}}^2}\,\text{var}\,\widehat{\beta}. \quad (10)$$

Here $\mathbf{E}|\bar{\mathbf{x}}|^{-2} \geq (\vec{\mu}^2 + \beta)$, and the two first terms of the right-hand side are not greater than $\beta^2\vec{\mu}^2/(\vec{\mu}^2 + \beta) = R^0$. Let us estimate the third term. It does not exceed

$$4\beta\mathbf{E}|\bar{\mathbf{x}}|^{-3}\sum_i \beta_i|x_i| \leq 4\beta\mathbf{E}|\bar{\mathbf{x}}|^{-3}\sqrt{\bar{\mathbf{x}}^2\sum_i \beta_i^2} = 4\beta^2\mathbf{E}|\bar{\mathbf{x}}|^{-2}\sigma. \quad (11)$$

In the last term of (10) for $N \geq 2$, we have var $\widehat{\beta} \leq 2\beta^2\sigma^2$; it follows that the sum of last two terms in (10) is not greater than $6\beta^2\sigma\mathbf{E}|\bar{\mathbf{x}}|^{-2}$. We substitute the upper estimate of $\mathbf{E}|\bar{\mathbf{x}}|^{-2}$ from Lemma 2.5 and come to the statement of our theorem. $\square$

Thus, the infinite-dimensional shrinkage estimator $\widehat{\mu} = (1 - \widehat{\beta}/\bar{\mathbf{x}}^2)\,\bar{\mathbf{x}}$ provides the quadratic risk the same as the best a priori estimator $\widehat{\mu} = \alpha_0\bar{\mathbf{x}}$ with the accuracy up to $O(\sigma)$. Small $\sigma^2 \leq \rho = \min_i d_i/d$ produce an essentially multiparametric effect: the quadratic risk of estimator (1) approaches the quadratic risk of unknown unimprovable a priori estimator.

Wide Class of Distributions

Let us prove now that the same estimator (1) provides the quadratic losses $(\vec{\mu} - \widehat{\mu})^2$ approaching R^0 for distributions different from normal.

We establish this fact for populations in which all variables have the fourth central moment in the scheme of series of infinite-dimensional observations $\mathbf{x}$ as sample size $N \to \infty$.

Consider the sequence

$$\mathfrak{P} = \{(\mathfrak{S}, \vec{\mu}, \Sigma, \mathfrak{X}, \widehat{\mu}, L)_N\}, \quad N = 1, 2, 3, \ldots, \tag{12}$$

of estimation problems for populations $\mathfrak{S}$ with expectation vectors $\vec{\mu} = \mathbf{E}\mathbf{x} = (\mu_1, \mu_2, \ldots)$ and infinite-dimensional covariance matrices $\Sigma = \mathrm{cov}(\mathbf{x}, \mathbf{x})$, in which an estimator $\widehat{\mu}$ is calculated over samples $\mathfrak{X} = \{\mathbf{x}_m\}$ of size $N \to \infty$ with quadratic loss function $L = (\vec{\mu} - \widehat{\mu})^2$ (we do not write out indexes for arguments of (12)). Denote $\overset{\circ}{\mathbf{x}} = \mathbf{x} - \vec{\mu} = (\overset{\circ}{x}_1, \overset{\circ}{x}_2, \ldots)$. Let d_i be eigenvalues of Σ, $i = 1, 2, \ldots$, ordered so that $d_1 \geq d_2 \geq d_3 \ldots$.

Suppose (12) satisfies the following requirements.

A. For each N, the series $\mu_1^2 + \mu_2^2 + \ldots = \vec{\mu}^2$ converges and as $N \to \infty$, we have $\lim \vec{\mu}^2 = \vec{\mu}_0^2$.

B. For each N, the series $d_1 + d_2 + \ldots = d = \mathrm{tr}\Sigma$ converges and $d > 0$.

C. For each N in the system of coordinates where Σ is diagonal, we have

$$\sup_{i=1,2,\ldots} \frac{\mathbf{E}\overset{\circ}{x}_i^4}{(\mathbf{E}\overset{\circ}{x}_i^2)^2} \leq \kappa, \quad \kappa \geq 1,$$

where the ratio is assumed to be 0 if the nominator is 0, and where κ does not depend on N.

Under condition C, we have $d^2 \leq \mathbf{E}(\overset{\circ}{\mathbf{x}}{}^2)^2 \leq \kappa d^2$.

Denote $\rho_i = d_i/d$, $i = 1, 2, \ldots$. Introduce the parameters

$$\sigma^2 = \frac{\mathrm{tr}\Sigma^2}{(\mathrm{tr}\Sigma)^2} = \sum_i \rho_i^2 \leq 1, \quad \gamma = \frac{\mathrm{var}(\overset{\circ}{\mathbf{x}}{}^2)}{\mathbf{E}(\overset{\circ}{\mathbf{x}}{}^2)^2} \leq 1 - \frac{1}{\kappa}.$$

The parameter σ^2 defines the multiparametric character of the problem: for special case 1, we have $\sigma^2 = 1/n$, where n is a number parameters of the statistical model.

The parameter γ restricts the dependence of variables. If components of $\mathbf{x}$ are independent, then

$$\gamma \leq \frac{(\kappa - 1)\,\sigma^2}{1 + (\kappa - 1)\,\sigma^2}.$$

If all components of $\mathbf{x}$ are proportional and $\overset{\circ}{x}_i = \sqrt{\rho_i/\rho_1}\ \overset{\circ}{x}_1$, $i = 1, 2, \ldots$, then the quantity $\gamma = 1 - (\mathbf{E}\overset{\circ}{x}_1^2)^2/\mathbf{E}|\overset{\circ}{x}_1|^4$ can approach 1.

    D. Let $d/N \to \beta > 0$ as $N \to \infty$.

    E. Suppose as $N \to \infty$, we have $\sigma = \sigma(N) \to 0$ and $\gamma = \gamma(N) \to 0$.

Asymptotic conditions A–E are the extension of the Kolmogorov "increasing dimension asymptotics" used first in [17] and [45].

Consider special case 1. Then, the Kolmogorov condition $n/N \to y > 0$ passes to condition D with $\beta = y$; condition C is provided by the requirements

$$0 < M = \sup_{|\mathbf{e}|=1} \mathbf{E}(\mathbf{e}^T\overset{\circ}{\mathbf{x}})^4 < c_1, \text{ and the spectral norm } \|\Sigma^{-1}\| \le c_2,$$

where $\mathbf{e}$ are nonrandom unity vectors, the quantities $\sigma^2 = 1/n \to 0$, and our quantity γ does not exceed the parameter

$$\sup_{\|\Omega\|=1} \text{var}(\overset{\circ}{\mathbf{x}}^T \Omega\ \overset{\circ}{\mathbf{x}}/n)/M \to 0,$$

where Ω are nonrandom, symmetrical, positively semidefinite matrices.

Denote (as in the above) the mean sample vector by $\bar{\mathbf{x}}$, $\overset{\circ}{\bar{\mathbf{x}}} = \bar{\mathbf{x}} - \vec{\mu} = (\overset{\circ}{\bar{x}}_1, \overset{\circ}{\bar{x}}_2, \ldots)$, $d = \text{tr}\,\Sigma$.

Remark 1. Under Assumptions A–E, we have

$$\mathbf{E}(\bar{\mathbf{x}}^2)^2 \le (\vec{\mu}^2)^2 + 2\vec{\mu}^2\beta + \kappa\beta^2 + O(N^{-1}) \text{ and } \vec{\mu}^T\Sigma^{-1}\vec{\mu} \to 0. \quad (13)$$

Indeed, we have

$$\mathbf{E}(\bar{\mathbf{x}}^2)^2 = \mathbf{E}(\vec{\mu}^2 + 2\mu^T\overset{\circ}{\bar{\mathbf{x}}} + \overset{\circ}{\bar{\mathbf{x}}}^2)^2 \le$$
$$\le \mathbf{E}\left(|\vec{\mu}|^4 + 2\vec{\mu}^2(\overset{\circ}{\bar{\mathbf{x}}})^2 + 8(\vec{\mu}^T\overset{\circ}{\bar{\mathbf{x}}})^2 + 2(\overset{\circ}{\bar{\mathbf{x}}})^2)^2\right).$$

Here the square $\vec{\mu}^2$ is bounded, $\mathbf{E}|\overset{\circ}{\mathbf{x}}|^2 \to \beta$, $\mathbf{E}(\vec{\mu}^T\overset{\circ}{\mathbf{x}})^2 = \vec{\mu}^T\Sigma\vec{\mu}/$ $N \to 0$, and thus $\mathbf{E}(\overset{\circ}{\mathbf{x}}^2)^2$ equals

$$\mathbf{E}\sum_{i,j}\|\overset{\circ}{\bar{x}}_i|^2\,|\overset{\circ}{\bar{x}}_j|^2 \le \sum_{i,j}N^{-2}\sqrt{\mathbf{E}|\overset{\circ}{\mathbf{x}}_i|^4\mathbf{E}|\overset{\circ}{\mathbf{x}}_j|^4} \le \kappa N^{-2}\sum_{i,j}d_id_j = \kappa\beta^2.$$

The Remark is grounded.

LEMMA 2.8. *Under Assumptions A–E as $N \to \infty$, we have*

$$\operatorname{var}\bar{\mathbf{x}}^2 \le d^2 N^{-2}(\kappa/N + \sigma^2) \to 0.$$

Proof. It is obvious that $\operatorname{var}\bar{\mathbf{x}}^2 \le 2\operatorname{var}(\vec{\mu}^T\overset{\circ}{\mathbf{x}}) + 2\operatorname{var}\overset{\circ}{\mathbf{x}}^2$. Here the first summand is no greater than $2\vec{\mu}^T\Sigma^{-1}\vec{\mu} \to 0$. The variance

$$\operatorname{var}\overset{\circ}{\mathbf{x}}^2 = N^{-4}\sum_{m=1}^{N}[\mathbf{E}(\overset{\circ}{\mathbf{x}}_m^2)^2 - (\mathbf{E}\overset{\circ}{\mathbf{x}}_m^2)^2] + N^{-2}(1 - N^{-1})\operatorname{tr}\Sigma, \quad (14)$$

where $\overset{\circ}{\mathbf{x}}_m = \mathbf{x}_m - \vec{\mu}$. Let $\vec{\mu} = 0$ and $\overset{\circ}{\mathbf{x}}_m = \mathbf{x} \overset{def}{=} (x_1, x_2, \ldots)$. By condition D, the first summand in (14) is not larger than

$$N^{-3}\mathbf{E}(\mathbf{x}^2)^2 \le N^{-3}\mathbf{E}\sum_{i,j}x_i^2x_j^2 \le$$

$$\le N^{-3}\sum_{i,j}\sqrt{\mathbf{E}x_i^4}\sqrt{\mathbf{E}x_j^4} \le N^{-3}\kappa d^2 \to 0.$$

The second summand of the right-hand side of (14) also vanishes. The proof is complete. $\square$

Consider the unbiased estimator

$$\widehat{\beta} = \frac{1}{N(N-1)}\sum_{m-1}^{N}(x_m - \bar{\mathbf{x}})^2$$

of the parameter $\beta = d/N$.

LEMMA 2.9. *Under Assumptions A–E as $N \to \infty$, we have* $\operatorname{var}\widehat{\beta} \to 0$.

Proof. It suffices to show that

$$N^{-2}(N-1)^{-2} \operatorname{var} \sum_{m=1}^{N} \overset{\circ}{\mathbf{x}}_m^2 \to 0.$$

Since the vectors $\overset{\circ}{\mathbf{x}}_m$ are independent, the sum here is a sum of variances of $\overset{\circ}{\mathbf{x}}_m^2$. But $\operatorname{var} \overset{\circ}{\mathbf{x}}_m^2 \leq \mathbf{E}\overset{\circ}{\mathbf{x}}_m^2 \leq \kappa d^2$. It follows that for $N > 1$, $\operatorname{var} \widehat{\beta} \leq \kappa d^2/N(N-1)^2 \to 0$. The lemma is proved. $\square$

THEOREM 2.10. *Under Assumptions A–E, the estimator $\widehat{\mu}$ of the form* (1) *has the quadratic losses such that*

$$\operatorname*{plim}_{N \to \infty} (\vec{\mu} - \widehat{\mu})^2 = R^0 \overset{def}{=} \frac{\vec{\mu}_0^2 \beta}{\mu_0^2 + \beta}.$$

Proof. In view of Lemmas 2.8 and 2.9, we have $\widehat{\beta} \to \beta$ and $\bar{\mathbf{x}}^2 \to \mu_0^2 + \beta$ in probability (i.p.). Let us write the quadratic losses in the form

$$L = (\vec{\mu} - \widehat{\mu})^2 = (\vec{\mu} - \bar{\mathbf{x}})^2 - 2\widehat{\beta} + 2(\vec{\mu}^T \bar{\mathbf{x}}) \frac{\widehat{\beta}}{\bar{\mathbf{x}}^2} + \frac{\widehat{\beta}^2}{\bar{\mathbf{x}}^2}. \qquad (15)$$

Here $\mathbf{E}(\vec{\mu} - \bar{\mathbf{x}})^2 = \operatorname{tr} \Sigma/N \to \beta$ and $(\vec{\mu} - \bar{\mathbf{x}})^2 \to \beta$ in probability since $\operatorname{var} \bar{\mathbf{x}}^2 \to 0$ by Lemma 2.8. The quantity $\widehat{\beta} \to \beta$ in probability by Lemma 2.9. In the third term of the right-hand side of (15), $\vec{\mu}^T \mathbf{x} \to \mu_0^2$ in probability since $\vec{\mu}^T \Sigma \vec{\mu}/N \to 0$. Further, $|\bar{\mathbf{x}}|^{-2} \to (\mu_0^2 + \beta)^{-1}$ and $\widehat{\beta}^2 \to (\mathbf{E}\widehat{\beta})^2 = \beta^2$ in probability. We obtain the theorem statement. $\square$

To prove the convergence $(\vec{\mu} - \widehat{\mu})^2 \to R^0$ in the square mean, it suffices to bound eight's moments and require the uniform boundedness of moments $\mathbf{E}|\bar{\mathbf{x}}|^{-4}$.

Conclusions

For a wide class of distributions restricted by bounded dependence condition $\gamma \to 0$, Theorem 2.10 establishes the quadratic loss decrease for shrinkage estimators (1) in problems with variables of arbitrary dimension (greater than 3) and different variances under correlations, decreasing on the average. Indeed, let r_{ij} be the correlation coefficient for variables x_i and x_j in the observer coordinate system. Then, the second of conditions E can be written as

$$\sigma^2 = \frac{\operatorname{tr}\Sigma^2}{(\operatorname{tr}\Sigma)^2} = \sum_{i,j} \rho_i \rho_j (r_{ij})^2 \to 0.$$

In case of finite dimension and identical variance of variables, we obtain the same effect as for the James–Stein estimator.

2.4. UNIMPROVABLE COMPONENT-WISE ESTIMATION

In this section, we extend the notion of shrinkage estimators to shrinking of separate components. This technique was investigated first in [64].

Denote by $\mathfrak{X}$ a sample from $\mathbf{N}(\vec{\mu}, I)$ of size N, and let $\bar{\mathbf{x}} = (\bar{x}_1, \bar{x}_2, \ldots, \bar{x}_n)$ be sample average vector.

Definition 1. The estimator

$$\widehat{\mu} = \{\varphi(\bar{x}_1), \varphi(\bar{x}_2), \ldots, \varphi(\bar{x}_n)\} \tag{1}$$

is called a *component-wise* estimator of the vector $\vec{\mu} = (\mu_1, \mu_2, \ldots, \mu_n)$, and the function $\varphi(\cdot)$ identical for all components function is called the *estimating function*.

Let us search for an estimator dominating the class $\mathfrak{K}$ with the accuracy up to remainder terms small for large n and N.

We restrict ourselves to normal n-dimensional distributions $\mathbf{N}(\vec{\mu}, I)$ with unit covariance matrix I. Denote by $y = n/N$ the quadratic risk of the standard estimator $\widehat{\mu} = \bar{\mathbf{x}}$.

We apply specifically multiparametric technique of studying relations between functions of unknown parameters and functions of observable variables. Define the density function

$$f(t) = n^{-1} \sum_{i=1}^{n} f_i(t) = \langle f_i(t) \rangle, \quad \text{where}$$

$$f_i(t) = \sqrt{\frac{N}{2\pi}} \exp\left(-\frac{N(t - \mu_i)^2}{2}\right), \quad i = 1, 2, \ldots n. \tag{2}$$

Here (and in the following) the subscript i in sums runs over $i = 1, 2, \ldots, n$, and let angular brackets denote averaging over i.

Let $\mathfrak{K}$ denote a class estimators of the form (1) with the differentiable estimating functions $\varphi(t)$ of the scalar argument.

THEOREM 2.11. *For the estimators $\widehat{\mu}$ from $\mathfrak{K}$, we have*

$$R = R(\varphi) = n^{-1} \mathbf{E} \sum_i (\mu_i - \widehat{\mu}_i)^2 = R_0 + \int (\varphi(t) - \varphi_0(t))^2, \tag{3}$$

where $\quad \varphi_0(t) = t + \dfrac{1}{N}\dfrac{f'(t)}{f(t)}, \quad R_0 = \dfrac{n}{N} - \dfrac{n}{N^2}\int \dfrac{[f'(t)]^2}{f(t)}\,dt,$ (4)

(the prime indicates the derivative in t).

Proof. We find that $R(\varphi)$ equals

$$n^{-1}\sum_i \mathbf{E}\big[(\mu_i - \bar{x}_i)^2 - 2(\mu_i - \varphi_0(\bar{x}_i))(\bar{x}_i - \varphi_0(\bar{x}_i)) + (\bar{x}_i - \varphi_0(\bar{x}_i))\big]^2$$

$$= \frac{n}{N} - \frac{2}{nN}\sum_i \int (\mu_i - t)\frac{f'(t)}{f(t)} f_i(t)\,dt + \frac{1}{N^2}\sum_i \int \frac{[f'(t)]^2}{f(t)}\,dt.$$

In the second term, we substitute $(\mu_i - t)f_i(t) = N^{-1}f'_i(t)$ and sum over i. The required expression for R_0 follows. The theorem is proved. $\square$

The estimator $\widehat{\mu} = \{\varphi_0(\bar{x}_i),\ i = 1, 2, \ldots, n\}$ may be called the best a priori component-wise estimator. Let us study its effect on the quadratic risk decrease.

From Theorem 2.11, it follows first that $R_0 \leq y = n/N$ and that

$$\frac{n}{N}\int \frac{[f'(t)]^2}{f(t)}\,dt \leq 1. \tag{5}$$

If the length of the vector $\vec{\mu}$ is known a priori, then the shrinkage estimator $\widehat{\mu} = \alpha\bar{\mathbf{x}}$ may be used with the shrinkage coefficient $\alpha = \vec{\mu}^2/(\vec{\mu}^2 + y)$ (here and in the following, the square of a vector denotes the square of its length). Obviously, it decreases the quadratic risk and leads to

$$R(\alpha) \overset{def}{=} \mathbf{E}(\vec{\mu} - \alpha\bar{\mathbf{x}})^2 = \frac{\vec{\mu}^2}{\vec{\mu}^2 + y}\,y.$$

Remark 1. The inequality holds $R_0 \leq R(\alpha)$.

Indeed, let us examine it. By the Cauchy–Buniakovsky inequality, we have

$$1 = \Big[\int tf'(t)\,dt\Big]^2 \leq \int t^2 f(t)\,dt \int \frac{[f'(t)]^2}{f(t)}\,dt = \frac{(\vec{\mu}^2 + y)(y - R_0)}{y^2}.$$

It follows immediately that $R_0 \leq R_1$.

In a special case when all components of $\vec{\mu}$ are identical, we have $\mu_1 = \mu_2 = \ldots \mu_n$ (and for $n = 1$) the function $\varphi_0(t) = \mu_1$ for all t, we have $f'(t)/f(t) = N(\mu_1 - t)$, and the quadratic risk of the best component-wise estimator is $R_0 = 0$ in contrast to the best shrinkage estimator that leads (for $\vec{\mu} \neq 0$) to the quadratic risk $R(\alpha) > 0$. In case of a large scattering of the quantities μ_i, the a priori best estimating function $\varphi_0(t)$ leads to $R_0 \approx 0$ again. Let us establish this fact.

THEOREM 2.12. *Let the set of components of the vector $\vec{\mu}$ be divided into two subsets A and B so that the distance between these subsets is not less than $\Delta > 0$. Denote*

$$R = \mathbf{E}(\vec{\mu} - \hat{\mu}^0)^2 = \mathbf{E} \sum_{i}^{n} (\mu_i - \varphi_0(\bar{x}_i))^2,$$

$$R_A = \mathbf{E} \sum_{i=1}^{k} (\mu_i - \varphi_0(\bar{x}_i))^2, \quad R_B = \mathbf{E} \sum_{i=k+1}^{n} (\mu_i - \varphi_0(\bar{x}_i))^2,$$

$0 < k < n$. *Then,*

$$0 \leq R - R_A - R_b \leq \Delta^2/4 \exp(-N\Delta^2/8) \, n/N. \tag{6}$$

Proof. We retain notation (2) for the functions $f_i = f_i(t)$, but numerate components from A by subscripts $i = 1, 2, \ldots k$, and components from B by subscripts $j = k+1, k+2, \ldots, n$. Denote $f_A = k/n \, \langle f_i \rangle$, $f_B = (n-k)/n \, \langle f_j \rangle$, where angular brackets denote averaging over i and j, and denote $f = f(t) = f_A + f_B$. We find that

$$R - R_A - R_B = \frac{n}{N^2} \int \frac{(f'_A f_B - f_A f'_B)^2}{f f_A f_B} \, dt.$$

Let $\mu_j > \mu_i$. Using the Cauchy–Buniakovsky inequality in view of $\mu_j - \mu_i \geq \Delta$, we find that

$$\frac{(f'_A f_B - f_A f'_B)^2}{f_A f_B} = \frac{\langle (\mu_j - \mu_i) \, f_i f_j \rangle^2}{y \langle f_i \rangle \, \langle f_j \rangle} \frac{k(n-k)}{n^2} \leq$$

$$\leq \langle (\mu_j - \mu_i)^2 f_i f_j \rangle \frac{k(n-k)}{yn^2} \leq f_A f_B \, \Delta^2/y,$$

where the angular brackets denote averaging over i and j. It is obvious that $\sqrt{f_A f_B} \leq f/2$. We obtain the inequality

$$R_0 - R_A - R_B \leq \frac{\Delta^2}{2N} \int \sqrt{f_A f_B}\, dt.$$

Since $k(n-k) \leq n^2/4$, the product $f_A f_B \leq \langle f_i f_j \rangle /4$.

We note that $f_i f_j \leq N/2\pi \exp[-N\Delta^2/4 - N(t-\mu_{ij})^2]$, where $\mu_{ij} = (\mu_i + \mu_j)/2$. Also it is obvious that

$$\langle \exp(-N(t-\mu_{ij})^2) \rangle \leq n^{-2} \Big[\sum_{i,j} \exp(N(t-\mu_{ij})^2/2) \Big]^2.$$

Integrating n^2 summands, we obtain

$$\int \sqrt{f_A f_B}\, dt \leq \frac{n}{2}\, \exp(-N\Delta^2/8).$$

It follows that (6) is valid. The theorem is proved. $\square$

Estimator for the Density of Parameters

We estimate the empiric density of unknown parameters. In order to explicate better the essentially multiparametric effects, we replace the problem of estimating vectors $\vec{\mu}$ of finite length by the problem of estimating the vector $\mathbf{v}$ of parameters $v_i = \sqrt{n}\mu_i$, $i = 1, 2, \ldots, n$, having the square average $\vec{\mu}^2$. We define the quantity $y = n/N$ and the function

$$f(t, y) = \langle f_i(t, y) \rangle, \quad \text{where} \quad f_i(t, y) = (2\pi y)^{-1/2} \exp(-(t-v_i)^2/2y),$$
$$(7)$$

$i = 1, 2, \ldots, n$, characterizing the set $\{v_i\}$.

Let a sample $\tilde{\mathfrak{X}} = \{\tilde{\mathbf{x}}_m\}$ of n-dimensional observations $\tilde{\mathbf{x}}_m = \sqrt{n}\bar{\mathbf{x}}_m$, $m = 1, 2, \ldots, N$, be given and let $\bar{\mathbf{u}} = (u_1, u_2, \ldots, u_n) = \sqrt{n}\,\bar{\mathbf{x}}$ be a vector of sample averages for $\tilde{\mathfrak{X}}$. We estimate the vectors $\mathbf{v} = \{v_i\}$ over $\tilde{\mathfrak{X}}$. Consider the class of component-wise estimators of $\mathbf{v}$ of the form $\hat{\mathbf{v}} = \{\varphi(u_i)\}$, where $\varphi(t)$ is the differentiable estimating function.

By Theorem 2.12, the best a priori component-wise estimator of $\mathbf{v}$ has the form $\widehat{\mathbf{v}}_0 \overset{def}{=} \{\varphi_0(u_i, y)\}$, where the estimating function

$$\varphi_0(t) = \varphi_0(t, y) = t + y\, \frac{f'(t, y)}{f(t, y)}. \tag{8}$$

The following statement immediately follows from Theorem 2.11.

Remark 2. Let $\widehat{\mathbf{v}} = \{\varphi(u_1), \varphi(u_2), \ldots, \varphi(u_n)\}$, where $\varphi(\cdot)$ is the differentiable function. Then,

$$R = R(\varphi) = \mathbf{E}(\mathbf{v} - \widehat{\mathbf{v}})^2 = \mathbf{E}\, \frac{1}{n} \sum_i (v_i - \varphi(u_i))^2 =$$

$$= R_0 + \int (\varphi(t) - \varphi_0(t, y))^2\, f(t, y)\, dt,$$

where

$$R_0 = \mathbf{E}(\mathbf{v} - \widehat{\mathbf{v}}_0)^2 = y - y^2 \int \frac{[f'(t, y)]^2}{f(t, y)}\, dt.$$

Note that for any $p > 0$, we have

$$p \int \frac{[f'(t, p)]^2}{f(t, p)} \leq 1. \tag{9}$$

Remark 3. Consider the Bayes distribution $\mathbf{N}(0, \beta)$ of independent quantities $\mu_1, \mu_2, \ldots, \mu_n$ identical for all μ_i. Then, the Bayes expectation $\mathbf{E}_B$ of the quadratic risk is

$$\mathbf{E}_B R_0 \geq y\beta/(y + \beta).$$

Let us prove this inequality. We have

$$\mathbf{E}_B R_0 = y - y^2 \int \frac{[f'(t, d)]^2}{f(t, d)}\, dt.$$

Here the function in the integrand is not greater

$$\frac{\langle f_i'(t, d)\rangle^2}{f_i(t, d)} = \frac{\langle (v_i - t) f_i(t, d)\rangle^2}{f_i(t, d)} \leq \langle (v_i - t)^2 f_i(t, d)\rangle$$

and the integral of the right-hand side is not greater d. It follows that $\mathbf{E}_B R_0 \geq y - y^2/d = y\beta/(y + \beta)$. Remark 3 is justified. $\square$

For $\beta = 0$, the quantity $R_0 = 0$, while for the great scattering of the component magnitudes and large β, the Bayes quadratic risk does not differ much from the quadratic risk y of the standard estimator.

Let $\varepsilon > 0$. Consider the statistics

$$\widehat{f}(t) = \widehat{f}(t, \varepsilon) = n^{-1} \sum_i \widehat{f}_i(t, \varepsilon),$$

$$\text{where} \quad \widehat{f}_i(t, \varepsilon) = \frac{1}{\sqrt{2\pi\varepsilon}} \exp[-(t - u_i)^2/2\varepsilon], \qquad (10)$$

$i = 1, 2, \ldots, n$. The function $\widehat{f}(t, \varepsilon)$ presents an ε-regularized density of empirical distribution of $u_1, u_2, \ldots, u_n$. Note that $\widehat{f}(t, \varepsilon)$ presents an unbiased estimator of the density $f(t, d)$, where $d = y + \varepsilon$ approximating $f(t, y)$ for small ε. We use this function for the approximation of the a priori best estimating function, substituting $d = y + \varepsilon$ instead of y. Let us obtain an upper estimate of the risk increase produced by this replacement.

LEMMA 2.10. *The quadratic risk of the a priori estimator* $\widehat{\mathbf{v}}_1 = \{\varphi_1(u_i)\}$, $i = 1, 2, \ldots, n$, *of the vector* $\mathbf{v}$ *equals*

$$\varphi_1(t) = t + y\, \frac{f'(t, d)}{f(t, d)}, \quad d = y + \varepsilon,$$

$$\text{equals} \quad R_1 = R_1(\varphi_1) = n^{-1} \sum_i \mathbf{E}(v_i - \varphi_1(u_i, d))^2 \leq R_0 + a\varepsilon^2/y,$$

where

$$\varphi_1(t) = t + \frac{f'(t, d)}{f(t, d)}, \quad d = y + \varepsilon, \quad \varepsilon > 0$$

and a is a numeric coefficient.

Proof. Denote for brevity $x_i = |t - v_i|/\sqrt{p}$,

$$f_i = f_i(t, p) = (2\pi p)^{-1/2p} \exp(-x_i^2/2),$$
$$f_{it} = \frac{\partial}{\partial p} f_1, \quad f_{ip} = \frac{\partial}{\partial p} f_i, \quad f_{itp} = \frac{\partial^2}{\partial p \, \partial t} f_i,$$

$p > 0, \; i = 1, 2, \ldots, n.$

Let us express the difference $R_1 - R_0$ in terms of the sum of squares of derivatives of the function $f(t, p)$ at the points p intermediate between y and $y + \varepsilon$. In view of Remark 3, we have

$$R_1 - R_0 = \int \mathbf{E}[\varphi_0(u_i, y) - \varphi_0(u_i, y + \varepsilon)]^2 \, f(t, y) \, dt =$$
$$= n^{-1}\varepsilon^2 \mathbf{E} \sum_i \left| \frac{\partial \varphi_0(u_i, p)}{\partial p} \right|^2 =$$
$$= n^{-1}\varepsilon^2 \, \mathbf{E} \sum_i \left(\frac{|f_{it}|}{f_i} + \frac{|f_{itp}|}{f_i} p_i + \frac{|f_{it} f_{ip}|}{f_i^2} p_i \right)^2,$$

where $p = p_i \in [y, y + \varepsilon]$ for all i. The calculation of expectation with respect to u_i reduces to the integration over $f_i(t, y)dt$. Leaving only squares of summands, we write

$$R - R_0 \le 3n^{-1}\varepsilon^2 \sum_i \mathbf{E} \int \left[\frac{f_{it}^2}{f_i^2} + p^2 \frac{f_{itd}^2}{f_i^2} + p^2 \frac{f_{it}^2 f_{ip}^2}{f_i^4} \right] f_i(t, y) \, dt,$$

where $p = p(i)$. Note that $f_i(t, y) \le p^{1/2} y^{-1/2} f_i(t, p)$, $p > 0$. We calculate the derivatives

$$f_{it} = \frac{x_i f_i}{\sqrt{p}}, \quad f_{ip} = \frac{(x_i^2 - 1) f_i}{2p}, \quad f_{itp} = \frac{x_i(x_i^2 - 3) f_i}{2p^{3/2}}.$$

The replacement $p = p_i$ by y only strengths the inequality. We calculate the integrals and obtain that $R - R_0 \le a\varepsilon^2 y$, where a is a number. Lemma 2.10 is proved. $\square$

Further, note that the statistics $\widehat{f}(t, \varepsilon)$ presents a sum of n independent addends. Let us estimate its variance.

LEMMA 2.11. *The variances*

$$\operatorname{var} \widehat{f}(t, \varepsilon) \le \frac{1}{2n\sqrt{\pi\varepsilon}} f(t, d), \quad \operatorname{var} \widehat{f}'(t, \varepsilon) \le \frac{1}{\sqrt{2\pi} \, n\varepsilon^{3/2}} f(t, d),$$

where $d = y + \varepsilon/2$.

Proof. Denoting averages over i by angular brackets and using the independence of addends, we obtain that

$$\operatorname{var} \widehat{f}(t) = n^{-1} \langle \operatorname{var} \widehat{f}_i(t) \rangle \leq n^{-1} \sum_i \mathbf{E} \widehat{f}_i^2(t)$$

$$= \frac{1}{2\pi\varepsilon} \frac{1}{2\pi y} \int \left\langle \exp[-\frac{(r - v_i)^2}{2y} - \frac{(t - r)^2}{\varepsilon}] \right\rangle dr = \frac{1}{2\sqrt{\pi\varepsilon}} f(t, d).$$

The first statement is proved. Next, we have

$$\operatorname{var} \widehat{f}'(t) = n^{-1} \langle \operatorname{var} \widehat{f}_i'(t) \rangle$$

$$\leq n^{-1} \langle \mathbf{E}[\widehat{f}_i'(t)]^2 \rangle = (2\pi\varepsilon n)^{-1} \, \mathbf{E} \langle \exp[-(u_i - t)^2/\varepsilon] \, (u_i - t)^2/\varepsilon^2 \rangle$$

$$\leq (2\pi\varepsilon^2 n)^{-1} \mathbf{E} \langle \exp[-(u_i - t)^2/2\varepsilon] \rangle$$

$$= (2\pi)^{-1/2} \, \varepsilon^{-3/2} n^{-1} \, \mathbf{E} \langle \widehat{f}_i(t) \rangle = (2\pi)^{-1/2} \, \varepsilon^{-3/2} n^{-1} \, f(t, d).$$

We obtain the lemma statement. $\square$

Estimator for the Best Estimating Function

Substituting $\widehat{f}(t)$ instead of $f(t, y)$ to (8) we obtain some estimating function, that is, however, not bounded from above. We consider some δ-regularized statistics approximating $\varphi_0(t)$ of the form

$$\widehat{\varphi}_0(t) = \widehat{\varphi}_0(t, \varepsilon, \delta) = \begin{cases} t + y \dfrac{\widehat{f}'(t)}{\widehat{f}(t)} & \text{for } \widehat{f}(t) > \delta \\ t & \text{for } \widehat{f}(t) \leq \delta, \qquad \delta > 0. \end{cases}$$

$$(11)$$

We choose $\widehat{\mathbf{v}}_0 = \widehat{\mathbf{v}}_0(\varepsilon, \delta) = (\widehat{\varphi}_0(u_1), \widehat{\varphi}_0(u_2), \ldots, \widehat{\varphi}_0(u_n))$ as an estimator of the vector $\mathbf{v} = (v_1, v_2, \ldots, v_n)$.

Remark 4. Let $n \to \infty$, $\varepsilon \to +0$, $\delta \to +0$, $y = n/N \to y_\infty$ and for each t the function $f(t, y) \to f(t, y_\infty) > 0$.

Then, for each t the difference $\widehat{\varphi}_0(t) - \varphi_0(t) \to 0$ in probability. We study now the quadratic risk $R = R(\widehat{\mathbf{v}}_0)$ of the estimator $\widehat{\mathbf{v}}_0$.

LEMMA 2.12. *If $\varepsilon \le y$, then the quadratic risk of the estimator* (11) *equals*

$$R = \mathbf{E}(\mathbf{v} - \widehat{\mathbf{v}}_0)^2 = \mathbf{E}n^{-1} \sum_i (v_i - \widehat{\varphi}_0(u_i))^2 =$$

$$= R_0 + \mathbf{E} \int \left(y \frac{f'(t,y)}{f(t,y)} - \widehat{\varphi}_0(t) \right)^2 f(t,y) \ dt.$$

Proof. Note that, by definition, the functions $\widehat{f}_i(u_i)$ and $\widehat{f}'_i(u_i)$ do not depend on u_i. Substitute $\varphi_0(\cdot)$ from (11). We find that

$$R = \mathbf{E}n^{-1} \sum_i \int (v_i - t - y\widehat{\varphi}_0(t))^2 f_i(t,y) \ dt =$$

$$= \frac{1}{n} \sum_i \int (v_i - t)^2 f_i(t,y) \ dt - \frac{2}{n} \mathbf{E} \sum_i \int (v_i - t)\widehat{\varphi}_0(t) f_i(t,y) \ dt +$$

$$+ y^2 \mathbf{E} \int [\widehat{\varphi}_0(t)]^2 \ f(t,y) dt.$$

Here the first summand equals y. In the second summand, we can substitute $v_i - t = y f'_i(t,y)/f_i(t,y)$, As a result we obtain

$$R = y - 2y\mathbf{E} \int \widehat{\varphi}_0(t) f'(t,y) \ dt + y^2 \mathbf{E} \int [\widehat{\varphi}_0(t)]^2 f(t,y) \ dt =$$

$$= R_0 + \mathbf{E} \int \left(y \frac{f'(t,y)}{f(t,y)} - \widehat{\varphi}_0(t) \right)^2 f(t,y) \ dt. \qquad (12)$$

The lemma is proved. $\square$

LEMMA 2.13. *If $y > 0$ then*

$$\Delta_1 \stackrel{\text{def}}{=} R_1 - R_0 = y^2 \int \left(\frac{f'(t,y)}{f(t,y)} - \frac{f'(t,d)}{f(t,d)} \right)^2 f(t,y) \ dt \le a\varepsilon^2/y. \qquad (13)$$

Proof. Denote $f_i = (2\pi p)^{-1/2} \exp(-(t - v_i)^2/2p)$,

$$f_{it} = \frac{\partial}{\partial t} f_i, \quad f_{ip} = \frac{\partial}{\partial p} f_i, \quad f_{ipt} = \frac{\partial^2}{\partial p \partial t} f_i.$$

Let us express the difference in (13) in terms of a derivative at an intermediate point $y \leq p \leq d = y + \varepsilon$. We obtain

$$\Delta_1 \leq \varepsilon^2 \int n^{-1} \sum_i \left(\frac{|f_{it}|^2}{f_i} + p \frac{|f_{ipt}|}{f_i} + p \frac{|f_{ip}f_{it}|}{f_i^2} \right) f(t,y) \, dt,$$

where the arguments of f_i and its derivatives are $p = p_i$, $y \leq p \leq d$, $i = 1, 2, \ldots, n$. Keeping only squares of addends, we can write

$$\Delta_1 \leq 3\varepsilon^2 \sqrt{\frac{d}{y}} \int n^{-1} \sum_i \left(\frac{f_{it}^2}{f_i^2} + d^2 \frac{f_{ipt}^2}{f_i^2} + d^2 \frac{f_{ip}f_{it}}{f_i^2} \right) f(t,p) \, dt.$$

We pass to the variables $x_i = (v_i - t)/\sqrt{p}$ and obtain that

$$f_{it}^2 \leq x_i^2 \, f_i^2, \qquad f_{ip}^2 \leq x_i^2 \, f_i^2/p \leq (x_i^2 + 1)^2 f_i^2/2p,$$
$$f_{ipt}^2 \leq (x_i^2 + 2|x_i| + 1)^2 \, f_i^2/4p^2, \quad p = p_i, \quad i = 1, 2, \ldots, n.$$

Here $p \geq y$, $d \leq 2y$. Integrating with respect to dx_i, we obtain that $\Delta_1 \leq a\varepsilon^2/y$, where a is a number. Lemma 2.13 is proved. $\square$

Now we pass from functions $f(t,y)$ to $f(t, y + \varepsilon) = \mathbf{E}\widehat{f}(t)$. The second addend of the right-hand side of (12) is not greater $2(\Delta_1 + \Delta_2)$, where Δ_1 is defined in Lemma 2.13 and

$$\Delta_2 = \mathbf{E} \int \left(y \frac{f'(t,d)}{f(t,d)} - \widehat{\varphi}_0(t) \right)^2 f(t,y) \, dt.$$

In the right-hand side, we isolate the contribution of t such that $\widehat{f}(t) < \delta$, for which $\widehat{\varphi}_0(t) = t$. Denote

$$\Delta_{21} = y^2 \mathbf{E} \int \left(\frac{f'(t,d)}{f(t,d)} - \frac{\widehat{f}'(t)}{\widehat{f}(t)} \right)^2 \mathrm{ind}(\widehat{f}(t) > \delta) f(t,y) \, dt.$$

and

$$\Delta_{22} = \int \left(y \frac{f'(t,d)}{f(t,d)} - t \right)^2 f(t,y) \, dt.$$

Then, $\Delta_2 = \Delta_{21} + \Delta_{22}$. Denote $\lambda = \sqrt{y} \, \delta$.

LEMMA 2.14. *For $\varepsilon \leq y, \Delta_{21} \leq 4y^{5/2}\varepsilon^{-3/2}\lambda^{-2}n^{-1}$.*

Proof. It is easy to see that

$$\Delta_{21} \leq y^2\delta^{-2} \int \mathbf{E}(\widehat{f}'(t) - r(t)\widehat{f}(t))^2 f(t, y) \, dt,$$

where $r(t) = f'(t, d)/f(t, d)$, $d = y + \varepsilon$. The quantity under the expectation sign is the variance

$$\operatorname{var}(\widehat{f}'(t) - r(t)\widehat{f}(t)) \leq 2\operatorname{var}\widehat{f}'(t) + 2r^2(t) \operatorname{var}\widehat{f}(t).$$

We use the inequality $f(t, y) \leq (2\pi y)^{-1/2}$ and estimate the variance using Lemma 2.11.

In view of (9) we obtain

$$\Delta_{21} \leq \frac{2y^2}{\sqrt{\pi}\, n\delta^2} \left[\frac{1}{eps^{1/2}} \int f(t, d) \, dt + \frac{1}{\varepsilon^{3/2}} \int \frac{[f'(t, d)]^2}{f(t, d)} \, dt \right] \leq$$
$$\leq \frac{4y}{\sqrt{\pi}\, n\lambda^2} \left(\frac{y}{\varepsilon}\right)^{3/2}.$$

Lemma 2.14 is proved. $\square$

Denote

$$V_1 = \frac{\vec{\mu}^2}{y}, \quad V_m = \frac{\langle v_i^{2m}\rangle^{1/m}}{y} = N\left[\sum_i (\mu_i^2)^m\right]^{1/m},$$

$m = 1, 2, \ldots$ The parameter V_m may be interpreted as a "signal-to-noise" ratio.

LEMMA 2.15. *The function*

$$h(\lambda) \stackrel{def}{=} y^2 \int \frac{[f'(t, d)]^2}{f(t, d)} \operatorname{ind}(f(t, d) \leq 2\delta) \, dt \; < \; \sqrt{6\pi\lambda}\, \sqrt{\vec{\mu}^2/y + 6} \cdot y.$$

For any integer $k \geq 2$, the inequality $h(\lambda) \leq 4\pi k y \, \lambda^{1-2/k}(A_k + V_{k-1})$ holds, where $A_k = 1/2 + [(2k - 3)!!]^{1/(k-1)}$.

Proof. To be concise, denote $f = f(t,d)$, $f_i = f_i(t,d)$ for all i. We apply the Cauchy–Buniakovsky inequality and find that

$$f'^2 = [f'(t,d)]^2 = \langle (v_i - t)f_i/d \rangle^2 \leq \langle f_i \rangle \langle (v_i - t)^2 f_i \rangle /d^2.$$

Consequently,

$$h(\lambda) \leq \frac{y^2}{d^2} \int \langle (v_i - t)^2 f_i \rangle \, \mathrm{ind}(f \leq 2\delta) \, dt. \tag{14}$$

Apply the Cauchy–Buniakovsky inequality once more and obtain the function $\langle f_i \rangle = f \leq \lambda/\sqrt{y}$ under the square root sign. We conclude that

$$h(\lambda) < \sqrt{2\lambda} \, \frac{y}{d^{5/4}} \int \sqrt{\langle (v_i - t)^4 f_i \rangle} \, dt.$$

Let us multiply and divide the integrand by the function $\rho(t) = \sqrt{d}\pi^{-1}(d+t^2)^{-1}$, $\int \rho(t) \, dt = 1$. The average with respect to $\rho(t) \, dt$ is not greater than the root from the square average; therefore the inequality holds

$$h(\lambda) \leq \sqrt{2\pi\lambda} \, \frac{y}{d^{3/2}} \left[\int \langle (v_i - t)^4 f_i \rangle (d + t^2) \, dt \right]^{1/2}. \tag{15}$$

We calculate the integrals in (15) using the substitution $t = v_i + x\sqrt{d}$ for each i and get the integrals with respect to the measure $(2\pi)^{-1/2} \exp(-x^2/2) \, dx$. We find that the right-hand side of (15) does not exceed

$$\sqrt{2\pi\lambda y} \, \sqrt{3\langle v_i^2 \rangle + 18y}.$$

The first lemma statement is proved.

Further, let some integer $k \geq 2$. Note that for $(v_i - t)^2 f_i^{1/k} < c = kd^{1-1/2k}$. From (14) it follows that

$$h(\lambda) \leq c\frac{y^2}{d^2} \int \langle f_i^{1-1/k} \rangle \, \mathrm{ind}(f \leq 2\delta) \, dt \leq c\frac{y^2}{d^2} \int \langle f_i \rangle^{1-1/k} \, dt$$

$$\leq 2c\frac{y^2}{d^2} \, \delta^{1-2/k} \int \langle f_i^{1/k} \rangle \, dt. \tag{16}$$

We multiply and divide the last integrand by the same function $\rho(t)$ and apply the integral Cauchy–Bunyakovskii inequality. It follows that the integral in the right-hand side of (16) is not greater

$$\pi^{1-1/k}d^{(1-1/k)/2}\left\langle\int(1+t^2/d)^{k-1}f_i\ dt\right\rangle^{1/k}.\qquad(17)$$

We calculate the integral in (17) for each i by the substitution $t=v_i+x\sqrt{d}$ and use the inequality $d+t^2\le d+2v_i^2+2dx^2$. When raising to powers $k-1$ in the integrand, the moments of $\mathbf{N}(0,1)$ and moments of empiric distribution of $\{v_i\}$ appear. We estimate them from the above by the corresponding higher moments $\mathbf{E}x^{2(k-1)}=(2k-3)!!$ and $\langle v_i^{2(k-1)}\rangle$ so that a constant remains under the integral with respect to the measure $(2\pi)^{-1/2}\ \exp(-x^2/2)\ dx$. This constant is not larger than

$$\left(1+2\ [(2k-3)!!]^{1/(k-1)}+2V_{k-1}\right)^{k-1}.\qquad(18)$$

Substituting (18) to (17), we obtain the lemma statement. $\square$

LEMMA 2.16.

$$\Delta_{22}\le\frac{1}{2\sqrt{\pi}}\left(\frac{d}{\varepsilon}\right)^{1/4}\frac{d}{\sqrt{n\lambda}}+h(\lambda).\qquad(19)$$

Proof. We find that Δ_{22} is not larger than

$$y^2\mathbf{E}\int\frac{[f'(t,d)]^2}{f(t,d)}\ \mathrm{ind}(\widehat{f}(t)<\delta)\ dt=y^2\int\frac{[f'(t,d)]^2}{f(t,d)}\ \mathbf{P}(\widehat{f}(t)<\delta)\ dt.$$

We divide the integration region into the subregions $\mathfrak{D}_1=\{t:f(t,d)>2\delta\}$ and $\mathfrak{D}_2=\{t:f(t,d)\le2\delta\}$. In the region $\mathfrak{D}_1$, the difference $|f(t,d)-\widehat{f}(t)|$ is not less than $\delta>0$, $f(t,d)=\mathbf{E}\widehat{f}(t)$, and by virtue of the Chebyshev inequality, $\mathbf{P}(\widehat{f}(t)\le\delta)\le\sigma/\delta$, where $\sigma^2=\mathrm{var}\widehat{f}(t)$. We estimate this variance by Lemma 2.11. In the region $\mathfrak{D}_2$, we have $f(t,d)>2\delta$, and by Lemma 2.14, the contribution of $\mathfrak{D}_2$ to (19) is not greater than $h(\lambda)$. We obtain the required statement. $\square$

THEOREM 2.13. *If* $0 \le \varepsilon \le y$, *then the quadratic risk of the estimator* (11) *is*

$$R = R(\widehat{\mathbf{v}}_0) = \mathbf{E}n^{-1} \sum_i (v_i - \widehat{\varphi}_0(u_i, \varepsilon, \lambda))^2$$

$$\le R_0 + ay(1 + |\vec{\mu}|/\sqrt{y})\, n^{-7/23},$$

where a is a numeric coefficient.

Proof. Denote $\theta = \varepsilon/y \le 1$. Summing upper estimates of remainder terms obtained in Lemmas 2.13, 2.14, 2.15, and 2.16, we obtain

$$R \le R_0 + \Delta_1 + \Delta_{21} + \Delta_{22}$$
$$\le ay[\theta^2 + \theta^{-3/2}\lambda^{-2}n^{-1} + \theta^{-1/4}\lambda^{-1}n^{-1/2} + \lambda^{1/2}(1 + |\vec{\mu}|)/\sqrt{y}].$$

Choose $\lambda = \theta^4$, $\theta = n^{-2/23}$ we arrive at the inequality in the theorem formulation. $\square$

For sufficiently large n, restricted ratios $y = n/N$, and for bounded moments $\mu_{(k)}^2$, the quadratic risk $R(\widehat{\mathbf{v}}_0)$ approaches $R_0 < y$.

It is not to be expected that the application of the multiparametric approach to the improvement of estimators without restrictions on parameters (see Remark 4) would succeed since in the case of a considerable scattering of parameters, the multiparametric problem reduces to a number of one-dimensional ones.

In conclusion, we note that the slow decrease in the magnitude of the remainder terms with the increase in n is produced by double regularization of the optimal estimator (11). Such regularization proves to be necessary also in other multiparametric problems, where the improvement effect is reached by an additional mixing of a great number of boundedly dependent variables that produces new specifically multiparametric regularities. The averaging interval must be small enough to provide good approximation to the best-in-the-limit weighting function, and it must be sufficiently large in order to get free from random distortions. Therefore, the improvement is guaranteed only with inaccuracy of the order of magnitude of n^α, where $0 < \alpha < 1$.

CHAPTER 3

SPECTRAL THEORY OF SAMPLE COVARIANCE MATRICES

Spectra of covariance matrices may be successfully used for the improvement of statistical treatment of multivariate data. Until recently, the extreme eigenvalues were of a special interest, and theoretical investigations were concerned mainly with the analysis of statistical estimation of the least and greatest eigenvalues. The development of essentially multivariate methods required more attention to problems of estimating spectral functions of covariance matrices that may be involved in the construction of new more efficient versions of most popular multivariate procedures. The progress of theoretical investigations in studying spectra of random matrices of increasing dimension (see Introduction) suggested new asymptotic technique and produced impressive results in the creation of improved statistical methods. We may say that the main success of multiparametric statistics is based on methods of spectral theory of large sample covariance matrices and their limit spectra. This chapter presents the latest achievements in this field.

The convergence of spectral distribution functions for random matrices of increasing dimension was established first by V. A. Marchenko and L. A. Pastur, and then V. L. Girko, Krishnaiah, Z. D. Bai and Silverstein et al. Different methods were used, but the most fruitful approach is based on study of resolvents as functions of complex parameter and their normed traces. The main idea is as follows. Let A be any real, symmetric, positive definite matrix of size $n \times n$ and $H(z) = (I - zA)^{-1}$ be its resolvent. Denote $h(z) = n^{-1}$ tr $H(z)$. Then, the empirical distribution function of eigenvalues λ_i of A

$$F(u) = \sum_{i=1}^{n} \operatorname{ind}(\lambda_i \leq u)$$

71

may be calculated as follows:

$$F(u) = \frac{1}{\pi} \lim_{\varepsilon \to +0} \operatorname{Im} \int_0^u h(z^{-1}) z^{-1} dv,$$

where $z = u - i\varepsilon$. If $h(z)$ converges as $n \to \infty$, the limit distribution function for eigenvalues of matrices A is obtained. We use the Kolmogorov increasing dimension asymptotics in which the dimension n of observation vectors increases along with sample size N so that $n/N \to y > 0$. This asymptotics was first introduced in another region in 1967 for studying spectra of random matrices by Marchenko and Pastur [43]. In this chapter, we present the results of investigations by the author of this book [63], [65], [67], [69], and [72] in developing spectral theory of large sample covariance matrices.

To prove the convergence of spectral functions, we choose the method of one-by-one exclusion of independent variables.

Let us illustrate the main idea and results of this approach in the following example. Let $\mathfrak{X}$ be a sample of identically distributed random vectors $\mathfrak{X} = \{\mathbf{x}_1, \mathbf{x}_2, \ldots, \mathbf{x}_n\}$ from $\mathbf{N}(0, \Sigma)$. Consider the matrix

$$S = \frac{1}{n} \sum_{m=1}^N \mathbf{x}_m \mathbf{x}_m^T,$$

which presents sample covariance matrix of a special form for the case when expectation of variables is known a priori. Let us single out an independent vector $\mathbf{x}_m$ from $\mathfrak{X}$, $m = 1, 2, \ldots, N$. Define

$$H_0 = H_0(t) = (I + tS)^{-1}, \quad h_0(t) = n^{-1} \operatorname{tr}(I + tS)^{-1},$$

$$S^m = S - N^{-1}\mathbf{x}_m\mathbf{x}_m^T, \quad H_0^m = (I + tS^m)^{-1}, \quad \psi_m = \mathbf{x}_m^T H_0 \mathbf{x}_m / N,$$

$$H_0 = H_0^m - tH_0^m\mathbf{x}_m\mathbf{x}_m^T H_0 / N, \quad H\mathbf{x}_m = (1 - t\psi_m)H_0^m\mathbf{x}_m,$$

$m = 1, 2, \ldots, N$. Obviously, $1 - t\mathbf{E}\psi_m = 1 - t\mathbf{E} \operatorname{tr}(H_0 S)/N = s_0(t) \overset{def}{=} 1 - y + yh_0(t)$ for each m.

For simplicity of notations, let the ratio $n/N = y$ be constant as $n \to \infty$. Assume that $\operatorname{var}(t\psi_m) \to 0$ as $n \to \infty$ (this fact will be proved later in this chapter).

PROPOSITION 1. *Let $\mathfrak{X}$ be a sample of size N from n-dimensional population $\mathbf{N}(0, \Sigma)$ and $n/N \to y > 0$ as $n \to \infty$. Then, for each $t \geq 0$,*

$$h_0(t) = \mathbf{E}n^{-1}\ \mathrm{tr}(I + tS)^{-1} = n^{-1}\ \mathrm{tr}\left(I + ts_0(t)\Sigma\right)^{-1} + \omega_n,$$

where $s_0(t) = 1 - t + yh_0(t)$ and $\omega_n \to 0$.

We present a full proof. Choose a vector $\mathbf{x}_m \in \mathfrak{X}$. For each m, we have

$$tH_0\mathbf{x}_m\mathbf{x}_m^T = t(1 - t\psi_m)\ H_0^m\mathbf{x}_m\mathbf{x}_m^T.$$

Here the expectation of the left-hand side is $t\mathbf{E}H_0S = I - \mathbf{E}H_0$. In the right-hand side, $1 - t\psi_m = s_0(t) - \Delta_m$, where Δ_m is the deviation of $t\psi_m$ from the expectation value, $\mathbf{E}\Delta_m^2 \to 0$. We notice that $\mathbf{E}H_0^m\mathbf{x}_m\mathbf{x}_m = \mathbf{E}H_0^m\Sigma$. It follows that

$$I - \mathbf{E}H_0 = ts_0(t)\mathbf{E}H_0^m\Sigma - t\mathbf{E}H_0^m\mathbf{x}_m\mathbf{x}_m^T\Delta_m.$$

Substitute the expression for H_0^m in terms of H_0. Our equation may be rewritten in the form

$$I = \mathbf{E}H_0\left(I + ts_0(t)\Sigma\right) + \Omega,$$

where $\Omega = t^2 s_0(t)\mathbf{E}H_0^m\mathbf{x}_m\mathbf{x}_m^T H_0\Sigma/N - t\mathbf{E}H_0^m\mathbf{x}_m\mathbf{x}_m^T\Delta_m$. We multiply this from the right-hand side by $R = \left(I + ts_0(t)\Sigma\right)^{-1}$, calculate the trace, and divide by n. It follows that $n^{-1}\ \mathrm{tr}\ R = h_0(t) + \omega_n$, where

$$|\omega_n| \leq t^2 s_0(t)\mathbf{E}\left(\mathbf{x}_m^T H_0\Sigma R H_0^m\mathbf{x}_m\right)/(nN) + t\mathbf{E}\left(\mathbf{x}_m^T\Delta_m R H_0^m\mathbf{x}_m\right)/n.$$

Let us estimate these matrix expressions in norm applying the Schwarz inequality. We conclude that $|\omega_n|$ is not greater than

$$t^2\|\Sigma\|\ \mathbf{E}\mathbf{x}_m^2/(nN) + t\left[\mathbf{E}(\mathbf{x}_m^2/n)^2\mathbf{E}\Delta_m^2\right]^{1/2} \leq \tau^2/N + \tau\sqrt{\mathrm{var}(t\psi_m)},$$

where $\tau = \sqrt{M}t \geq 0$. The proposition is proved. $\square$

The obtained asymptotic expression for $h_0(t)$ is remarkable in that it states the convergence of spectral functions of S and shows that their principal parts are functions of only Σ, that is, of only two moments of variables. Relations between spectra of sample covariance matrices and true covariance matrices will be called *dispersion equations*.

3.1. SPECTRAL FUNCTIONS OF LARGE SAMPLE COVARIANCE MATRICES

Following [67], we begin with studying matrices of the form

$$S = N^{-1} \sum_{m=1}^{N} \mathbf{x}_m \mathbf{x}_m^T,$$

where random $\mathbf{x}_m \in \mathbb{R}^n$ are of interest in a variety of applications different from statistics (see Introduction). Let us call them Gram matrices in contrast to sample covariance matrices $C = S - \bar{\mathbf{x}}\bar{\mathbf{x}}^T$, which depend also on sample averages $\bar{\mathbf{x}}$ and are more complicated.

Gram Matrices

We restrict distributions $\mathfrak{S}$ of $\mathbf{x}$ with the only requirement that all components of $\mathbf{x}$ have fourth moments and $\mathbf{E}\mathbf{x} = 0$. Denote $\Sigma = \text{cov}(\mathbf{x}, \mathbf{x})$. Define the resolvent

$$H_0 = H_0(z) = (I - zS)^{-1}$$

as functions of a complex parameter z (with the purpose to use the analytical properties of $H_0(z)$).

For measuring remainder terms as $n \to \infty$, we define two parameters: the maximum fourth moment of a projection of $\mathbf{x}$ onto nonrandom axes (defined by vectors $\mathbf{e}$ of unit length)

$$M = \sup_{|\mathbf{e}|=1} \mathbf{E}(\mathbf{e}^T \mathbf{x})^4 > 0 \qquad (1)$$

and special measures of the quadratic from variance

$$\nu = \sup_{\|\Omega\|=1} \text{var}(\mathbf{x}^T \Omega \mathbf{x}/n), \quad \text{and} \quad \gamma = \nu/M, \qquad (2)$$

where Ω are nonrandom, symmetric, positive semidefinite matrices of unit spectral norm. For independent components of $\mathbf{x}$, the parameter $\nu \leq M/n$.

Let us solve the problem of isolating principal parts of spectral functions as $n \to \infty$ when $n/N \to y > 0$.

Define the region of complex plane $\mathfrak{G} = \{z : \operatorname{Re} z < 0$ or $\operatorname{Im} z \neq 0\}$ and the function

$$\alpha = \alpha(z) = \begin{cases} 1 & \text{if } \operatorname{Re} z \leq 0, \\ |z|/|\operatorname{Im} z| & \text{if } \operatorname{Re} z > 0 \text{ and } \operatorname{Im} z \neq 0. \end{cases}$$

To estimate expressions involving the resolvent, we will use the following inequalities.

Remark 1. Let A be a real symmetric matrix, $\mathbf{v}$ be a vector with n complex components, and $\mathbf{v}^H$ be the Hermitian conjugate vector (here and in the following, the superscript H denotes the Hermitian conjugation). If $u \geq 0$ and $z \in \mathfrak{G}$, then

$$|1 - zu|^{-1} \leq \alpha, \quad \|(I - zA)^{-1}\| \leq \alpha,$$
$$\|I - (I - zA)^{-1}\| \leq \alpha, \quad |1 - \mathbf{v}^H (I - zA)^{-1} \mathbf{v}|^{-1} \leq \alpha.$$

For $z \in \mathfrak{G}$, we have $\|H_0(z)\| \leq \alpha$.

We will use the method of alternative elimination of independent sample vectors. In this section, we assume that the sample size $N > 1$. Let $\mathbf{e}$ be a nonrandom complex vector of length 1, $\mathbf{e}^H \mathbf{e} = 1$.

Denote

$$h_0(z) = \mathbf{E} n^{-1} \operatorname{tr} H_0(z), \quad y = n/N, \quad s_0(z) = 1 - y + y h_0(z),$$
$$S^m = S - N^{-1} \mathbf{x}_m \mathbf{x}_m^T, \quad H_0^m = (I - zS^m)^{-1},$$
$$\varphi_m = \varphi_m(z) = \mathbf{x}_m^T H_0^m \mathbf{x}_m / N, \quad \psi_m = \psi_m(z) = \mathbf{x}_m^T H_0 \mathbf{x}_m / N,$$
$$v_m = v_m(z) = \mathbf{e}^H H_0^m \mathbf{x}_m, \quad u_m = u_m(z) = \mathbf{e}^H H_0 \mathbf{x}_m, \quad (3)$$

$m = 1, 2, \ldots, N$.

Remark 2. If $z \in \mathfrak{G}$, the following relations are valid

$$H_0 = H_0^m + z H^m \mathbf{x}_m \mathbf{x}_m^T H_0 / N, \quad H_0 \mathbf{x}_m = (1 + z \psi_m) H_0^m \mathbf{x}_m,$$
$$u_m = v_m + z \psi_m v_m = v_m + z \varphi_m u_m,$$
$$\psi_m = \varphi_m + z \varphi_m \psi_m, \quad (1 + z \psi_m)(1 - z \varphi_m) = 1,$$
$$|1 - z \varphi_m|^{-1} \leq \alpha, \quad |u_m| \leq \alpha |v_m|, \quad m = 1, 2, \ldots, N. \quad (4)$$

LEMMA 3.1. *If $z \in \mathfrak{G}$, then*

$$\mathbf{E}v_m = 0, \quad \mathbf{E}v_m^4 \le M\alpha^4, \quad 1 + z\mathbf{E}\psi_m = s_0(z),$$
$$|\mathbf{E}u_m|^2 \le \sqrt{M}|z|^2\alpha^2 \operatorname{var}\psi_m, \quad m = 1, 2, \ldots, N. \tag{5}$$

Proof. The first two inequalities immediately follow from the independence of $\mathbf{x}_m$ and H^m. Next, we have

$$z\mathbf{E}\psi_m = \mathbf{E}z\mathbf{x}_m^T H_0 \mathbf{x}_m/N = \mathbf{E}z \operatorname{tr}(SH_0)/N = \mathbf{E} \operatorname{tr}(H_0 - I)/N =$$
$$= y(h_0(z) - 1) = s_0(z) - 1.$$

We note that $\mathbf{E}u_m = \mathbf{E}v_m\Delta_m$, where $\Delta_m = \psi_m - \mathbf{E}\psi_m$. The last lemma statement follows form the Schwarz inequality. $\square$

Define the variance of a complex variable as

$$\operatorname{var}(z) = \mathbf{E}(z - \mathbf{E}z)(z^* - \mathbf{E}z^*),$$

where (and in the following) the asterisk denotes complex conjugation, Introduce the parameters

$$\tau = \sqrt{M}|z| \quad \text{and} \quad \delta = 2\alpha^2 y^2(\gamma + \tau^2\alpha^4/N).$$

To estimate variances of functionals uniformly depending on a large number of independent variables, we use the technique of expanding in martingale differences. It would be sufficient to cite the Burkholder inequality (see in [91]). However, we present the following statement with the full proof.

LEMMA 3.2. *Given a set $\mathfrak{X} = \{X_1, X_2, \ldots, X_N\}$ of independent variables, consider a function $\varphi(\mathfrak{X})$ such that $\varphi(\mathfrak{X}) = \varphi^m(\mathfrak{X}) + \Delta_m(\mathfrak{X})$, where $\varphi^m(\mathfrak{X})$ does not depend on X_m. If second moments exist for $\varphi(\mathfrak{X})$ and $\Delta_m = \Delta_m(\mathfrak{X})$, $m = 1, 2, \ldots, N$, then*

$$\operatorname{var} \varphi(\mathfrak{X}) \le \sum_{m=1}^{N} \mathbf{E}(\Delta_m - \mathbf{E}_m\Delta_m)^2,$$

where $\mathbf{E}_m$ is the expectation calculated by integration with respect to the distribution of the variable X_m only.

Proof. Denote by F_m the distribution function of independent X_m, $m = 1, 2, \ldots, N$, and let dF^m denote the product $dF_1 dF_2 \ldots dF_m$, $m = 1, 2, \ldots, N$. Consider the martingale differences

$$\beta_1 = \varphi - \int \varphi dF_1; \quad \beta_m = \int \varphi dF^{m-1} - \int \varphi dF^m, \quad m = 2, \ldots, N,$$

where $\varphi = \varphi(\mathfrak{X})$. In view of the independence of $X_1, X_2, \ldots, X_N$, it can be readily seen that $\mathbf{E}\beta_i \beta_j = 0$ if $i \neq j$, $i, j = 1, 2, \ldots, N$. Majoring the square of the first moment by second moment, we obtain

$$\mathbf{E}\beta_m^2 = \mathbf{E}[\int (\varphi - \int \varphi dF_m)\, dF^{m-1}]^2 \leq$$
$$\leq \mathbf{E} \int (\varphi - \int \varphi dF_m)^2 dF^{m-1} = \mathbf{E}(\varphi - \mathbf{E}_m \varphi)^2.$$

We have $\mathbf{E}\beta_m^2 \leq \mathbf{E}\,(\Delta_m - \int \Delta_m dF_m)^2$. The statement of the lemma follows. $\square$

LEMMA 3.3. *If $z \in \mathfrak{G}$, then we have*

$$\mathrm{Var}(\mathbf{e}^H H_0 \mathbf{e}) \leq \tau^2 \alpha^6 / N, \quad \mathrm{Var}\, \varphi_m \leq M\delta/2,$$
$$\text{and} \quad \mathrm{Var}\, \psi_m \leq a M \alpha^4 \delta, \quad m = 1, 2, \ldots, N,$$

where a is a numerical constant.

Proof. From (4), it follows that $\mathbf{e}^H H_0 \mathbf{e} = \mathbf{e}^H H_0^m \mathbf{e} + z v_m u_m / N$, $m = 1, 2, \ldots, N$. Using Remark 2, we find that

$$\mathrm{Var}(\mathbf{e} H_0 \mathbf{e}) \leq |z|^2 \sum_{m=1}^{N} \mathbf{E}|v_m u_m|^2 / N^2 \leq$$
$$\leq |z|^2 \alpha^2 \mathbf{E}|v_m|^4 / N \leq \tau^2 \alpha^6 / N.$$

Now we fix some integer $m = 1, 2, \ldots, N$. Denote $\Omega = \mathbf{E}H_0^m$, $\Delta H_0^m = H_0^m - \Omega$. Since Ω is nonrandom, we have

$$\text{Var } \varphi_m = \mathbf{E}|\mathbf{x}_m^T \Delta H_0^m \mathbf{x}_m|^2/N^2 + \text{Var}(\mathbf{x}_m^T \Omega \mathbf{x}_m)/N^2. \qquad (6)$$

Note that H_0^m is a matrix of the form H_0 with N less by 1 if we replace the argument t by $t' = (1 - N^{-1})\, t$. We apply the first statement of this lemma to estimate the conditional variance $\text{Var}\,(\mathbf{e}_m^H H_0^m \mathbf{e}_m)$ under fixed $\mathbf{x}_m$, where $\mathbf{e}_m$ is a unit vector directed along $\mathbf{x}_m$, and find that the first summand in (6) is not greater than

$$\mathbf{E}|\mathbf{x}_m|^4 \tau^2 \alpha^6/N^3 \le M\tau^2 \alpha^6 y^2/N.$$

To estimate the second summand, we substitute the parameter γ from (2) to (6) and obtain that the second addend in (6) is not greater than $\|\Omega\|^2 M n^2 \gamma/N^2 \le M\alpha^2 y^2 \gamma$. Summing both addends, we obtain the right-hand side of the second inequality in the statement of the lemma.

Further, the equation connecting φ_m and ψ_m in Remark 2 may be rewritten in the form

$$(1 - z\varphi_m)\Delta\psi_m = (1 + z\mathbf{E}\psi_m)\Delta\varphi_m - z\mathbf{E}\,\Delta\varphi_m\Delta\psi_m,$$

where $\Delta\varphi_m = \varphi_m - \mathbf{E}\varphi_m$ and $\Delta\psi_m = \psi_m - \mathbf{E}\psi_m$. We square the absolute values of both parts of this equation and take into account that $|1 - z\varphi_m|^{-1} \le \alpha$ and $|1 + z\mathbf{E}\psi_m| \le \alpha$. It follows that

$$\text{Var } \psi_m \le \alpha^4 \,\text{Var } \varphi_m + \alpha^2 |z|^2 \,\text{Var } \varphi_m \,\text{Var } \psi_m.$$

Here in the second summand of the right-hand side,

$$|z| \,\text{Var } \psi_m \le \mathbf{E}|z\psi_m|^2 = \mathbf{E}|z\varphi_m(1 - z\varphi_m)^{-1}|^2 \le (1 + \alpha)^2.$$

But $\alpha \ge 1$. It follows that $\text{Var } \psi_m \le 5\alpha^4 \,\text{Var } \varphi_m$. The last statement of our lemma is proved. $\square$

Remark 3. If $z \in \mathfrak{G}$ and $u \geq 0$, then $|1 - zs_0(z)u|^{-1} \leq \alpha$.

Indeed, first let Re $z \leq 0$. We single out a sample vector $\mathbf{x}_m$. Using (5), we obtain

$$s_0(z) = 1 + z\mathbf{E}\,\psi_m = \mathbf{E}(1 - z\varphi_m)^{-1},$$
$$zs_0(z) = \mathbf{E}(z^{-1} - \varphi_m^{-1})^{-1} = \mathbf{E}r(z^{-1} - \varphi_m^{-1})^*,$$

where $r \geq 0$. We examine that Re $\varphi_m \geq 0$ for Re $z \leq 0$. It follows that, in this case, Re $zs_0(z) \leq 0$ and $|1 - zs_0(z)u| \geq 1$.

Now, let Re $z > 0$ and Im $z \neq 0$. The sign of Im z coincides with the sign of Im $h_0(z)$ and with the sign of Im $s_0(z)$. Therefore,

$$|1 - zs_0(z)u| \geq |z|\,\left|\mathrm{Im}\,z/|z|^2 + u\,\mathrm{Im}\,s_0(z)\right| \geq |\mathrm{Im}\,z/z| = \alpha^{-1}.$$

Our remark is grounded for the both cases. $\square$

THEOREM 3.1. *For any population in which all four moments of all variables exist, for any $z \in \mathfrak{G}$, we have*

$$\mathrm{Var}(\mathbf{e}^H H_0(z)\mathbf{e}) \leq \tau^2 \alpha^6 / N,$$
$$\mathbf{E}H_0(z) = (I - zs_0(z)\Sigma)^{-1} + \Omega_0,$$

where $\|\Omega_0\| \leq o_N \overset{def}{=} a\tau^2 \alpha^4(\sqrt{\delta} + \alpha/N)$ *and a is a numerical constant.*

Proof. The first statement of the theorem is proved in Lemma 3.3. To prove the second one, we fix an integer m, $m = 1, 2, \ldots, N$, and multiply both sides of the first relation in (4) by $\mathbf{x}_m \mathbf{x}_m^T$. It follows

$$H_0 \mathbf{x}_m \mathbf{x}_m^T = H_0^m \mathbf{x}_m \mathbf{x}_m^T + z H_0^m \mathbf{x}_m \mathbf{x}_m^T H_0 \mathbf{x}_m \mathbf{x}_m^T / N.$$

Multiplying by z, calculate the expectation values:

$$z\mathbf{E}H_0 \mathbf{x}_m \mathbf{x}_m^T = \mathbf{E}z H_0 S = \mathbf{E}(H_0 - I) =$$
$$= z\mathbf{E}H_0^m \Sigma + z\mathbf{E}H_0^m \mathbf{x}_m \mathbf{x}_m^T (1 - z\psi_m).$$

We substitute $z\psi_m = s_0(z) - 1 + z\Delta\psi_m$, where $\Delta\psi_m = \psi_m - \mathbf{E}\psi_m$, and obtain that

$$\mathbf{E}H_0 = I + zs_0(z)\mathbf{E}H_0^m\Sigma + \Omega_1,$$

where $\Omega_1 = z^2\mathbf{E}H_0^m\mathbf{x}_m\mathbf{x}_m^T\Delta\psi_m$. Using (4) once more to replace H_0^m, we find that

$$(I - zs_0(z)\Sigma)\mathbf{E}H_0 = I + \Omega_1 + \Omega_2,$$

where $\Omega_2 = zs_0(z)\mathbf{E}(H_0^m - H_0)\Sigma$. Denote $R = (I - zs_0(z)\Sigma)^{-1}$. By Remark 3, $\|R\| \leq \alpha$. Multiplying by R, we obtain $\mathbf{E}H_0 = R + \Omega$, where $\Omega = R\Omega_1 + R\Omega_2$. We notice that Ω is a symmetric matrix and, consequently, its spectral norm equals $|\mathbf{e}^H\Omega\mathbf{e}|$, where $\mathbf{e}$ is one of its eigenvalues.

Denote $\mathbf{f} = R\mathbf{e}$, $\mathbf{f}^H\mathbf{f} \leq \alpha$. We have $\|\Omega\| = |\mathbf{f}^T\Omega_1\mathbf{e} + \mathbf{f}^T\Omega_2\mathbf{e}|$. Now,

$$|\mathbf{f}^H\Omega_1\mathbf{e}| = |z| \; \mathbf{E}|\mathbf{f}^H H_0^m\mathbf{x}_m(\mathbf{x}_m^T\mathbf{e}) \; \Delta\psi_m| \leq$$
$$\leq |z|^2 \left(\mathbf{E}|\mathbf{f}^H H_0^m\mathbf{x}_m|^4 \; \mathbf{E}|\mathbf{x}_m^T\mathbf{e}|^4\right)^{1/4} \sqrt{\operatorname{Var}\psi_m}. \qquad (7)$$

Here $\mathbf{E}|\mathbf{f}^H H_0^m\mathbf{x}_m|^4 \leq M\mathbf{E}|\mathbf{f}^H H_0^m \; H_0^{m*}\mathbf{f}|^2 \leq M\alpha^8$, $\mathbf{E}|\mathbf{x}_m^T\mathbf{e}|^4 \leq M$, $\operatorname{Var}\psi_m \leq aM\alpha^4\sqrt{\delta}$. It follows that the left-hand side of (7) is not greater than $M|z|^2\alpha^4\sqrt{\delta}$. Then,

$$\|\Omega_2\| \leq |\mathbf{f}^H\Omega_2\mathbf{e}| = |z|^2 \; |s_0(z)| \; |\mathbf{E}\mathbf{f}^H H_0\mathbf{x}_m(\mathbf{x}_m^T H_0^m\Sigma\mathbf{e})/N| \leq$$
$$\leq |z|^2 \; |s_0(z)| \left(\mathbf{E}|\mathbf{f}^H H_0\mathbf{x}_m|^2 \; \mathbf{E}|\mathbf{x}_m^T H_0^m\Sigma\mathbf{e}|^2\right)^{1/2} /N.$$

Here we have that $|s_0(z)| = |\mathbf{E}(1 - z\varphi_m)^{-1}| \leq \alpha$; using (3) and (4), we find that

$$\mathbf{E}|\mathbf{f}^H H_0\mathbf{x}_m|^2 \leq |\mathbf{f}|^2 \; \alpha^2\mathbf{E}|\mathbf{e}_1^H H_0^m\mathbf{x}_m|^2 \leq \sqrt{M}\alpha^6,$$

where $\mathbf{e}_1 = \mathbf{f}/|\mathbf{f}|$. Obviously,

$$\mathbf{E}|\mathbf{x}_m^T H_0^m\Sigma\mathbf{e}|^2 = \sqrt{M}(\mathbf{e}^T\Sigma H_0^m\Sigma H_0^{m*}\Sigma\mathbf{e}) \leq M^{3/2}\alpha^2.$$

Therefore, $|\mathbf{f}^T \Omega_2 \mathbf{e}|$ is not greater than $M|z|^2 \alpha^5 / N$. We obtain the required upper estimate of $\|\Omega\|$. This completes the proof. $\square$

Corollary. For any $z \in \mathfrak{G}$, we have

$$h_0(z) = n^{-1} \mathrm{tr}(I - zs_0(z)\Sigma)^{-1} + \omega, \qquad (8)$$

where $|\omega| \leq o_N$, and o_N is defined in Theorem 3.1.

Parameters restricting the dependence

Let us investigate the requirements of parameters (1) and (2). Note that the boundedness of the moments M is an essential condition restricting the dependence of variables. Indeed, let Σ be a correlation matrix with the Bayes distribution of the correlation coefficients that is uniform on the segment $[-1, 1]$. Then, the Bayes mean $\mathbf{E}M \geq \mathbf{E}n^{-1}\mathrm{tr}\Sigma^2 \geq (n+2)/3$. In case of $N(0, \Sigma)$ with the matrix Σ with all entries 1, the value $M = 3n^2$.

Let us prove that relation (8) can be established with accuracy to terms, in which $M \to \infty$ and moments of variables are restricted only in a set. Denote

$$\Lambda_k = n^{-1}\mathrm{tr}\,\Sigma^k, \quad Q_k = \mathbf{E}(\mathbf{x}^2/n)^k, \quad W = n^{-2}\sup_{\|\Omega\|=1} \mathbf{E}(\mathbf{x}^T \Omega \mathbf{x}')^4,$$

$k \geq 0$, where $\mathbf{x}$ and $\mathbf{x}'$ are independent vectors and Ω are nonrandom, symmetric, positive semidefinite matrices of unit spectral norm.

Remark 4. If $t \geq 0$, then relation (8) holds with the remainder term ω such that $\omega^2/2 \leq \left[Q_2 y^2(\nu + Wt^2/N) + W/N^2\right]t^4$.

Using this inequality, we can show that if M is not bounded but Q_1 and Q_2 are bounded, then there exists a case when $\nu \to 0$ as $n \to \infty$ and $\omega \to 0$ in (8).

Indeed, let $\mathbf{x} \sim \mathbf{N}(0, \Sigma)$. Denote $\Lambda_k = n^{-1}\mathrm{tr}\,\Sigma^k$, $k = 1, 2, \ldots$. For normal $\mathbf{x}$, we have $M = 3\|\Sigma\|^2$, $Q_2 = \Lambda_1^2 + 2\Lambda_2/n$, $W = 3(\Lambda_2^2 + 2\Lambda_4/n)$, $\nu = 2\Lambda_2/n$. Consider a special case when $\Sigma = I + \rho E$, where E is a matrix all of whose entries are 1, and $0 \leq \rho \leq 1$. Then, $M = 3(1+n\rho)^2$, $\Lambda_1 = 1+\rho$, $\Lambda_2 = 1+2\rho+n\rho^2$, $\Lambda_k \leq a_k + b_k \rho^k n^{k-1}$, where a_k and b_k are positive numbers independent of n, and all

$Q_k < c$, where c does not depend on n. If $\rho = \rho(n) = n^{-3/4}$ as $n \to \infty$, then $M \to \infty$, whereas the quantities Λ_3, Λ_4, and Q_3 remain finite. Nevertheless, the quantities $\nu = O(n^{-1})$ and $\omega \to 0$.

Sample Covariance Matrices

The traditional (biased) estimator of the true covariance matrix Σ is

$$C = \frac{1}{N} \sum_{m=1}^{N} (\mathbf{x}_m - \bar{x})(\mathbf{x}_m - \bar{\mathbf{x}}_m)^T.$$

To pass to matrix C, we use the relation $C = S - \bar{\mathbf{x}}\bar{\mathbf{x}}^T$, where $\bar{\mathbf{x}}$ are sample mean vectors, and show that this difference does not influence the leading parts of spectral equations and affects only remainder terms.

Consider the resolvent $H = H(z) = (I - zC)^{-1}$ of matrix C and define $h(z) = n^{-1}\text{tr}H(z)$ and $s(z) = 1 - y + yh(z)$.

Denote also

$$V = V(z) = \mathbf{e}^H H_0(z)\bar{\mathbf{x}}, \quad U = \mathbf{e}^H H(z)\bar{\mathbf{x}},$$
$$\Phi = \Phi(z) = \bar{\mathbf{x}}^T H_0(z)\ \bar{\mathbf{x}}, \quad \Psi = \Psi(z) = \bar{\mathbf{x}}^T H(z)\bar{\mathbf{x}},$$

where $\mathbf{e}$ is a nonrandom complex vector with $\mathbf{e}^H\mathbf{e} = 1$, the superscript H stands for the Hermitian conjugation.

Remark 5. If $z \in \mathfrak{G}$, then

$$H(z) = H_0(z) - zH_0(z)\bar{\mathbf{x}}\bar{\mathbf{x}}^T H(z),$$
$$U = V - z\Phi U = V - z\Psi V,$$
$$(1 + z\Phi)(1 - z\Psi) = 1, \quad |1 - z\Psi| \leq \alpha.$$

Indeed, the first three identities may be checked straightforwardly. The fourth statement follows from Remark 1.

Remark 6. If $z \in \mathfrak{G}$, then

$$q \stackrel{def}{=} |z|\ \bar{\mathbf{x}}^T H_0(z)H_0^*(z)\bar{\mathbf{x}} \leq \alpha^2.$$

Let us derive this inequality. Let $\mathbf{w}$ be a complex vector. We denote complex scalars $\mathbf{w}^T\mathbf{w}$ by $\mathbf{w}^2$ and the real product $\mathbf{w}^H\mathbf{w}$ by $|\mathbf{w}|^2$. Denote the matrix product $Z^H Z$ by $|Z|^2$. Denote $\Omega^2 = I - zC$, $\bar{\mathbf{y}} = \Omega^{-1}\bar{\mathbf{x}}$, $\mathbf{a} = |H(z)||\bar{\mathbf{x}}$. Then,

$$H_0(z) = (I - zC - z\,\bar{\mathbf{x}}\,\bar{\mathbf{x}}^T)^{-1} = \Omega^{-1}(I - z\,\bar{\mathbf{y}}\bar{\mathbf{y}}^H)^{-1}\Omega^{-1},$$
$$\bar{\mathbf{y}}^H\bar{\mathbf{y}} = \bar{\mathbf{x}}^T(I - zC)^{-1}\bar{\mathbf{x}} = \mathbf{a}^T(I - zC^*)\mathbf{a}.$$

Therefore,

$$q = |z|\bar{\mathbf{x}}^T H_0(z)H_0^*(z)\bar{\mathbf{x}} = |z| \cdot |\Omega^{-1}(I - z\bar{\mathbf{y}}\bar{\mathbf{y}}^H)^{-1}\bar{\mathbf{y}}|^2 =$$
$$= |z| \cdot |\bar{\mathbf{y}}^H\Omega^{-2}\bar{\mathbf{y}}|^{-2}|1 - z\bar{\mathbf{y}}^2|^{-2} = |z|\mathbf{a}^2\left|1 - z\mathbf{a}^2 + |z|^2\mathbf{a}^T C\mathbf{a}\right|^{-2}.$$

Denote $t = \mathbf{a}^2/(1+|z|^2\mathbf{a}^T C\mathbf{a})$. Let $z \neq 0$. If $\operatorname{Re} z < 0$, then we have $q \leq |z|\mathbf{a}^2 \leq 1$. If $\operatorname{Re} z \geq 0$, then the quantity $q \leq |z|t^2/|1 - zt|^2$. The maximum of the right-hand side of this inequality is attained for $t = 1/\operatorname{Re} z$ and equals $q = q_{\max} = |z|^2|\operatorname{Im} z|^{-2} = \alpha^2(z)$. This is our assertion. $\square$

LEMMA 3.4. *If $z \in \mathfrak{G}$, then we have $|z| \operatorname{var} V \leq 2\tau\alpha^4(1+\tau\alpha^2)/N$.*

Proof. To use Lemma 3.2, we single out one of sample vectors, say, $\mathbf{x}_m$. Denote $H_0 = H_0(z)$, $H = H(z)$, $H_0^m = H_0^m(z)$, $\tilde{\mathbf{x}} = \bar{\mathbf{x}} - N^{-1}\mathbf{x}_m$. We have

$$V = \mathbf{e}^H H_0\mathbf{x} = \mathbf{e}^H H_0\tilde{\mathbf{x}} + \mathbf{e}^H H^m\mathbf{x}_m/N + z\mathbf{e}^H H_0^m\mathbf{x}_m\mathbf{x}_m^T H_0\tilde{\mathbf{x}}/N,$$

where the first summand in the left-hand side does not depend on $\mathbf{x}_m$. Denote $w_m = \mathbf{x}_m^T H_0^m\tilde{\mathbf{x}}$. By Lemma 3.2, we have

$$|z| \operatorname{var} V \leq |z| N^{-2} \sum_{m=1}^{N} \mathbf{E}|u_m(1 + zw_m)|^2.$$

But

$$\mathbf{E}|u_m|^2 \leq \sqrt{M}\alpha^4, \quad \mathbf{E}|zu_mw_m|^2 \leq \sqrt{M}\alpha^4|z|^2(\mathbf{E}|w_m|^4)^{1/2}.$$

It follows that

$$|z|^2 \mathbf{E}|w_m|^4 \le M|z|^2 \mathbf{E}(\tilde{\mathbf{x}}^T H_0^m \; H_0^{m*}\tilde{\mathbf{x}})^2 \le Mq',$$

where q' may be reduced to the form of the expression for q in Remark 6, with the number N less by unit with the argument $z' = (1 - N^{-1})z$. From Remark 6, it follows that $q' \le \alpha^2$ and that $|z|^2 \mathbf{E}|w_m|^4 \le M\alpha^4$. We obtain the required upper estimate of $|z|$ var V. Lemma 3.4 is proved. $\square$

LEMMA 3.5. *If $z \in \mathfrak{G}$, then $\|\mathbf{E}H(z) - \mathbf{E}H_0(z)\| \le a\omega$, where $\omega^2 = \tau^2 \alpha^6 y(\tau^2 \alpha^2 \delta + (1 + \tau\alpha^2)/N)$, and a is numerical coefficient.*

Proof. In view of the symmetry of matrices $H = H(z)$ and $H_0 = H_0(z)$, we have

$$\|\mathbf{E}H - \mathbf{E}H_0\| \le \|z\mathbf{E}H\bar{\mathbf{x}}\bar{\mathbf{x}}^T H_0\| = \mathbf{E}|zVU| \le$$
$$\le \left(\mathbf{E}|zV^2|\mathbf{E}|zU^2|\right)^{1/2}$$

with some nonrandom unit vectors $\mathbf{e}$ in definitions of V and U. We have

$$\mathbf{E}|zV^2| = |z\mathbf{E}V^2| + |z| \text{ Var } V,$$
$$\mathbf{E}|zU^2| = |z|\alpha^2\mathbf{E}\bar{\mathbf{x}}^2 \le \sqrt{M}|z| \; \alpha^2 y.$$

But for any $m = 1, 2, \ldots, N$, we have $\mathbf{E}V = \mathbf{E}\mathbf{e}^T H_0 \; \bar{\mathbf{x}} = |\mathbf{E}u_m|$. From Lemma 3.1, it follows that $|\mathbf{E}u_m|^2 \le aM^{3/2}|z|^2 \alpha^6 \delta$, where a is a number. Gathering up these estimates, we obtain the lemma statement. $\square$

LEMMA 3.6. *If $z \in \mathfrak{G}$ and $u > 0$, then $|1 - zs(z)u|^{-1} \le \alpha$.*

Proof. Denote

$$\tilde{C} = C - N^{-1}(\mathbf{x}_m - \bar{\mathbf{x}})(\mathbf{x}_m - \bar{\mathbf{x}})^T, \quad \tilde{H} = (I - z\tilde{C})^{-1},$$
$$\tilde{\Phi} = (\mathbf{x}_m - \bar{\mathbf{x}})^T \tilde{H}(\mathbf{x}_m - \bar{\mathbf{x}})^T/N,$$
$$\tilde{\Psi} = (\mathbf{x}_m - \bar{\mathbf{x}})^T H(\mathbf{x}_m - \bar{\mathbf{x}})^T/N.$$

We examine the identities

$$H = \widetilde{H} + zH(\mathbf{x}_m - \bar{\mathbf{x}})(\mathbf{x}_m - \bar{\mathbf{x}})^T \widetilde{H}/N, \quad (1 - z\widetilde{\Phi})(1 + z\widetilde{\Psi}) = 1.$$

It follows that

$$s(z) = 1 + y(h(z) - 1) = 1 + \mathbf{E}N^{-1}\mathrm{tr}(H(z) - I) =$$
$$= 1 + z\mathbf{E}N^{-1}\mathrm{tr}\, H(z)\, C = 1 + z\mathbf{E}N^{-1}(\mathbf{x}_m - \bar{\mathbf{x}})^T H(z)\, (\mathbf{x}_m - \bar{\mathbf{x}}) =$$
$$= 1 + z\mathbf{E}\widetilde{\Psi} = \mathbf{E}(1 - z\widetilde{\Phi})^{-1}.$$

Suppose that $\mathrm{Re}\, z \leq 0$ and $u > 0$. Then, $\mathrm{Re}\, zs(z) = r(z^{-1} - \Phi)^*$, where $r > 0$ and $\mathrm{Re}\, zs(z) \leq 0$. Therefore, $|1 - zs(z)u| \geq 1$.

Suppose $\mathrm{Im}\, z \neq 0$. The sign of z coincides with the sign of $h(z)$ and with the sign of $s(z)$. Therefore,

$$|1 - zs(z)u| = |z| \cdot |\mathrm{Im}\, z/|z|^2 + u\mathrm{Im}\, s(z)| \geq \alpha^{-1}.$$

This proves the lemma. $\square$

THEOREM 3.2. *If $z \in \mathfrak{G}$ and $N > 1$, then*

$$\mathrm{Var}(n^{-1}\mathrm{tr}\, H(z)) \leq a\tau^2\, \alpha^4/N,$$
$$\mathbf{E}H(z) = (I - zs(z)\Sigma)^{-1} + \Omega, \quad h(z) = n^{-1}\mathrm{tr}(I - zs(z)\Sigma)^{-1} + \omega,$$

$$(9)$$

where $s(z) = 1 + y(h(z) - 1)$, and

$$\|\Omega\| \leq a\tau \max(1, \lambda)\, \alpha^3[\tau\alpha\, \sqrt{\delta} + (1 + \tau^2\alpha)/\sqrt{N}],$$
$$|\omega| \leq a\tau\, \alpha^2 \max(1, \lambda)[\tau\alpha^2\sqrt{\delta} + (1 + \tau\alpha^3)/\sqrt{N}],$$

and a are (different) numerical constants.

Proof. From the matrix C, we single out the summand C^m independent of $\mathbf{x}_m$:

$$C = C^m + \Delta_m, \quad \Delta_m = (1 + N^{-1})\, \mathbf{x}_m\, \mathbf{x}_m^T - \mathbf{x}_m\, \bar{\mathbf{x}}^T - \bar{\mathbf{x}}\, \mathbf{x}_m^T.$$

Denote $H = H(z)$, $H^m = (I - zC^m)^{-1}$. We have the identity $H^m = H + zH^m\Delta_m H$. Applying Lemma 3.2, we obtain

$$\text{Var}(n^{-1}\text{tr}H) = \sum_{m=1}^{N} \mathbf{E}|z\, n^{-1}\text{tr}(H\Delta_m H^m)|^2 \leq$$

$$\leq 3|z|^2\, n^{-2}N^{-1}\mathbf{E}((1 + N^{-1}) \cdot |\mathbf{x}_m^T H^m H\mathbf{x}_m|^2 +$$

$$+ |\mathbf{x}_m^T H^m H\mathbf{x}_m|^2 + |\bar{\mathbf{x}}^T H^m H\bar{\mathbf{x}}_m|^2),$$

where $\|H^m\| \leq \alpha$, $\|H\| \leq \alpha$, $\mathbf{E}(\mathbf{x}_m^2)^2 \leq Mn^2$, and $\mathbf{E}|\bar{\mathbf{x}}|^4 \leq My^2$. We conclude that $\text{Var}(n^{-1}\text{tr } H) \leq 3\tau^2\alpha^4 (1 + 3/N)/N$. The first statement of our theorem is proved.

Now, we start from Theorem 3.1. Obviously,

$$\mathbf{E}H(z) = (I - zs(z)\Sigma)^{-1} + (\mathbf{E}H(z) - \mathbf{E}H_0(z)) +$$
$$+(I - zs(z)\Sigma)^{-1}z(s_0(z) - s(z))\, \Sigma\, (I - zs_0(z)\Sigma)^{-1} + \Omega_0.$$
$$(10)$$

Here

$$|s_0(z) - s(z)| = y|h_0(t) - h(t)| \leq \tau y\alpha^2/N.$$

Note that last three summands in the right-hand side of (10) do not exceed $\tau^2\alpha^4\, y/N + a\omega + \|\Omega_0\|$, where ω is from Lemma 3.5. Substituting ω and Ω_0, we obtain that

$$|\omega| \leq \tau^2\alpha^4\, y/N + \tau\alpha^2/N + a\tau^2\alpha^4\, (\sqrt{\delta} + \alpha/N).$$

This gives the required upper estimate of $|\omega|$. The proof is complete. $\square$

The equation (9) for $h(z)$ was first derived for normal distribution in paper [63] in the form of a limit formula. In [65] and [67], these limit formulas were obtained for a wide class of populations. In [71], relations (9) were established for a wide class of distributions with fixed n and N.

Limit Spectra

We investigate here the limiting behavior of spectral functions for the matrices S and C under the increasing dimension asymptotics. Consider a sequence $\mathfrak{P} = \{\mathfrak{P}_n\}$ of problems

$$\mathfrak{P}_n = (\mathfrak{S}, \Sigma, N, \mathfrak{X}, S, C)_n, \quad n = 1, 2, \ldots \tag{11}$$

in which spectral functions of matrices C and S are investigated over samples $\mathfrak{X}$ of size N from populations $\mathfrak{S}$ with $\mathrm{cov}(\mathbf{x}, \mathbf{x}) = \Sigma$ (we do not write out the subscripts in arguments of $\mathfrak{P}_n$). For each problem $\mathfrak{P}_n$, we consider functions

$$h_{0n}(t) = n^{-1}\mathrm{tr}(I - zS)^{-1}, \quad h_n(t) = n^{-1}\mathrm{tr}(I - zC)^{-1},$$

$$F_{nS}(u) = \frac{1}{n}\sum_{i=1}^{n} \mathrm{ind}(\lambda_i^S \leq u), \quad F_{nC}(u) = \frac{1}{n}\sum_{i=1}^{n} \mathrm{ind}(\lambda_i^C \leq u),$$

where λ_i^S and λ_i^C are eigenvalues of S and C, respectively, $i = 1, 2, \ldots, n$.

We restrict (11) by the following conditions.

A. For each n, the observation vectors in $\mathfrak{S}$ are such that $\mathbf{Ex} = 0$ and the four moments of all components of $\mathbf{x}$ exist.

B. The parameter M does not exceed a constant c_0, where c_0 does not depend on n. The parameter γ vanishes as $n \to \infty$ in $\mathfrak{P}$.

C. In $\mathfrak{P}$, $n/N \to \lambda > 0$.

D. In $\mathfrak{P}$ for each n, the eigenvalues of matrices Σ are located on a segment $[c_1, c_2]$, where $c_1 > 0$ and c_2 does not depend on n, and $F_{n\Sigma}(u) \to F_\Sigma(u)$ as $n \to \infty$ almost for any $u \geq 0$.

Corollary (of Theorem 3.1). Under Assumptions A–D for any $z \in \mathfrak{G}$, the limit exists $\lim_{n\to\infty} h_n(z) = h(z)$ such that

$$h(z) = \int (1 - zs(z)u)^{-1} dF_\Sigma(u), \qquad s(z) = 1 - \lambda + \lambda h(z), \tag{12}$$

and for each z, we have

$$\lim_{n\to\infty} \|\mathbf{E}(I - zC)^{-1} - (I - zs(z)\Sigma)^{-1}\| \to 0.$$

Let us investigate the analytical properties of solutions to (12).

THEOREM 3.3. *If $h(z)$ satisfies (12), $c_1 > 0$, $\lambda > 0$, and $\lambda \neq 1$,
then*
  1. $|h(z)| \leq \alpha(z)$ and $h(z)$ is regular near any point $z \in \mathfrak{G}$;
  2. for any $v = \operatorname{Re} z > 0$ such that $v < v_2 = c_1^{-1}(1 - \sqrt{\lambda})^{-2}$ or
$v > v_1 = c_2^{-1}(1 + \sqrt{\lambda})^{-2}$, we have

$$\lim_{\varepsilon\to+0} \operatorname{Im} h(v + i\varepsilon) = 0;$$

  3. if $v_1 \leq v \leq v_2$, then $0 \leq \operatorname{Im} h(v + i\varepsilon) \leq (c_1\lambda v)^{-1/2} + \omega$,
where $\omega \to 0$ as $\varepsilon \to +0$;
  4. if $v = \operatorname{Re} z < 0$ then $s(-v) \geq (1 + c_2\lambda |v|)^{-1}$;
  5. if $|z| \to \infty$ on the main sheet of the analytical function $h(z)$,
then we have

$$\text{if } 0 < \lambda < 1, \text{ then } \ zh(z) = -(1 - \lambda)^{-1}\Lambda_{-1} + O(|z|^{-1}),$$
$$\text{if } \lambda = 1, \text{ then } \ zh^2(z) = -\Lambda_{-1} + O(|z|^{-1/2}),$$
$$\text{if } \lambda > 1, \text{ then } \ zs(z) = -\beta_0 + O(|z|^{-1}),$$

where β_0 is a root of the equation

$$\int (1 + \beta_0 u)^{-1} dF_\Sigma(u) = 1 - \lambda^{-1}.$$

Proof. The existence of the solution to (12) follows from
Theorem 3.1. Suppose $\operatorname{Im}(z) > 0$, then $|h(z)| \leq \alpha = \alpha(z)$. To
be concise, denote $h = h(z)$, $s = s(z)$. For all $u > 0$ and z outside
the beam $z > 0$, we have $|1 - zsu|^{-1} \leq \alpha$. Differentiating $h(z)$ in
(12), we prove the regularity of $h(z)$. Define

$$b_\nu = b_\nu(z) = \int |1 - zs(z)u|^{-2} u^\nu dF_\Sigma(u), \quad \nu = 1, 2.$$

Let us rewrite (12) in the form

$$(h-1)/s = z \int u(1-zsu)^{-1} dF_\Sigma(u). \tag{13}$$

It follows that

$$\text{Im}[(h-1)/s] = |s|^{-2}\text{Im}\,h = b_1\text{Im}\,z + b_2\lambda|z|^2\text{Im}\,h.$$

Dividing by b_2, we use the inequality $b_1/b_2 \leq c_1^{-1}$. Fix some $v = \text{Re}\,z > 0$ and tend $\text{Im}\,z = \varepsilon \to +0$. It follows that the product

$$(|s|^{-2}b_2^{-1} - \lambda v^2)\text{Im}\,h \to 0.$$

Suppose that $\text{Im}\,h$ does not tend to 0 (v is fixed). Then, there exists a sequence $\{z_k\}$ such that, for $z_k = v + i\varepsilon_k$, $h = h(z_k)$, $s = s(z_k)$, we have $\text{Im}\,h \to a$, where $a \neq 0$. For these z_k, we obtain $|s|^{-2}b_2^{-1} \to \lambda v^2$ as $\varepsilon_k \to +0$. We apply the Cauchy–Bunyakovskii inequality to (5). It follows that $|h-1|^2/|z_k s|^2 \leq b_2$. We obtain that $|h-1|^2 \leq \lambda^{-1}+o(1)$ as $\varepsilon_k \to +0$. It follows that $|s-1|^2 \leq \lambda+o(1)$. So the values s are bounded for $\{z_k\}$. On the other hand, it follows from (12) that $\text{Im}\,h = b_1\,\text{Im}(zs) = b_1(\text{Re}\,s \cdot \text{Im}\,z + \lambda v\,\text{Im}\,h)$. We find that $(b_1^{-1} - \lambda v)\text{Im}\,h \to 0$ as $\text{Im}\,z \to 0$. But $\text{Im}\,h \to a \neq 0$ for $\{z_k\}$. It follows that $b_1^{-1} \to \lambda v$. Combining this with the inequality $|s|^{-2}b_2^{-1} \to \lambda v^2$, we find that $|s|^{-2}b_2^{-1} - b_1^{-1}v \to 0$. Note that b_1 is finite for $\{z_k\}$ and $c_2^{-2} \leq b_1 b_2^{-1} \leq c_1^{-1}$. Substitute the boundaries $(1 \pm \sqrt{\lambda}) + o(1)$ for $|s|$. We obtain that $v_1 + o(1) \leq v \leq v_2 + o(1)$ as $\varepsilon_k \to +0$. We can conclude that $\text{Im}\,h \to 0$ for any positive v outside the interval $[v_1, v_2]$. This proves the second statement of our theorem.

Now suppose $v_1 \leq v \leq v_2$. From (12), we obtain the inequality $\lambda\text{Im}\,h \cdot \text{Im}(zh) \leq c_1^{-1}$. But h is bounded. It follows that the quantity $(\text{Im}\,h)^2 \leq (c_1 v \lambda)^{-1}$. The third statement of our theorem is proved.

Further, let $v = \text{Re}\,z < 0$. Then, the functions h and s are real and non-negative. We multiply both parts of (12) by λ. It follows that $(h-1)/zs \leq b_1 \leq c_2$. We obtain $s \geq (1 + c_2\lambda|z|)^{-1}$.

Let us prove the fifth theorem statement. Let $\lambda < 1$. For real $z \to -\infty$, the real value of $1 - zsu$ in (12) tends to infinity.

Consequently, $h \to 0$ and $s \to 1 - \lambda$. For sufficiently large $|\operatorname{Re} z|$, we have

$$h(z) = \sum_{k=1}^{\infty} \Lambda_{-k}(zs)^{-k},$$

where $\Lambda_k = \int u^k dF_{\Sigma}(u)$. We conclude that

$$h(z) = -(1 - \lambda)^{-1} \Lambda_{-1} z^{-1} + O(|z|^{-2})$$

for real $z < 0$ and for any $z \in \mathfrak{G}$ as $|z| \to \infty$ in view of the properties of the Laurent series. Now let $\lambda = 1$. Then $h = s$. From (12), we obtain that $h \to 0$ as $z \to -\infty$ and $h^2 = \Lambda_{-1}|z|^{-1} + O(|z|^{-2})$. Now suppose that $\lambda > 1$, $z = -t < 0$, and $t \to \infty$. Then, by Lemma 3.6, we have $s \geq 0$, $h \geq 1 - 1/\lambda$, and $s \to 0$. Equation (12) implies $ts \to \beta_0$ as is stated in the theorem formulation. This completes the proof of Theorem 3.3. $\square$

Remark 7. Under Assumptions A–D for each $u \geq 0$, the limit exists

$$F(u) = \operatorname*{plim}_{n \to \infty} F_{nC}(u)$$

such that
$$\int (1 - zu)^{-1} dF(u) = h(z) . \tag{14}$$

Indeed, to prove the convergence, it is sufficient to cite Corollary 3.2.1 from [22] that states the convergence of $\{h_{nS}(z)\}$ and $\{h_{nC}(z)\}$ almost surely. By Lemma 3.5, both these sequences converge to the same limit $h(z)$. To prove that the limits of $F_{nS}(u)$ and $F_{nC}(u)$ coincide, it suffices to prove the uniqueness of the solution to (12). It can be readily proved if we perform the inverse Stieltjes transformation.

THEOREM 3.4. *Under Assumptions A–D,*

1. if $\lambda = 0$, then $F(u) = F_{\Sigma}(u)$ almost everywhere for $u \geq 0$;
2. if $\lambda > 0$ and $\lambda \neq 1$, then $F(0) = F(u_1-0) = \max(0, 1-\lambda^{-1})$, $F(u_2) = 1$, where $u_1 = c_1(1 - \sqrt{\lambda})^2$, $u_2 = c_2(1 + \sqrt{\lambda})^2$, and c_1 and c_2 are bounds of the limit spectra Σ;

3. if $y > 0$, $\lambda \neq 1$, and $u > 0$, then the derivative $F'(u)$ of the function $F(u)$ exists and $F'(u) \leq \pi^{-1}(c_1 \lambda u)^{-1/2}$;

Proof. Let $\lambda = 0$. Then $s(z) = 1$. In view of (12), we have

$$h(z) = \int (1 - zu)^{-1} dF(u) = \int (1 - zu)^{-1} dF_\Sigma(u).$$

At the continuity points of $F_\Sigma(u)$, the derivative

$$F'_\Sigma(u) = \frac{1}{\pi} \lim_{\varepsilon \to +0} \operatorname{Im} \frac{1}{z} h\left(\frac{1}{z}\right) = F'(u),$$

where $z = u - i\varepsilon$, $u > 0$.

Let $\lambda > 0$. By Theorem 3.2 for $u < u_1$ and for $u > u_2$ (note that $u_1 > 0$ if $\lambda > 0$), the values $\operatorname{Im}[(u - i\varepsilon)^{-1} h((u - i\varepsilon)^{-1})] \to 0$ as $\varepsilon \to +0$. But we have

$$\operatorname{Im} \frac{h((u - i\varepsilon)^{-1})}{u - i\varepsilon} > (2\varepsilon)^{-1}[F(u + \varepsilon) - F(u - \varepsilon)]. \qquad (15)$$

It follows that $F'(u)$ exists and $F'(u) = 0$ for $0 < u < u_1$ and for $u > u_2$. The points of the increase of $F(u)$ can be located only at the point $u = 0$ or on the segment $[u_1, u_2]$. If $\lambda < 1$ and $|z| \to \infty$, we have $\int (1 - zu)^{-1} dF(u) \to 0$ and, consequently, $F(0) = 0$. If $\lambda > 1$ and $|z| \to \infty$, then $h(z^{-1})z^{-1} \approx (1 - \lambda^{-1})/z$ and $F(0) = 1 - \lambda^{-1}$. The second statement of our theorem is proved.

Now, let $z = v + i\varepsilon$, where $v > 0$ is fixed and $\varepsilon \to +0$. Then, using (12) we obtain that $\operatorname{Im} h = b_1 \operatorname{Im}(zs)$. Obviously,

$$|\operatorname{Im} h| \leq \int |1 - zsu|^{-1} dF_\Sigma(u) \leq \frac{1}{c_1 \operatorname{Im}(zs)} = \frac{b_1}{c_1 \operatorname{Im} h}.$$

If $\operatorname{Im} h$ remains finite, then $b_1 \to (\lambda v)^{-1}$. Performing limit transition in (15), we prove the last statement of the theorem. $\square$

THEOREM 3.5. *If Assumptions A–D hold and $0 < \lambda < 1$, then for any complex z, z' outside of the half-axis $z > 0$, we have*

$$|h(z) - h(z')| < c_3 |z - z'|^\zeta,$$

where c_3 and $\zeta > 0$ do not depend on z and z'.

Proof. From (12), we obtain

$$|h(z)| \leq \lambda^{-1} \max(\lambda, |1 - \lambda| + 2c_1^{-1}|z|^{-1}).$$

By definition, the function $h(z)$ is differentiable for each z outside the segment $\mathfrak{V} = [v_1, v_2]$, $v_1 > 0$.

Denote a δ-neighborhood of the segment $\mathfrak{V}$ by $\mathfrak{V}_\delta$. If z is outside of $\mathfrak{V}_\delta$, then the derivative $h'(z)$ exists and is uniformly bounded. It suffices to prove our theorem for $v \in \mathfrak{V}_1$, where $\mathfrak{V}_1 = \mathfrak{V}_\delta - \{z : \text{Im } z = 0\}$. Choose $\delta = \delta_1 = v_1/2$. Then $\delta_1 < |z| < \delta_2$ for $z \in \mathfrak{V}_1$, where δ_2 does not depend on z. We estimate the absolute value of the derivative $h'(z)$. For $\text{Im } z \neq 0$, from (15) by the differentiation we obtain

$$\left(z^{-1}y^{-1} - \int X^{-2}u \, dF_\Sigma(u) \right) h'(z) = \frac{s(z)}{z\lambda} \int X^{-2}u \, dF_\Sigma(u), \quad (16)$$

where $X = (1 - zs(z)u) \neq 0$. Denote

$$\varphi(z) = \frac{1}{zy} - \int X^{-2}u \, dF_\Sigma(u), \quad b_1 = \int |X|^{-2}u \, dF_\Sigma(u),$$
$$h_1 = \text{Im } h(z), \quad z_0 = \text{Re } z, \quad z_1 = \text{Im } z, \quad s_0 = \text{Re } s(z),$$

and let α with subscripts denote constants not depending on z. The right-hand side of (16) is not greater than $\alpha_1 b_1$ for $z \in \mathfrak{V}_1$ and therefore $|h'(z)| < \alpha_2 b_1 |\varphi(z)|^{-1}$.

We consider two cases. Denote $\alpha_3 = (2\delta_2 c_2)^{-1}$.

At first, let $\text{Re } s(z) = s_0 \leq \alpha_3$. Using the relation $h_1 = b_1$ $\text{Im } (zs(z))$, we obtain that the quantity $-\text{Im } \varphi(z)$ equals

$$z_1 |z|^{-2}\lambda^{-1} + 2b_1^{-1} \int |X|^{-4}u^2(1 - z_0 s_0 u + z_1\lambda \, h_1 u)h_1 dF_\Sigma(u).$$

In the integrand here, we have $z_0 > 0$, $1 - z_0 s_0 u \geq 1/2$, $z_1 h_1 > 0$. From the Cauchy–Bunyakovskii inequality, it follows that

$$\int |X|^{-4} u^2 dF_\Sigma(u) \geq b_1^2.$$

Hence $|\operatorname{Im} \varphi(z)| \geq b_1 h_1$ and $|h'(z)| \leq \alpha_2 h_1^{-1}$. Let

$$\operatorname{Re} \varphi(z) = \lambda^{-1} z_0 |z|^{-2} - b_1 + 2[\operatorname{Im} zs(z)]^2 \int |X|^{-4} u^3 dF_\Sigma(u).$$

Define $p = \lambda^{-1} z_0 |z|^{-2} - b_1$. We have

$$p = \lambda^{-1} |z|^{-2} z_0 z_1 |\operatorname{Im} zs(z)|^{-1} (s_0 - \lambda h_1 z_1 / z_0).$$

Here $|h_1| < \alpha_4$, $z_0 \geq \delta_1 > 0$, $s_0 > \alpha_3 > 0$, and we obtain that $p > 0$ if $z_1 < \alpha_6$, where $\alpha_6 = \alpha_3 \alpha_5 / \lambda \alpha_4$. If $z \in \mathfrak{V}_1$ and $z_1 > \alpha_6$, then the Hölder inequality follows from the existence of a uniformly bounded derivative of the analytic function $h(z)$ in a closed domain.

Now let $z \in \mathfrak{V}_1$, $z_1 < \alpha_6$, $p > 0$, and $s_0 > \alpha_3 > 0$. Then, $|h'(z)| \leq \alpha_7 b_1 |\operatorname{Re} \varphi(z)|^{-1}$, where

$$\operatorname{Re} \varphi(z) \geq 2(\operatorname{Im} zs(z))^2 c_1 \int |X|^{-4} u^2 dF_\Sigma(u) \geq$$
$$\geq 2(\operatorname{Im} zs(z))^2 \, c_1 b_1 = 2 c_1 h_1^2.$$

Substituting $b_1 = h_1 / \operatorname{Im}(zs(z))$ and taking into account that $s_0 > 0$, we obtain that $|h'(z)| \leq \alpha_7 h_1^{-2}$. Thus, for $v \in \mathfrak{V}_\delta$ and $0 < z_1 < \alpha_6$ for any s_0, it follows that $|h'(z)| \leq \alpha_8 \max (h_1^{-1}, h_1^{-2}) \leq \alpha_9 h_1^{-2}$. Calculating the derivative along the vertical line we obtain the inequality $h_1^2 \, |dh_1/dz| \leq \alpha_9$, whence

$$h_1^3(z) \leq h_1^3(z') + 3\alpha_9 |z - z'| \leq \left(h_1(z') + \alpha_{10} \, |z - z'|^{1/3} \right)^3$$

if $\operatorname{Im} z \cdot \operatorname{Im} z' > 0$. The Hölder inequality for $h_1 = \operatorname{Im} h(z)$ with $\zeta = 1/3$ follows. This completes the proof of Theorem 3.5. $\square$

Example. Consider limit spectra of matrix Σ of a special form of the "ρ-model" considered first in [63]. It is of a special interest since it admits an analytical solution to the dispersion equation (12). For this model, the limit spectrum of Σ is located on a segment $[c_1, c_2]$, where $c_1 = \sigma^2(1 - \sqrt{\rho})^2$ and $c_2 = \sigma^2(1 + \sqrt{\rho})^2$, $\sigma > 0$, $0 \leq \rho < 1$. Its limit spectrum density is

$$\frac{dF_\Sigma(u)}{du} = \begin{cases} (2\pi\rho)^{-1}(1 - \rho)u^{-2}\sqrt{(c_2 - u)(u - c_1)}, & c_1 \leq u \leq c_2, \\ 0 & \text{for} \quad u < c_1 \quad \text{and for} \quad u > c_2. \end{cases}$$

The moments $\Lambda_k = \int u^k dF_\Sigma(u)$ for $k = 0, 1, 2, 3, 4$ are

$$\Lambda_0 = 1, \quad \Lambda_1 = \sigma^2(1 - \rho), \quad \Lambda_2 = \sigma^4(1 - \rho),$$
$$\Lambda_3 = \sigma^6(1 - \rho^2), \quad \Lambda_4 = \sigma^8(1 - \rho)(1 + 3\rho + \rho^2).$$

If $\rho > 0$, the integral

$$\eta(z) = \int (1 - zu)^{-1} dF_\Sigma(u) = \frac{1 + \rho - \kappa z - \sqrt{(1 + \rho - \kappa z)^2 - 4\rho}}{2\rho},$$

where $\kappa = \sigma^2(1 - \rho)^2$. The function $\eta = \eta(z)$ satisfies the equation $\rho\eta^2 + (\kappa z - \rho - 1)\eta + 1 = 0$. The equation $h(z) = \eta(zs(z))$ can be transformed to the equation $(h - 1)(1 - \rho h) = \kappa z h s$, which is quadratic with respect to $h = h(z)$, $s = 1 - \lambda + \lambda h$. If $\lambda > 0$, its solution is

$$h = \frac{1 + \rho - \kappa(1 - \lambda)z - \sqrt{(1 + \rho - \kappa(1 - \lambda)z)^2 - 4(\rho + \kappa z\lambda)}}{2(\rho + \kappa\lambda z)}.$$

The moments $M_k = (k!)^{-1}h^{(k)}(0)$ for $k = 0, 1, 2, 3$ are

$$M_0 = 1, \quad M_1 = \sigma^2(1 - \rho), \quad M_2 = \sigma^4(1 - \rho)(1 + \lambda(1 - \rho)),$$
$$M_3 = \sigma^6(1 - \rho)(1 + \rho + 3\lambda(1 - \rho) + \lambda^2(1 - \rho)^2).$$

Differentiating the functions of the inverse argument, we find that, in particular, $\Lambda_{-1} = \kappa^{-1}$, $\Lambda_{-2} = \kappa^{-2}(1 + \rho)$, $M_{-1} = \kappa^{-1}(1 - \lambda)^{-1}$,

$M_{-2} = \kappa^{-2}(\rho + \lambda(1-\rho))(1-\lambda)^{-3}$. The continuous limit spectrum of the matrix C is located on the segment $[u_1, u_2]$, where

$$u_1 = \sigma^2(1 - \sqrt{\lambda + \rho(1-\lambda)})^2, \quad u_2 = \sigma^2(1 + \sqrt{\lambda + \rho(1-\lambda)})^2$$

and has the density

$$f(u) = \begin{cases} \dfrac{(1-\rho)\sqrt{(u_2-u)(u-u_1)}}{2\pi u\,(\rho u + \sigma^2(1-\rho)^2 y)} & \text{if } u \in [u_1, u_2], \\ 0 & \text{otherwise.} \end{cases}$$

If $\lambda > 1$, then the function $F(u)$ has a jump $1 - \lambda^{-1}$ at the point $u = 0$. If $\lambda = 0$, then $F(u) = F_\Sigma(u)$ has a form of a unit step at the point $u = \sigma^2$. The density $f(u)$ satisfies the Hölder condition with $\zeta = 1/2$.

In a special case when $\Sigma = I$ and $\rho = 0$, we obtain the limit spectral density $F'(u) = (2\pi)^{-1}u^{-2}\sqrt{(u_2 - u)(u - u_1)}$ for $u_1 \le u \le u_2$, where $u_{2,1} = (1 \pm \sqrt{\lambda})^2$. This "semicircle" law of spectral density was first found by Marchenko and Pastur [43].

3.2. SPECTRAL FUNCTIONS OF INFINITE SAMPLE COVARIANCE MATRICES

The technique of efficient essentially multivariate statistical analysis presented in the review [69] and in book [71] is based on using spectral properties of large-dimensional sample covariance matrices. However, this presentation is restricted to dimensions not much greater than sample size. To treat efficiently the case when the sample size is bounded another asymptotics is desirable. In this section, we assume that the observation vectors are infinite dimensional and investigate spectra of infinite sample covariance matrices under restricted sample size that tends to infinity.

Let $\mathfrak{F}$ be an infinite-dimensional population and $\mathbf{x} = (x_1, x_2, \ldots)$ be an infinite-dimensional vector from $\mathfrak{F}$. Let $\mathfrak{X} = \{\mathbf{x}_m\}$, $m = 1, 2, \ldots$ be a sample of size N from $\mathfrak{F}$.

We restrict ourselves with the following two requirements.

1. Assume that fourth moments exist for all components of vector $\mathbf{x} \in \mathfrak{F}$, and let the expectation $\mathbf{E}\mathbf{x} = 0$.

Denote $\Sigma = \mathrm{cov}(\mathbf{x}, \mathbf{x})$ and let $d_1 \geq d_2 \geq d_3 \geq \ldots$ be eigenvalues of the matrix Σ.

2. Let the series of eigenvalues of Σ $d_1 + d_2 + d_3 + \cdots$ converge, and let $d \overset{def}{=} \mathrm{tr}\Sigma > 0$.

We introduce three parameters characterizing the populations. Denote

$$M = \sup_{|\mathbf{e}|=1} \mathbf{E}(\mathbf{e}^T \mathbf{x})^4 > 0, \tag{1}$$

where (and in the following) $\mathbf{e}$ are nonrandom, infinite-dimensional unity vectors.

Define the parameter

$$k = \mathbf{E}(\mathbf{x}^2)^2 / (\mathbf{E}\mathbf{x}^2)^2 \geq 1,$$

where (and in the following) squares of vectors denote squares of their lengths.

Denote

$$\gamma = \sup_{\|\Omega\|=1} \text{var}(\mathbf{x}^T \Omega \mathbf{x})/\mathbf{E}(\mathbf{x}^2)^2 \leq 1, \tag{2}$$

where Ω are nonrandom, positive, semidefinite, symmetrical infinite-dimensional matrices of spectral norm 1 (we use only spectral norms of matrices). Denote $\rho_i = d_1/d$, $i = 1, 2, \ldots$, and $\rho = \rho_1$.

In the particular case of normal distributions $\mathbf{x} \sim \mathbf{N}(0, \Sigma)$, we have $M = 3\|\Sigma\|^2$, $k = 3$, and

$$\gamma = 2\frac{\text{tr}\Sigma^2}{\text{tr}\Sigma^2 + 2\,\text{tr}^2\Sigma} = 2\sum_{i=1}^{\infty} \frac{\rho_i^2}{1 + 2\rho_i^2} \leq 2\rho.$$

Let n components of the vector $\mathbf{x} \sim \mathbf{N}(0, \Sigma)$ have the variance 1, while the remaining components have variance 0. In this case, $\rho = 1/n$ and

$$\gamma = \frac{2\overline{r_{ij}^2}}{1 + 2\overline{r_{ij}^2}}, \quad \overline{r_{ij}^2} = \frac{1}{n}\sum_{i,j=1}^{n} r_{ij}^2,$$

where $r_{ij} = \Sigma_{ij}/\sqrt{d_i d_j}$ are correlation coefficients.

To apply methods developed in Section 3.1 to infinite-dimensional vectors, we introduce an asymptotics, preserving the basic idea of Kolmogorov: as $n \to \infty$, the summary variance d of variables has the same order of magnitude as the sample size N. We consider the quantities N and d as large, M, $d_1 \leq \sqrt{M}$, $b = d/N$ as bounded, and γ and ρ as small.

Dispersion Equations for Infinite Gram Matrices

First we investigate spectral properties of simpler random matrices of the form

$$S = N^{-1}\sum_{m=1}^{N} \mathbf{x}_m \mathbf{x}_m^T, \tag{3}$$

where $\mathbf{x}_m \in \mathfrak{X}$ are random, independent, identically distributed infinite-dimensional vectors, $m = 1, 2, \ldots, N$. The matrix S is sample covariance matrices of a special form when the expectation of variables is known a priori.

Obviously,

$$d_1/N = b\rho, \quad N^{-1}\mathrm{tr}\Sigma = b, \quad N^{-2}\mathrm{tr}\Sigma^2 \le b^2\rho, \quad \Sigma = \mathbf{E}S,$$
$$\mathbf{E}(\mathbf{x}^2)^2 \le kd^2, \quad \mathrm{var}(\mathbf{x}^2/N) \le kb^2\gamma,$$

and for $\mathbf{x} \sim \mathbf{N}(0,\Sigma), \quad \mathrm{var}(\mathbf{x}^2/N) \le 2b^2\rho$.

Consider the matrices $R_0 = R_0(t) = (S + tI)^{-1}$ and $H_0 = H_0(t) = SR_0(t), \ t > 0$. Let us apply the method of one-by-one exclusion of independent vectors $\mathbf{x}_m$. Let $t > 0$. Denote

$$S^m = S - N^{-1}\mathbf{x}_m\mathbf{x}_m^T, \qquad R_0^m = (S^m + tI)^{-1},$$
$$v_m = v_m(t) = \mathbf{e}^T R_0^m(t)\mathbf{x}_m, \qquad u_m = u_m(t) = \mathbf{e}^T R_0(t)\mathbf{x}_m,$$

$$\phi_m = \phi_m(t) = \mathbf{x}_m^T(S^m + tI)^{-1}\mathbf{x}_m/N,$$
$$\psi_m = \psi_m(t) = \mathbf{x}_m^T(S + tI)^{-1}\mathbf{x}_m/N,$$

$m = 1, 2, \ldots.$ We have the identities

$$R_0 = R_0^m - R_0^m\mathbf{x}_m\mathbf{x}_m^T R_0/N, \quad \psi_m = \phi_m - \phi_m\psi_m, \quad u_m = v_m - \psi_m v_m$$

and the inequalities

$$|\psi_m| \le 1, \quad |u_m| \le |v_m|, \quad \|R_0\| \le \|R_0^m\| \le 1/t, \quad \mathbf{E}v_m^4 \le M/t^4$$

$m = 1, 2, \ldots.$ To obtain upper estimations of the variance for functionals depending on large number of independent variables, we apply Lemma 3.2.

LEMMA 3.7. *If $t > 0$, the variance* $\mathrm{var}(\mathbf{e}^T R_0(t)\mathbf{e}) \le M/Nt^4$.

Proof. For each $m = 1, 2, \ldots, N$ in view of (4), we have the identity $\mathbf{e}^T R_0(t)\mathbf{e} = \mathbf{e}^T R_0^m(t)\mathbf{e} - u_m v_m/N$. Using the expansion in martingale-differences from Lemma 3.2, we find that

$$\mathrm{var}(\mathbf{e}^T R_0(t)\mathbf{e}) \le \sum_{m=1}^{N} N^{-2}|\mathbf{E}u_m v_m|^2 \le N^{-1}\mathbf{E}v_m^4.$$

We arrive at the statement of Lemma 3.7. $\square$

For normal observations, var $(\mathbf{e}^T R_0(t)\mathbf{e}) \leq 3bd_1\rho/t^4$.

LEMMA 3.8. *If $t > 0$, then for each $m = 1, 2, \ldots, N$, we have* var $\phi_m(t) \leq a\omega$, var $\psi_m(t) \leq a\omega$, *where*

$$\omega = ak\frac{b^2}{t^2}\left(\gamma + \frac{M}{Nt^2}\right), \tag{4}$$

and a are absolute constants.

Proof. Denote $\Omega = \mathbf{E}R_0^m(t)$, $\Delta_m = R_0^m(t) - \Omega$. Then,

$$1/2 \operatorname{var} \phi_m = \operatorname{var}(\mathbf{x}_m^T\Omega\mathbf{x}_m/N) + \mathbf{E}(\mathbf{x}_m^T\Delta_m\mathbf{x}_m)^2/N^2. \tag{5}$$

The first summand is not greater than $kb^2\gamma/t^2$. The second summand in (4) equals $\mathbf{E}(\mathbf{x}_m^2)^2/N^2\,\mathbf{E}(\mathbf{e}^T\Delta_m\mathbf{e})^2$, where random $\mathbf{e} = \mathbf{x}_m/|\mathbf{x}_m|$.

Note that $\mathbf{E}(\mathbf{e}^T\Delta_m\mathbf{e})^2 = \operatorname{var}(\mathbf{e}^T R_0^m\mathbf{e})$, where we can estimate the right-hand side using Lemma 3.2 with N less by 1 and some different $t > 0$. We find that

$$\mathbf{E}(\mathbf{x}_m^T\Delta_m\mathbf{x}_m)^2/N^2 \leq M\mathbf{E}(\mathbf{x}_m^2)^2/N^3t^4 \leq ak\,Mb^2/Nt^4$$

and consequently var $\phi_m \leq a\omega$, where ω is defined by (4). The first inequality is obtained. The second inequality follows from the identity $\psi_m = (1 + \phi_m)^{-1}\phi_m$. The lemma is proved. $\square$

For normal observations, the remainder term in (4) is equal to

$$\omega = a\left(1 + \frac{bd_1}{t^2}\right)\frac{b^2\rho}{t^2}.$$

Denote $s = s(t) = 1 - N^{-1}\operatorname{tr}\mathbf{E}H_0(t)$, $k = \mathbf{E}(\mathbf{x}^2)^2/(\mathbf{E}\mathbf{x}^2)^2$. It is easy to see that $k \geq 1$, and for $t > 0$ $|1 - s| \leq b/t$.

THEOREM 3.6. *If $t > 0$, we have*

$$\mathbf{E}H_0(t) = s(t)\,(s(t)\Sigma + tI)^{-1}\Sigma + \Omega,$$
$$1 - s(t) = s(t)/N\,\operatorname{tr}\left[(s(t)\Sigma + tI)^{-1}\Sigma\right] + o, \tag{6}$$

where Ω is a symmetric, positive, semidefinite infinite-dimensional matrix such that

$$\|\Omega\| \le a\sqrt{k}\tau^2 \left(g\sqrt{\gamma} + \frac{1+g\tau}{\sqrt{N}}\right),$$

$$|o| \le a\ k\tau^2 \left[g^2\sqrt{\gamma} + \frac{1+g\tau+g^2\tau}{\sqrt{N}}\right],$$

where (and in the following) $\tau = \sqrt{M}/t$, $g = b/\sqrt{M}$, and a is a numeric coefficient.

Proof. Let us multiply the first identity in (3) by $\mathbf{x}_m\mathbf{x}_m^T$ from the right and calculate the expectation. We find that

$$\mathbf{E}R_0 S = \mathbf{E}R_0^m\mathbf{x}_m\mathbf{x}_m^T - \mathbf{E}R_0^m\mathbf{x}_m\ \mathbf{x}_m R_0\mathbf{x}_m\ \mathbf{x}_m^T/N =$$
$$= \mathbf{E}R_0^m\Sigma - \mathbf{E}\psi_m R_0^m\mathbf{x}_m\mathbf{x}_m^T. \tag{7}$$

The quantity $\mathbf{E}\,\psi_m = \mathbf{E}\,\mathrm{tr}\,R_0\,S/N = \mathbf{E}N^{-1}\mathrm{tr}\,H_0 = 1 - s$. Denote $\delta_m = \psi_m - \mathbf{E}\psi_m$. We substitute in (7) $R_0 S = H_0$, $\psi_m = 1 - s + \delta_m$ and obtain

$$\mathbf{E}H_0 = s\mathbf{E}R_0^m\Sigma - \mathbf{E}R_0^m\mathbf{x}_m\mathbf{x}_m^T\delta_m. \tag{8}$$

Now in the first term of the right-hand side of (8), we replace the matrix R_0^m by R_0. Taking into account that $|s| \le 1 + b/t$, we conclude that the contribution of the difference $R_0^m - R_0$ in the first term of the right-hand side of (8) for some $\mathbf{e}$ is not greater in norm than

$$|s\mathbf{e}^T R_0^m\mathbf{x}_m\mathbf{x}_m^T R_0\Sigma\mathbf{e}|/N \le |s|d_1\mathbf{E}|v_m u_m|/N \le |s|\mathbf{E}v_m^2/N \le$$
$$\le (1 + b/t)\sqrt{M}d_1/Nt^2.$$

The second term of right-hand side of (8) in norm is not greater than

$$\mathbf{E}|v_m(\mathbf{x}_m^T\mathbf{e})\delta_m| \le \sqrt{\mathbf{E}v_m^2(\mathbf{x}_m^T\mathbf{e})^2\mathbf{E}\delta_m^2} \le a\sqrt{M}\omega/t,$$

where ω is defined by (4). We obtain the upper estimate in norm

$$\|\mathbf{E}H_0 - s\mathbf{E}R_0\Sigma\| \le a\sqrt{k}\ \tau^2(g\sqrt{\gamma} + \frac{1+g\tau}{\sqrt{N}}).$$

The first statement of our theorem is proved.

Further, the first term of the right-hand side of (8) equals $s\mathbf{E}R_0\Sigma$ plus the matrix Ω_1, for which

$$|N^{-1}\mathrm{tr}\,\Omega_1| = a|\mathbf{Ex}_m^T R_0^m \Sigma R_0\mathbf{x}_m/N| \le$$
$$\le |s|d_1\mathbf{E}|u_m v_m|/N \le (1+b/t)d_1\sqrt{M}/Nt^2.$$

The second term of the right-hand side of (8) is the matrix Ω_2, for which similarly

$$|N^{-1}\mathrm{tr}\,\Omega_2| \le N^{-1}\mathbf{E}|(\mathbf{x}_m^T R_0^m \mathbf{x}_m)\delta_m| \le$$
$$\le N^{-1}\sqrt{t^{-2}\mathbf{E}(\mathbf{x}_m^2)^2\mathbf{E}\delta_m^2} \le ab\,\sqrt{k\omega}/t.$$

Combining these two upper estimates we obtain the theorem statement. $\square$

Special case 1: pass to finite-dimensional variables

Let for some n, $d_1 \le d_2 \le \ldots \le d_n > 0$, and for all $m > n, d_m = 0$. Denote by y the Kolmogorov ratio $y = n/N$. Consider the spectral functions

$$\widetilde{H}_0 = \widetilde{H}_0(z) = (I+zS)^{-1}, \quad \widetilde{h} = \widetilde{h}(z) = n^{-1}\mathrm{tr}\,\widetilde{H}_0(z).$$

Note that the matrix $R_0(z) = z^{-1}\widetilde{H}_0(z^{-1})$, and $\widetilde{s}(z) \overset{\text{def}}{=} 1 - y + yh(z) = s(t)$, where $t = 1/z > 0$. The parameter γ introduced in Section 3.1 is not greater than the parameter γ in Section 3.2. By Lemma 3.3 from [11] (in the present notations), we obtain that $\mathrm{var}\,\psi_m \le 2b^2/t^2(\gamma + M/Nt^2)$, that strengthens the estimates in Lemma 3.8 owing to the absence of the coefficient $\sqrt{k} \ge 1$.

The dispersion equations in Sections 3.1 and 3.2 have the form

$$\mathbf{E}\widetilde{H}_0(z) = (I+z\widetilde{s}(z)\Sigma)^{-1} + \widetilde{\Omega} =$$
$$= \widetilde{s}(z) = 1 - y + N^{-1}\mathrm{tr}\,(I+zS)^{-1} =$$
$$= 1 - y + N^{-1}\mathrm{tr}(I+z\widetilde{s}(z)\Sigma)^{-1} + \widetilde{o},$$

$\|\widetilde{\Omega}\| \le a\tau^2(y\sqrt{\gamma}+1/N)$ and $|\widetilde{o}| = y\|\widetilde{\Omega}\|$ with the remainder terms different by lack of the coefficient $k \ge 1$.

Thus, Theorem 3.6 is a generalization of Theorem 3.1 and also of limit theorems proved in monographs [21–25].

Dispersion Equations for Sample Covariance Matrices

Let the standard sample covariance matrix be of the form

$$C = N^{-1} \sum_{m=1}^{N} (\mathbf{x}_m - \bar{\mathbf{x}})(\mathbf{x}_m - \bar{\mathbf{x}})^T, \qquad \bar{\mathbf{x}} = N^{-1} \sum_{m=1}^{N} \mathbf{x}_m,$$

where $\mathbf{x}_m$ are infinite-dimensional sample vectors and $\bar{\mathbf{x}}$ is the sample average vector. Note that $C = S - \bar{\mathbf{x}}\,\bar{\mathbf{x}}^T$. We use this relation to pass from spectral functions of S to spectral functions of C.

Remark 1. $\mathbf{E}\bar{\mathbf{x}}^2 = b$, $\mathbf{E}(\bar{\mathbf{x}}^2)^2 \leq (2k+1)\,b^2$.
Indeed, the first equality is obvious. Next we find

$$\mathbf{E}(\bar{\mathbf{x}}^2)^2 = b^2 + \operatorname{var}\bar{\mathbf{x}}^2 = b^2 + 2N^{-2}(1 - N^{-1})\operatorname{tr}\Sigma^2.$$

Here

$$\operatorname{tr}\Sigma^2 = \sum_{i,j=1}^{\infty} [\mathbf{E}(\mathsf{x}_i\mathsf{x}_j)^2] \leq \sum_{i,j=1}^{\infty} \mathbf{E}(\mathsf{x}_i\mathsf{x}_j)^2 = \mathbf{E}(\mathbf{x}^2)^2 \leq kd^2.$$

The required inequality follows.

Denote

$$R = R(t) = (C + tI)^{-1}, \qquad H(t) = R(t)C,$$
$$\Phi = \Phi(t) = \bar{\mathbf{x}}^T(S + tI)^{-1}\bar{\mathbf{x}},$$
$$\Psi = \Psi(t) = \bar{\mathbf{x}}^T(C + tI)^{-1}\bar{\mathbf{x}},$$
$$V = V(t) = \mathbf{e}^T R_0\,\bar{\mathbf{x}}, \qquad U = U(t) = \mathbf{e}^T R\,\bar{\mathbf{x}}.$$

We have the identities $(1 + \Psi)(1 - \Phi) = 1$, $U = V + U\Phi$.
It is easy to check that the positive $\Phi \leq 1$.

LEMMA 3.9. *If $t > 0$, then*

$$|\mathbf{E}V|^2 \leq \sqrt{M}\omega/t^2, \qquad \operatorname{var}V \leq a\,\frac{\sqrt{M}}{Nt^2}\left(1 + \sqrt{kM}\,b/t^2\right),$$

where a is a numeric coefficient.

Proof. Note that for each $m = 1, 2, \ldots, N$, we have $\mathbf{E}v_m = 0$ and $\mathbf{E}V = \mathbf{E}u_m = -\mathbf{E}v_m(\psi_m - \mathbf{E}\psi_m)$. It follows that

$$|\mathbf{E}V|^2 \leq \mathbf{E}v_m^2 \quad \operatorname{var} \psi_m \leq a\sqrt{M}\omega/t^2.$$

Further, we estimate the variance of V using Lemma 3.2. Denote $\widetilde{\mathbf{x}} = \overline{\mathbf{x}} - N^{-1}\mathbf{x}_m$, where $\mathbf{x}_m$ is the vector excluded from the sample. Using the first of equations (4), we rewrite V in the form

$$V = \mathbf{e}^T R_0^m \widetilde{\mathbf{x}} + \mathbf{e}^T R_0 \mathbf{x}_m/N - \mathbf{e}^T R_0 \mathbf{x}_m \ \mathbf{x}_m^T R_0^m \ \widetilde{\mathbf{x}}/N,$$

where the first summand does not depend on $\mathbf{x}_m$. In view of Lemma 3.2,

$$\operatorname{var} V \leq N^{-2} \sum_{m=1}^{N} \mathbf{E}v_m^2(1 - w_m)^2 \leq 2N^{-1}\sqrt{\mathbf{E}v_m^4(1 + \mathbf{E}w_m^4)},$$

where by definition $w_m = \mathbf{x}_m^T R_0^m \widetilde{\mathbf{x}}$. Note that $\mathbf{E}w_m^4 \leq M\mathbf{E}(\widetilde{\mathbf{x}}^2)^2/t^4 \leq 3Mkb^2/t^4$. We obtain the statement of Lemma 3.9. $\square$

LEMMA 3.10. *If $t > 0$, the spectral norm*

$$\|\mathbf{E}(R(t) - R_0(t))\| \leq a\frac{k\tau}{t}\left[A\gamma + \frac{B}{\sqrt{N}}\right],$$

$$\mathbf{E}\frac{1}{N} \operatorname{tr}(H(t) - H_0(t)) \leq b/Nt,$$

where $A = g^2\tau^2$, $B = 1 + g + g\tau + g^2\tau^4$, and the coefficient a is a number.

Proof. By definition $R - R_0 = R \ \overline{\mathbf{x}} \ \overline{\mathbf{x}}^T R_0$, and there exists a nonrandom vector $\mathbf{e}$ such that $\|\mathbf{E}(R - R_0)\| = \mathbf{E}(\mathbf{e}^T R \overline{\mathbf{x}} \ \overline{\mathbf{x}}^T R_0 \mathbf{e}) = \mathbf{E}UV$. Substitute $U = V + U\Phi$ and find that

$$|\mathbf{E}UV| = |\mathbf{E}V^2 + \mathbf{E}UV\Phi| \leq \mathbf{E}V^2 + \sqrt{\operatorname{var} V}\sqrt{\mathbf{E}U^2\Phi^2}.$$

Here $\Phi \leq 1$, $\mathbf{E}U^2 \leq \mathbf{E}\overline{\mathbf{x}}^2/t^2$. By Lemma 3.9, the expression in the right-hand side is not greater than

$$(\mathbf{E}V)^2 + \operatorname{var} V + \sqrt{\operatorname{var} V}\ \sqrt{b/t^2}.$$

Substituting upper estimates of V, and var V from Lemma 3.9, and ω from (4), we arrive at the statement of Lemma 3.5. $\square$

THEOREM 3.7. *If $t > 0$, then*

$$\mathbf{E}H(t) = s(t)(s(t)\Sigma + tI)^{-1}\Sigma + \Omega,$$

where $\|\Omega\| \le k\tau(A\sqrt{\gamma} + B/\sqrt{N})$, $A = g\tau(1 + g\tau)\sqrt{\gamma}$, $B = (1 + g)(1 + \tau) + g^2\tau^4$, and $\tau = \sqrt{M}/t$, $g = b/\sqrt{M}$.

Proof. We have $H_0(t) - H(t) = t(R(t) - R_0(t))$, where the latter difference was estimated in Lemma 3.10. Combining upper estimates of the remainder terms in Lemma 3.10 and Theorem 3.10, we obtain the statement of Theorem 3.7. $\square$

One can see that the principal parts of expectation of matrices $H(t)$ and $H_0(t)$ do not differ in norm asymptotically.

Limit Spectral Equations

Now we formulate special asymptotic conditions preserving the main idea of the Kolmogorov asymptotics: the sum of variances of all variables increases with the same rate as sample size.

Consider a sequence

$$\mathfrak{P} = (\mathfrak{F}, M, \Sigma, k, \gamma, \mathfrak{X}, S)_N, \quad N = 1, 2, \ldots$$

of problems of the investigation of infinite-dimensional matrices $\Sigma = \mathrm{cov}(\mathbf{x}, \mathbf{x})$ by sample covariance matrices S of the form (3) calculated over samples $\mathfrak{X}$ of size N from infinite-dimensional populations with the parameters M and γ defined above (we omit the subscripts N in arguments of $\mathfrak{P}$). Consider the empiric distribution function of eigenvalues d_i of the matrix Σ

$$G_N^0(v) = N^{-1}\sum_{i=1}^{\infty} d_i \cdot \mathrm{ind}(d_i < v). \tag{9}$$

Assume the following.

A. For each N, the parameter $M < c_1$, where c_1 does not depend on N.

B. As $N \to \infty$, the parameter $k < c_2$, where c_2 does not depend on N.

C. As $N \to \infty$, $b = b_N \overset{def}{=} \mathrm{tr}\Sigma/N \to \beta > 0$.

D. As $N \to \infty$, the quantities $\gamma = \gamma_N \to 0$.

E. Almost for all $v > 0$, the limit exists

$$G^0(v) = \lim_{N \to \infty} G_N^0(v).$$

Obviously (as in the above), $G_N^0(0) = 0$, and for almost all N, $G_N^0(c_3) = G^0(c_3) = \beta$, where $c_3 = \sqrt{M}$.

Denote (as in the above) $R_0(t) = (S + tI)^{-1}$, $H_0(t) = R_0(t)S$, $s(z) = s_N(z) = 1 - \mathbf{E}N^{-1} \mathrm{tr}H_0(z)$, and $H(t) = R(t)C$.

THEOREM 3.8. *Under Assumptions A–E for $z = t > 0$ and for* Im $z = t > 0$ *as $N \to \infty$, the convergence holds*

$$s(z) = s_N(z) \overset{def}{=} 1 - \mathbf{E}N^{-1} \mathrm{tr}H_0(z) \to \sigma(z),$$
$$1 - \mathbf{E}N^{-1} \mathrm{tr}H(z) \to \sigma(z), \tag{10}$$

where the function

$$\sigma(z) = 1 - \sigma(z) \int \frac{1}{\sigma(z)v + z} \, dG^0(v). \tag{11}$$

Proof. Let us rewrite the second of the equations (5) introducing the distribution function $G_N^0(v)$:

$$1 = s(z) + s(z) \int (s(z)v + t)^{-1} dG_N^0(v) + o_N, \tag{12}$$

where the remainder term o_N vanishes as $N \to \infty$. Here the function under the sign of the integral is continuous and bounded, and the limit transition is possible $G_N^0(v) \to G^0(v)$. The right-hand side of (12) with the accuracy up to o_N depends monotonously on $s = s_N$ and for each N uniquely defines the quantity $s_N(t)$;

one can conclude that for each $t > 0$ as $N \to \infty$, the limit exists $\sigma(t) = \lim s_N(t)$. We obtain the first assertion of the theorem.

The second assertion of Theorem 3.8 follows from Lemma 3.10.

□

Now consider empirical spectral functions of the matrices S and C:

$$G_{0N}(u) \stackrel{def}{=} \mathbf{E}N^{-1} \sum_{i=1}^{\infty} \lambda_i \cdot \mathrm{ind}(\lambda_i < u), \quad u > 0,$$

where λ_i are eigenvalues of S, and

$$G_N(u) \stackrel{def}{=} \mathbf{E}N^{-1} \sum_{i=1}^{\infty} \lambda_i \cdot \mathrm{ind}(\lambda_i < u), \quad u > 0,$$

where λ_i are eigenvalues of C. Obviously, $G_{0N}(0) = G_N(0) = 0$, and for all $u > 0$, we have $G_{0N}(u) \leq b$, $G_N(u) \leq b$, and we find that

$$s(z) = s_N(z) = 1 - \int (u + z)^{-1} dG_{0N}(u). \qquad (13)$$

From (13) it follows that the function $\sigma(z)$ is regular for $z > 0$ and allows analytical continuation in the region outside of the half-axis $z < 0$.

We assume that for all z not lying on the half-axis $z \leq 0$, the Hölder condition is valid

$$|\sigma(z) - \sigma(z')| \leq a \, |z - z'|^{\delta}, \qquad (14)$$

where a, $\delta > 0$.

THEOREM 3.9. *Let Assumptions A–E and (14) be valid. Then,*

1. there exists a monotonously increasing function $G(u) \leq \beta$ such that as $N \to \infty$ almost everywhere for $u > 0$, the convergence to a common limit holds $G_{0N}(u) \to G(u)$, $G_N(u) \to G(u)$;

2. for all z not lying on the half-axis $z \le 0$,

$$1 - \sigma(z) = \int \frac{1}{u+z} \, dG(u); \tag{15}$$

3. for $u > 0$ almost everywhere

$$G'(u) = \pi^{-1} \lim_{\varepsilon \to +0} \sigma(-u + i\varepsilon), \quad u > 0. \tag{16}$$

Proof. Set $G_{0N}(u) = bF_N(u)$, where $F_N(u)$ is some monotonous function with the properties of a distribution function. Note that for fixed $z = -u + i\varepsilon$, where $u > 0$, $\varepsilon > 0$, the quantity Im $s(z)/b\pi$ presents an ε-smoothed density $F_N(u)$. From the sequence of functions $\{F_N(u)\}$, we extract any two convergent subsequences $F_N^\nu(u) \to F^\nu(u)$, $\nu = 1, 2$. Let $u > 0$ be a continuity point of $F^1(u)$ and of $F^2(u)$ as well. Then, smoothed densities of distributions $F_N^1(u)$ and $F_N^2(u)$ converge as $N \to \infty$ and converge to a common limit

$$F(u, \varepsilon) = \frac{1}{\pi\beta} \int \text{Im } \sigma(-u' + i\varepsilon) \, du'. \tag{17}$$

In view of the Hölder condition for each $u > 0$ and $\varepsilon \to +0$, there exists the limit $\sigma(-u) = \lim \sigma(-u + i\varepsilon)$ and the limit

$$F(u) = \lim_{\varepsilon \to +0} F(u, \varepsilon) = \lim_{\varepsilon \to +0} \lim_{N \to \infty} F_N(u). \tag{18}$$

Denote $G(u) = \beta F(u)$. This is just the required limit for $G_N(u)$.

Further, for $z = -u + i\varepsilon$, $u, \varepsilon > 0$, then the functions $f(z) = -\mathbf{E}$ Im N^{-1} tr $H(z)/b\pi$ present smoothed densities for $G_N(u)/b$. By Assumption E and (9), these functions converge as $N \to \infty$ and converge to the derivative of $F(u, \varepsilon)$. As $\varepsilon \to +0$, Im $f(z) \to G'(u)$. Equation (15) follows from (13). Theorem 3.9 is proved. $\square$

Equations (13) and (15) establish a functional dependence between spectra of the matrices Σ and of S and C.

Let the region $\mathfrak{S} = \{u : u > 0 \ \& \ G'(u) > 0\}$ be called the region of (limit) spectrum of matrices S.

THEOREM 3.10. *Let Assumptions A–E hold, the condition* (14) *is valid and, moreover, as* $N \to \infty$, *the quantities* $d_1 \to \alpha > 0$.

Then in the spectrum region for real $z = -u > 0$, $u \in \mathfrak{S}$, *the function* $\sigma(z)$ *is defined and the following relations are valid.*

1. $\displaystyle u \int \frac{1}{|\sigma(z)v - u|^2} \, dG^0(v) = 1.$

2. $\displaystyle |\sigma(z)|^2 \int \frac{v}{|\sigma(z)v - u|^2} \, dG^0(v) = 1.$

3. $u \leq \alpha \, |\sigma(z)|^2.$

4. $\displaystyle 1 - \sqrt{\frac{\beta}{u}} \leq \mathrm{Re} \, \frac{1}{\sigma(z)} \leq 1 + \sqrt{\frac{\beta}{u}}.$

5. $u \leq (\sqrt{\alpha} + \sqrt{\beta})^2.$

Proof. Let $z = -u + i\varepsilon$, $u, \varepsilon > 0$. Denote $\sigma = \sigma(z)$, $\sigma_0 = \sigma_0(z) = \mathrm{Re} \, \sigma(z)$, $\sigma_1 = \sigma_1(z) = \mathrm{Im} \, \sigma(z)$,

$$I_k = I_k(z) = \int \frac{v^k}{|v\sigma(z) + z|^2} \, dG^0(v). \tag{19}$$

In view of the Hölder condition, the functions $\sigma(z)$ and $I_k(z), k = 0, 1$ are bounded and continuous for each $u > 0$ and $\varepsilon \to +0$, and are defined and continuous for $z = -u < 0$. From (15), one can see that the sign of $\sigma_1(z)$ coincides with the sign of $\varepsilon > 0$. Let us fix some $u \in \mathfrak{S}$, $\sigma_1 = \sigma_1(-u) > 0$. The denominator under the integral sign in (19) is not smaller than $u^2 \sigma_1^2 / |\sigma|^2$ and does not vanish for any fixed $u > 0$ and $\varepsilon \to +0$. We equate the imaginary parts in the equality (11). It follows that $1 - uI_0 = I_0 \sigma_0 \varepsilon / \sigma_1$. As $\varepsilon \to +0$, the quantity $uI_0 \to 1$. We obtain the limit relation in the first assertion of the theorem.

Let us divide both parts of (11) by σ and compare the imaginary parts. We find that $I_1 - |\sigma|^{-2} = I_0 \varepsilon / \sigma_1 \to 0$. The second statement of our theorem follows.

Note that the integration region in (19) is bounded by a segment $[0, \alpha]$. Consequently, $I_1 \leq \alpha I_0$. Substituting limit values of I_0 and I_1, we obtain Statement 3.

Let us apply the Cauchy–Bunyakovskii inequality to equation (11). It follows that for $z = -u < 0$

$$\left| \frac{1 - \sigma(z)}{\sigma(z)} \right|^2 \leq \beta \int \frac{1}{|\sigma(z)v + z|^2} \, dG^0(v) = \frac{\beta}{u},$$

We get Statement 4 of our theorem.

Consider these inequalities on the boundary at the point $u \in \mathfrak{S}$, where $\sigma_1 = 0$. At this point for $u \geq \beta$, we have the following inequality $\sqrt{\alpha/u} \geq |\sigma|^{-1} \geq |\sigma_0/\sigma^2| \geq 1 - \sqrt{\beta/u}$.

We arrive at the fifth statement of the theorem. Theorem 3.10 is proved. $\square$

Model 1. Consider infinite-dimensional vectors $\mathbf{x}$ and let sample size $N = 1, 2, \ldots$, while only n eigenvalues of Σ are different from 0 so that the ratio $n/N \to y > 0$ as $N \to \infty$. This case presents the increasing-"dimension asymptotics" that is the basic approach of the well-known spectral theory of random matrices by V. L. Girko and of the theory of essentially multivariate analysis [71].

Now we consider a special case when for each N (and n), all n nonzero eigenvalues of the matrix Σ coincide: $\alpha = d_1 = d_2 = \cdots = d_n$. Then, the function $G^0(v) = y\alpha \, \text{ind}(v < \alpha)$, and we obtain from (15) as $\varepsilon \to +0$:

$$(1 - \sigma(z))(\sigma(z)\alpha - u) = y\alpha\sigma(z),$$

where $z = -u < 0$. This is a quadratic equation in $\sigma = \sigma(z)$. Let us rewrite it in the form $A\sigma^2 + B\sigma + C = 0$, where $A = \alpha$, $B = \alpha(1 - y) + u$, $C = u$.

Starting from (15), we find the limit derivative

$$G'(u) = \pi^{-1} \text{Im } \sigma(z) = \frac{1}{2\pi\alpha} \text{Im} \sqrt{B^2 - 4AC} =$$
$$= \frac{1}{2\pi\alpha} \sqrt{(u_2 - u)(u - u_1)}, \quad u > 0,$$

where $u_{2,1} = \alpha(1 \pm \sqrt{y})^2$. For $y \neq 1$, the density of the limit distribution $F(u)$ of nonzero eigenvalues of S equals $F'(u) = G'(u)/yu$, while for $y > 1$, the function $F(u)$ has a jump $(1 - y)\alpha$ at the point $u = 0$. This limit spectrum was first found under a finite-dimensional setting by Marchenko and Pastur in [43], 1967.

Model 2. We make Model 1 more complicated by adding an infinite "tail" of identical infinitesimal eigenvalues to $n = n_1$ nonzero eigenvalues of Σ. For fixed N and increasing $n = n(N)$, $m = m(N)$, let $d_1 = d_2 = \ldots, d_n = \alpha > 0$, and $d_{n+1} = d_{n+2} = d_{n+m} = \delta > 0$, so that the ratios $n/N \to y_1 > 0$, $m/N \to y_2 > 0$, and $b_1 = y_1\alpha \to \beta_1 > 0$ and $b_2 = y_2\delta \to \beta_2 > 0$ while $\delta \to 0$. For $s = s(t)$, $t > 0$, from equation (15), we obtain the relation

$$\frac{1 - s(t)}{s(t)} = \frac{\beta_1}{\alpha s(t) + z} + \frac{\beta_2}{\delta s(t) + z} + o_N. \qquad (20)$$

As $N \to \infty$, we have $s = s(t) \to \sigma(t)$, the remainder term o_N vanishes, and equation (20) passes to a quadratic equation with respect to $\sigma(t)$. Rewrite it in the form $A\sigma^2 + B\sigma + C = 0$, where $A = \alpha(\beta_2 + z)$, $B = z(\beta_1 + \beta_2 + z - \alpha)$, $C = -z^2$, $z > 0$. We perform the analytical continuation of $\sigma(z)$ to complex arguments $z = -u + i\varepsilon$, $u, \varepsilon > 0$. Solving the Stilties equation we obtain

$$G(u_2) - G(u_1) = \lim_{\varepsilon \to +0} \pi^{-1} \int_{u_1}^{u_2} \operatorname{Im} \sigma(-u + i\varepsilon)\, du =$$

$$= \lim_{\varepsilon \to +0} \pi^{-1} \int_{u_1}^{u_2} \frac{\sqrt{B^2 - 4AC}}{2|A|}\, du. \qquad (21)$$

First, we study possible jumps of the function $G(u)$. They are produced when the coefficient A vanishes, i.e., when $z = -\beta_2$. The integration in the vicinity of $u = \beta_2 > 0$ in (21) reveals a jump

$$(2\pi d_1)^{-1}\sqrt{\operatorname{Re}(B^2 - 4AC)}$$

that equals $\beta_2(y_1 - 1)/2$ for $y_1 > 1$ and 0 for $y_1 \leq 1$. We conclude that for $y_1 < 1$, the function $G(u)$ is continuous everywhere for $u \geq 0$, and for $y_1 > 1$, there is a discontinuity at the point $u = \beta_2$ with a jump equal to $\beta_2(y_1 - 1)$.

The continuous part of $G(u)$ is defined by equation (21). In the case of Model 2, the limit spectral density of S equals

$$
u^{-1}G'(u) = \begin{cases} \dfrac{1}{2\pi\alpha|u - \beta_2|}\sqrt{(u_2 - u)(u - u_1)}, & \text{if } u_1 \le u \le u_2, \\ 0, & \text{if } u < u_1, \ u > u_2, \end{cases}
$$

where $u_{2,1} = \beta_2 + \alpha(1 \pm \sqrt{y})^2$, $y = \beta_1/\alpha$. The influence of a "background" of small variances shifts the spectrum by the quantity β_2. It is easy to examine that these boundaries stay within the upper bound, established by Statement 5 of Theorem 3.10.

Model 3. Let eigenvalues of the matrix Σ be $\lambda_k = a^2/(a^2 + N^{-2}k^2)$, $a > 0$, $k = 0, 1, 2, \ldots$. As $N \to \infty$

$$
b = \frac{1}{N}\sum_{k=0}^{\infty}\lambda_k \to \int_0^{\infty}\frac{a^2}{a^2 + t^2}\,dt = \frac{\pi a}{2} = \beta.
$$

The function

$$
G_N^0(v) \overset{def}{=} \frac{1}{N}\sum_{k=0}^{\infty}\lambda_k \operatorname{ind}(\lambda_k < v) \to G^0(v) \overset{def}{=} a \cdot \operatorname{arctg}\sqrt{\frac{v}{1-v}},
$$
$$
\text{and} \quad \frac{dG^0(v)}{dv} = \frac{a}{2\sqrt{v(1-v)}}, \qquad 0 < v < 1.
$$

Let $u > 0$, $\sigma = \sigma(-u)$. The dispersion equation (15) is transformed to

$$
\frac{\sigma - 1}{\sigma} = a^2\int\frac{1}{t^2 + a^2(u - \sigma)}\,dt = \frac{\beta}{\sqrt{u(u-\sigma)}}. \tag{22}
$$

The quantity $\operatorname{Im}\sigma(z)$ is connected with the spectral function $G'(u)$ by (15). Therefore, outside of the spectrum region, the function $\sigma(-u)$ is real. This function monotonously decreases with the increase of $u > 0$ and $\sigma(-u) \ge 1$. From (22), it follows that $u \ge \sigma(-u)$.

For small $u > 0$, equation (22) has no solution. But this equation is equivalent to the cubic equation

$$
\varphi(\sigma, u) = a_3\sigma^3 + a_2\sigma^2 + a_1\sigma^1 + a_0 = 0, \tag{23}
$$

where $a_3 = u$, $a_2 = \beta^2 - u^2 - 2u$, $a_1 = 2u^2 + u$, and $a_0 = -u^2$, solvable in complex numbers. We conclude that for small $u > 0$, the equation (23) has only complex roots. For $\operatorname{Im} \sigma > 0$, $G'(u) > 0$. Consequently, small $u > 0$ are included in the spectrum region $\mathfrak{S}$. Denote $\sigma_0 = \operatorname{Re} \sigma$, $\sigma_1 = \operatorname{Im} \sigma$.

In the spectrum region, the imaginary part of $\varphi(\sigma, u)$ is proportional to $\sigma_1 > 0$ with the coefficient

$$a_3(3\sigma_0^2 - \sigma_1^2) + 2a_2\sigma_0 + a_1 = 0. \tag{24}$$

By Theorem 3.10, the spectrum region is bounded, and there exists an upper spectrum boundary $u_B > 0$. This quantity may be calculated by solving simultaneously the system of two equations (23) and (24) with $\sigma_1 = 0$. From Theorem 3.10 and equation (22), we obtain the inequality

$$\sqrt{1 + \beta^2} \leq u_B \leq (1 + \sqrt{\beta})^2.$$

One can prove that the spectral density $G'(u)$ has vertical derivatives at the boundary $u = u_B$.

3.3. NORMALIZATION OF QUALITY FUNCTIONS

In this section, following [72], we prove that in case of high dimension of variables, most of standard rotation invariant functionals measuring the quality of regularized multivariate procedures may be approximately, but reliably, evaluated under the hypothesis of population normality. This effect was first described in 1995 (see [72]). Thus, standard quality functions of regularized procedures prove to be approximately distribution free. Our purpose is to investigate this phenomenon in detail. We

1. study some classes of functionals of the quality function type for regularized versions of mostly used linear multivariate procedures;

2. single out the leading terms and show that these depend on only two moments of variables;

3. obtain upper estimates of correction terms accurate up to absolute constants.

We restrict ourselves with a wide class of populations with four moments of all variables.

Let $\mathbf{x}$ be an observation vector in n-dimensional population $\mathfrak{S}$ from a class $\mathfrak{K}_4^0(M)$ of populations with $\mathbf{Ex} = 0$ and the maximum fourth moment

$$M = \sup \ \mathbf{E}(\mathbf{e}^T\mathbf{x})^4 > 0, \qquad (1)$$

where the supremum is calculated over all $\mathbf{e}$, and $\mathbf{e}$ (here and in the following) are nonrandom vectors of unit length. Denote $\Sigma = \mathrm{cov}(\mathbf{x}, \mathbf{x})$. We consider the parameter

$$\gamma = \sup_{\|\Omega\|=1} \ [\mathbf{x}^T\Omega \ \mathbf{x}/n]/M, \qquad (2)$$

where Ω are nonrandom, symmetrical, positive semidefinite $n \times n$ matrices.

Let $\mathfrak{X} = (\mathbf{x}_1, \ldots, \mathbf{x}_N)$ from $\mathfrak{S}$ be a sample from $\mathfrak{S}$ of size N. We consider the statistics

$$\bar{\mathbf{x}} = \tfrac{1}{N} \sum_{m=1}^{N} \mathbf{x}_m, \quad S = \tfrac{1}{N} \sum_{m=1}^{N} \mathbf{x}_m \mathbf{x}_m^T,$$

$$C = \tfrac{1}{N} \sum_{m=1}^{N} (\mathbf{x}_m - \bar{\mathbf{x}})(\mathbf{x}_m - \bar{\mathbf{x}})^T.$$

Here S and C are sample covariance matrices (for known and for unknown expectation vectors).

Normalization Measure

Definition 1. We say that function $f : \mathbb{R}^n \to \mathbb{R}^1$ of a random vector $\mathbf{x}$ is ε-*normal evaluable* (in the square mean) in a class of n-dimensional distributions $\mathfrak{S}$, if for each $\mathfrak{S}$ with expectation vector $\mathbf{a} = \mathbf{E}\mathbf{x}$ and covariance matrix $\Sigma = \text{cov}(\mathbf{x}, \mathbf{x})$, we can choose a normal distribution $\mathbf{y} \sim \mathbf{N}(\mathbf{a}, \Sigma)$ such that

$$\mathbf{E}(f(\mathbf{x}) - f(\mathbf{y}))^2 \leq \varepsilon.$$

We say that function $f : \mathbb{R}^n \to \mathbb{R}^1$ is *asymptotically normal evaluable* if it is ε-evaluable with $\varepsilon \to 0$.

Example 1. Let $n = 1$, $\xi \sim \mathbf{N}(0, 1)$, $x = \xi^3/\sqrt{15}$. Denote the distribution law of x by $\mathfrak{S}$. Then, the function $f(x) = x$ is ε-normal evaluable (by normal $y = \xi$) in $\mathfrak{S}$ with $\varepsilon = 0.45$.

Example 2. Let $\mathfrak{S}$ be an n-dimensional population with four moments of all variables and let $f(t)$ be a continuous differentiable function of $t \geq 0$ that has a derivative not greater b in absolute value. Then, $f(\bar{\mathbf{x}}^2)$ is ε-normal evaluable with $\varepsilon = c/N$.

Example 3. In Section 3.1, the assumptions were formulated, which provide the convergence of entries for the resolvents of sample covariance matrices S and C as $n \to \infty$, $N \to \infty$, and $n/N \to \lambda > 0$. Limit entries of these resolvents prove to be depending only on limit values of spectral functions of the matrix Σ. By the above definition, the entries of these resolvents are asymptotically normal evaluable as $n \to \infty$.

Spectral Functions of Sample Covariance Matrices

We consider the resolvent-type matrices $H_0 = H_0(t) = (I + A + tS)^{-1}$ and $H = H(t) = (I + A + tC)^{-1}$, where (and in the following) I denotes the identity matrix and A is a positively semidefinite symmetric matrix of constants. Define spectral functions

$$h_0(t) = n^{-1} \operatorname{tr} H_0(t), \quad h(t) = n^{-1} \operatorname{tr} H(t),$$
$$s_0(t) = 1 - y + y h_0(t), \quad s(t) = 1 - y + y h(t), \quad \text{where } y = n/N.$$

The phenomenon of normalization arises as a consequence of properties of spectral functions that were established in Section 3.1. We use theorems that single out the leading parts of spectral functions for finite n and N.

The upper estimates of the remainder terms are estimated from the above by functions of n, N, M, γ, and $t \geq 0$.

Our resolvents $H_0(\mathrm{t})$ and $H(\mathrm{t})$ differ from those investigated in Section 3.1 by the addition of nonzero matrix A. The generalization can be performed by a formal reasoning as follows. Note that the linear transformation $\mathbf{x}' = B\mathbf{x}$, where $B^2 = (I + A)^{-1}$, takes H_0 to the form $H_0' = B(I + tS')^{-1}B$, where

$$S' = N^{-1} \sum_{m=1}^{N} \mathbf{x}'\mathbf{x}'^{T}, \quad \mathbf{x}'_m = B\,\mathbf{x}_m, \quad m = 1, 2, \ldots, N.$$

It can be readily seen that $M' = \sup \mathbf{E}(\mathbf{e}^T\mathbf{x}')^4 \leq M$, and the products $M'\gamma' \leq M\gamma$, where γ' is the quantity (2) calculated for $\mathbf{x}'$. Let us apply results of Section 3.1 to vectors $\mathbf{x}' = B\mathbf{x}$. The matrix elements of our H_0 can be reduced to matrix elements of the resolvents of S' by the linear transformation B with $\|B\| \leq 1$. The remainder terms for $\mathbf{x}' = B\mathbf{x}$ are not greater than those for $\mathbf{x}$. The same reasoning also holds for the matrix H. A survey of upper estimates obtained in Sections 3.1 shows that all these remain valid for the new H_0 and H.

Let us formulate a summary of results obtained in Section 3.1, which will be a starting point for our development below. To be more concise in estimates, we denote by $p_n(t)$ polynomials of a fixed degree with respect to t with numeric coefficients and $\omega = \sqrt{\gamma + 1/N}$.

LEMMA 3.11. (corollary of theorems from Section 3.1). *If* $t \geq 0$ *and* $\mathfrak{S} \in \mathfrak{K}_4(M)$, *then*

1. $\mathbf{E}H_0(t) = (I + A + ts_0(t)\Sigma) + \Omega_0$,

   where $\|\Omega_0\| \leq p_3\omega$;

2. $\mathrm{var}(\mathbf{e}^T H_0 \mathbf{e}) \leq a \, Mt^3/N$;

3. $t\mathbf{E}\bar{\mathbf{x}}^T H_0(t)\bar{\mathbf{x}} = 1 - s_0(t) + o, \quad |o|^2 \leq p_5(t) \, \omega$;

4. $\mathrm{var}(t\bar{\mathbf{x}}^T H_0(t)\bar{\mathbf{x}}) \leq a \, Mt^2/N$;

5. $\mathbf{E}H(t) = (I + A + ts_0(t)\Sigma)^{-1} + \Omega$,

   where $\|\Omega\|^2 \leq p_6(t) \, \omega$;

6. $\mathrm{var}\,(\mathbf{e}^T H(t)\mathbf{e}) \leq Mt^2/N$;

7. $ts_0(t) \, \mathbf{E}\bar{\mathbf{x}}^T H(t)\bar{\mathbf{x}} = 1 - s_0(t) + o, \qquad |o| \leq p_2(t) \, \omega$;

8. $\mathrm{var}[t\bar{\mathbf{x}}^T H(t)\bar{\mathbf{x}}] \leq p_6(t)/N$.

Normal Evaluation of Sample-Dependent Functionals

We study rotation invariant functionals including standard quality functions that depend on expectation value vectors, sample means, and population and sample covariance matrices. For sake of generality, let us consider a set of k n-dimensional populations $\mathfrak{S}_1, \mathfrak{S}_2, \ldots, \mathfrak{S}_k$, with expectation vectors $\mathbf{E}\mathbf{x} = \mathbf{a}_i$, covariance matrices $\mathrm{cov}(\mathbf{x}, \mathbf{x}) = \Sigma_i$ for $\mathbf{x}$ form $\mathfrak{S}_i$, moments M_i of the form (1), and the parameters γ_i of the form (2), $i = 1, 2, \ldots, k$.

To be more concise, we redefine the quantities N, y, τ, and γ:

$$N = \min_i N_i, \quad y = n/N, \quad \gamma = \max_i \nu_i/M_i, \quad \tau = \max_i \sqrt{M_i}t_i, \quad (3)$$

$i = 1, 2, \ldots, k$. Definition 1 will be used with these new parameters.

Let $\mathfrak{X}_i$ be independent samples from $\mathfrak{S}_i$ of size N_i, $i = 1, 2, \ldots, k$. Denote $y_i = n/N_i$,

$$\bar{\mathbf{x}}_i = N_i^{-1} \sum_m \mathbf{x}_m, \quad S_i = N_i^{-1} \sum_m (\mathbf{x}_m - \mathbf{a}_i)(\mathbf{x}_m - \mathbf{a}_i)^T,$$

$$C_i = N_i^{-1} \sum_m (\mathbf{x}_m - \bar{\mathbf{x}})(\mathbf{x}_m - \bar{\mathbf{x}})^T,$$

where m runs over all numbers of vectors in $\mathfrak{X}_i$, and $i = 1, 2, \ldots, k$.
We introduce more general resolvent-type matrices

$$H_0 = (I + t_0 A + t_1 S_1 + \cdots + t_k S_k)^{-1},$$
$$H = (I + t_0 A + t_1 C_1 + \cdots + t_k C_k)^{-1},$$

where $t_0, t_1, \ldots, t_k \geq 0$ and A are symmetric, positively semidefinite matrices of constants.

We consider the following classes of functionals depending on $A, \bar{\mathbf{x}}_i, S_i, C_i$, and t_i, $i = 1, 2, \ldots, k$.

The class $\mathfrak{L}_1 = \{\Phi_1\}$ of functionals $\Phi_1 = \Phi_1(t_0, t_1)$ of the form $(k = 1)$

$$n^{-1} \operatorname{tr} H_0, \quad \mathbf{e}^T H_0 \mathbf{e}, \quad t_1 \bar{\mathbf{x}}_1^T H_0 \bar{\mathbf{x}}_1, \quad n^{-1} \operatorname{tr} H, \quad \mathbf{e}^T H \mathbf{e}, \quad t_1 \bar{\mathbf{x}}_1 H \bar{\mathbf{x}}_1.$$

Note that matrices $tH = C_\alpha \stackrel{def}{=} (C + \alpha I)^{-1}$ with $\alpha = 1/t$ may be considered as regularized estimators of the inverse covariance matrix Σ^{-1}.

The class $\mathfrak{L}_2 = \{\Phi_2\}$ of functionals $\Phi_2 = \Phi_2(t_0, t_1, \ldots, t_k)$ of the form

$$n^{-1} \operatorname{tr} H_0, \quad \mathbf{e}^T H_0 \mathbf{e}, \quad t_i \bar{\mathbf{x}}_i H_0 \bar{\mathbf{x}}_j,$$
$$n^{-1} \operatorname{tr} H, \quad \mathbf{e}^T H \mathbf{e}, \quad t_i \bar{\mathbf{x}}_i H \bar{\mathbf{x}}_j, \quad i, j = 1, 2, \ldots, k.$$

The class $\mathfrak{L}_3 = \{\Phi_3\}$ of functionals of the form $\Phi_3 = D_m \Phi_2$ and $\partial/\partial t_0 \, D_m \Phi_2$, where $\Phi_3 = \Phi_3(t_0, t_1, t_2, \ldots, t_k)$ and D_m is the partial differential operator of the mth order

$$D_m = \frac{\partial^m}{\partial z_{i_1} \ldots \partial z_{i_m}},$$

where $z_j = \ln t_j$, $t_j \geq 0$, $j = 0, 1, 2, \ldots, k$, and $i_1, i_2, \ldots, i_m$ are numbers from $\{0, 1, 2, \ldots, k\}$.

Note that by such differentiation of resolvents, we can obtain functionals with the matrices A, S, and C in the numerator. This class includes a variety of functionals which are used as quality functions of some multivariate procedures, for example:

$$\bar{\mathbf{x}}_i^T A \bar{\mathbf{x}}_j, \quad t_i \bar{\mathbf{x}}_i^T H \bar{\mathbf{x}}_j, \quad n^{-1} \operatorname{tr}(AH), \quad t_i n^{-1} \operatorname{tr}(HC_iH), \quad \mathbf{e}^T HAH\mathbf{e},$$
$$t_i \bar{\mathbf{x}}_i HC_i H \bar{\mathbf{x}}_j, \quad t_0 \mathbf{e}^T HAHAH\mathbf{e}, \quad i, j = 1, 2, \ldots, k, \text{ etc.}$$

The class $\mathfrak{L}_4 = \{\Phi_4\}$ of functionals of the form

$$\Phi_4 = \Phi_4(\eta_0, \eta_1, \ldots, \eta_k) =$$
$$= \int\int \Phi_3(t_0, t_1, \ldots, t_k)\, d\eta_0(t_0)\, d\eta_1(t_1)\ldots d\eta_k(t_k),$$

where $\eta_i(t)$ are functions of $t \geq 0$ with the variation not greater than 1 on $[0, \infty)$, $i = 0, 1, \ldots, k$, having a sufficient number of moments

$$\beta_j = \int t^j |d\eta_i(t)|, \quad i = 1, 2, \ldots, k,$$

and the functions Φ_3 are extended by continuity to zero values of arguments.

This class presents a number of functionals constructed by arbitrary linear combinations and linear transformations of regularized ridge estimators of the inverse covariance matrices with different ridge parameters, for example, such as sums of $\alpha_i(I + t_iC)^{-1}$ and functions $n^{-1} \operatorname{tr}(I + tC)^{-k}$, $k = 1, 2, \ldots$, $\exp(-tC)$ and other. Such functionals will be used in Chapter 4 to construct regularized approximately unimprovable statistical procedures.

The class $\mathfrak{L}_5 = \{\Phi_5\}$ of functionals $\Phi_5 = \Phi_5(z_1, z_2, \ldots, z_m)$, where the arguments $z_1, z_2, \ldots, z_m$ are functionals from $\mathfrak{L}_4$, and the Φ_5 are continuously differentiable with respect to all arguments with partial derivatives bounded in absolute value by a constant $c_5 \geq 0$.

Obviously, $\mathfrak{L}_1 \in \mathfrak{L}_2 \in \mathfrak{L}_3 \in \mathfrak{L}_4 \in \mathfrak{L}_5$.

THEOREM 3.11. *Functionals* $\Phi_1 \in \mathfrak{L}_1$ *are* ε*-normal evaluable in the class of populations with four moments of all variables with* $\varepsilon = \varepsilon_1 \stackrel{\text{def}}{=} p_{10}\,\omega.$

Proof. Let $\tilde{\mathfrak{S}}$ denote a normal population $\mathbf{N}(0, \Sigma)$ with a matrix $\Sigma = \Sigma_1 = \operatorname{cov}(\mathbf{x}, \mathbf{x})$ that is identical in $\mathfrak{S}$ and $\tilde{\mathfrak{S}}$. We set $N = N_1$, $y = y_1 = n/N$, $t_0 = 1$, $t_1 = t$.

Let $\mathbf{E}$ and $\widetilde{\mathbf{E}}$ denote the expectation operators for $\mathbf{x} \sim \mathfrak{S}$ and $\widetilde{\mathfrak{S}}$, respectively, and by definition, let

$$h_0(t) = \mathbf{E}n^{-1}\mathrm{tr}H_0, \quad s_0(t) = 1 - y + \mathbf{E}N^{-1}\mathrm{tr}(I + A)H_0,$$
$$\widetilde{h}_0(t) = \widetilde{\mathbf{E}}n^{-1}\mathrm{tr}H_0, \quad \widetilde{s}_0(t) = 1 - y + \widetilde{\mathbf{E}}N^{-1}\mathrm{tr}(I + A)H_0,$$
$$G_0 = (I + A + ts_0(t)\Sigma)^{-1}, \quad \widetilde{G}_0 = (I + A + t\widetilde{s}_0(t)\Sigma)^{-1}.$$

Statement 1 of Lemma 3.11 implies that

$$|h_0(t) - \widetilde{h}_0(t)|(1 + tN^{-1}\mathrm{tr}G_0\Sigma\widetilde{G}_0) \leq o_1,$$

where o_1 is defined by Lemma 3.11. The trace in the parentheses is non-negative. From Statement 2 of Lemma 3.11, it follows that $\mathrm{var}(n^{-1}\mathrm{tr}\ H_0) \leq o_2$ both in $\mathfrak{S}_1$ and in $\widetilde{\mathfrak{S}}_1$. We conclude that $n^{-1}\ \mathrm{tr}\ H_0$ is ε-normal evaluable with $\varepsilon \leq 4\sigma_1^2 + 2\sigma_2 \leq p_6(t)\ \omega^2$. Theorem 3.2 implies that

$$\|\mathbf{E}H_0 - \widetilde{\mathbf{E}}H_0\| \leq t|s_0(t) - \widetilde{s}_0(t)|\|G_0\Sigma\widetilde{G}_0\| + 2\sigma_1,$$

where $|s_0(t) - \widetilde{s}_0(t)| \leq 2o_1 y$ and $\|G_0\Sigma\widetilde{G}_0\| \leq \sqrt{M}$. Thus, the norm in the left-hand side is not greater than $2(1 + \tau y)o_1$. From Lemma 3.11, it follows that $\mathrm{var}(\mathbf{e}^T H_0\mathbf{e}) \leq o_2$ both for $\mathfrak{S}_1$ and for $\widetilde{\mathfrak{S}}_1$. We conclude that $\mathbf{e}^T H_0\mathbf{e}$ is ε-normal evaluable with $\varepsilon = p_8(t)\ \omega^2$.

Further, by the Statement 8 of Lemma 3.11, we have

$$|t\mathbf{E}\bar{\mathbf{x}}^T H_0\bar{\mathbf{x}}| - |t\widetilde{\mathbf{E}}\bar{\mathbf{x}}^T H_0\bar{\mathbf{x}}| \leq |s_0(t) - \widetilde{s}_0(t)| + 2|o_3|,$$

where the summands in the right-hand side are not greater than $p_3(t)\ \omega$ and $p_3(t)\ \omega$, respectively. In view of Statement 7 of Lemma 3.11, we have $\mathrm{var}(t\bar{\mathbf{x}}^T H_0\bar{\mathbf{x}}) \leq p_2(t)/N$ both for $\mathfrak{S}_1$ and for $\widetilde{\mathfrak{S}}_1$. We conclude that $t\bar{\mathbf{x}}^T H_0\bar{\mathbf{x}}$ is ε-normal evaluable with $\varepsilon = p_6(t)\ \omega^2$.

Now we define

$$h(t) = n^{-1}\text{tr } H, \quad s(t) = 1 - y + \mathbf{E}N^{-1}\text{tr}(I + A)H,$$
$$\widetilde{h}(t) = \widetilde{\mathbf{E}}n^{-1}\text{tr}H, \quad \widetilde{s}(t) = 1 - y + \widetilde{\mathbf{E}}N^{-1}\text{tr}H,$$
$$G = (I + A + ts(t)\Sigma)^{-1}, \quad \widetilde{G} = (I + A + t\widetilde{s}(t)\Sigma)^{-1}.$$

From Lemma 3.11, it follows that

$$|h(t) - \widetilde{h}(t)| \le tN^{-1}\text{tr}(G\Sigma\widetilde{G}) \; |s_0(t) - \widetilde{s}_0(t)| + \|\Omega\| \le p_3(t) \; \omega.$$

By Statement 6 of Lemma 3.11, we have $\text{var}(n^{-1}\text{tr}H) \le o_2/N$ both in $\mathfrak{S}_1$ and in $\widetilde{\mathfrak{S}}_1$. We conclude that $n^{-1}\text{tr}H$ is ε-normal evaluable with $\varepsilon = p_6(t) \; \omega$.

From Statement 5 of Lemma 3.11, it follows that

$$\|\mathbf{E}H - \widetilde{\mathbf{E}}H\| \le t|s_0(t) - \widetilde{s}_0(t)|\|G\Sigma\widetilde{G}\| + 2\|\Omega\| \; \le p_3(t) \; \omega.$$

By Statement 6 of Lemma 3.11, we have $\text{var}(\mathbf{e}^T H\mathbf{e}) \le p_5(t)/N$ both in $\mathfrak{S}_1$ and in $\widetilde{\mathfrak{S}}_1$. It follows that $\mathbf{e}^T H\mathbf{e}$ is ε-normal evaluable with $\varepsilon = p_6(t) \; \omega^2$.

Further, using Statements 1 and 7 of Lemma 3.11, we obtain that $\min(s_0(t), \widetilde{s}_0(t)) \ge (1 + \tau y)^{-1}$, and

$$|t\mathbf{E}\bar{\mathbf{x}}^T H\bar{\mathbf{x}}| - |t\widetilde{\mathbf{E}}\bar{\mathbf{x}}^T H\bar{\mathbf{x}}| \le |1/s_0(t) - 1/\widetilde{s}_0(t)| + o(t),$$

where $|o(t)| \le p_5(t)\omega$. The first summand in the right-hand side is not greater than $p_5(t)\omega$. From Lemma 3.11, it follows that $\text{var}(t\bar{\mathbf{x}}^T H\bar{\mathbf{x}}) \le p_6(t)/N$. We conclude that $t\bar{\mathbf{x}}^T H\bar{\mathbf{x}}$ is ε-normal evaluable with $\varepsilon = p_{10}(t)\omega$. This completes the proof of Theorem 3.11. $\square$

THEOREM 3.12. *Functionals* $\Phi_2 \in \mathcal{L}_2$ *are ε-normal evaluable with* $\varepsilon = \varepsilon_2 = k^2\varepsilon_1$.

Proof. We consider normal populations $\widetilde{\mathfrak{S}}_i = \mathbf{N}(\mathbf{a}_i, \Sigma_i)$ with $\mathbf{a}_i = \mathbf{E}\mathbf{x}$ and $\Sigma_i = \text{cov}(\mathbf{x}, \mathbf{x})$, for $\mathbf{x}$ in $\mathfrak{S}_i$, $i = 1, 2, \ldots, k$. Let $\mathbf{E}_i$ be the expectation operator for the random vectors

$$\mathbf{x}_1 \sim \mathfrak{S}_1, \ldots, \quad \mathbf{x}_i \sim \mathfrak{S}_i, \quad \mathbf{x}_{i+1} \sim \widetilde{\mathfrak{S}}_{i+1}, \ldots, \quad \mathbf{x}_k \sim \widetilde{\mathfrak{S}}_k,$$

where the tilde denotes the probability distribution in the corresponding population, $i = 1, 2, \ldots, k - 1$. Let $\mathbf{E}_0$ denote the expectation when all populations are normal, $\mathbf{x}_i \sim \widetilde{\mathfrak{S}}_i$, $i = 1, 2, \ldots, k$, and let $\mathbf{E}_k$ be the expectation operator for $\mathbf{x}_i \sim \mathfrak{S}_i$, $i = 1, 2, \ldots, k$.

Clearly, for each random f having the required expectations,

$$\mathbf{E}_0 f - \mathbf{E} f = \sum_{i=1}^{k} (\mathbf{E}_{i-1} f - \mathbf{E}_i f).$$

Let us estimate the square of this sum as a sum of k^2 terms. We set $f = \Phi_2$ for the first three forms of the functionals Φ_2 (depending on H_0). Choose some $i : 1 \leq i \leq k$. In view of the independence of $\mathbf{x}_i$ chosen from different populations, each summand can be estimated by Theorem 3.11 with $H_0 = H_{0i} = (I + B^i + t_i S_i)^{-1}$, where

$$B^i = I + t_0 A + \sum_{j=1}^{i-1} t_j S_j + \sum_{j=i+1}^{k} t_j S_j$$

is considered to be nonrandom for this i, $i = 1, 2, \ldots, k$. By Theorem 3.11 each summand is ε-normal evaluable with $\varepsilon = \varepsilon_1$. We conclude that $(\mathbf{E} f_0 - \mathbf{E} f)^2 \leq k^2 \varepsilon_1$. Similar arguments hold for f depending on H. This completes the proof of Theorem 3.12. $\square$

THEOREM 3.13. *Functionals* $\Phi_3 \in \mathfrak{L}_3$ *are ε-normal evaluable with* $\varepsilon_3 = a_{mk}(1+\|A\|)^2(1+\tau y)\, \varepsilon_2^{1/2(m+1)}$, *where* a_{mk} *are numerical constants and* τ *and* y *are defined by* (3).

Proof. Let $\mathfrak{X}$ denote a collection of samples from populations $(\mathfrak{S}_1, \mathfrak{S}_2, \ldots, \mathfrak{S}_k)$ and $\widetilde{\mathfrak{X}}$ a collection of samples from normal populations $(\widetilde{\mathfrak{S}}_1, \widetilde{\mathfrak{S}}_2, \ldots, \widetilde{\mathfrak{S}}_k)$ with the same first two moments, respectively. Let us compare $\Phi_3(\mathfrak{X}) = D_m \Phi_2(\mathfrak{X})$ and $\widetilde{\Phi}_3(\mathfrak{X}) = D_m \widetilde{\Phi}_2(\widetilde{\mathfrak{X}})$, where Φ_2 and $\widetilde{\Phi}_2$ are functionals from $\mathfrak{L}_2$. Note that

$$\frac{\partial H_0}{\partial z_i} = -H_0 t_i S_i H_0 \quad \text{and} \quad \frac{\partial H}{\partial z_i} = -H t_i C_i H,$$

where $z_i = \ln t_i$, $t_i > 0$, $i = 1, 2, \ldots, k$. The differential operator D_m transforms H_0 into sums (and differences) of a number of matrices $T_r = H_0 t_i S_i H_0 \ldots t_j S_j H_0$, $i, j = 0, 1, 2, \ldots, k$ with different numbers r of the multiples H_0, $1 \leq r \leq m+1$, $T_1 = H_0$. Note that $\|T_r\| \leq 1$, as is easy to see from the fact that the inequalities $\|H_0 \Omega_1\| \leq 1$ and $\|H_0\| \leq 1$ hold for $H_0 = (I + \Omega_1 + \Omega_2)^{-1}$ and for any symmetric, positively semidefinite matrices Ω_1 and Ω_2. Now, $\partial/\partial z_i \, T_r$ is a sum of r summands of the form T_{r+1} plus $r-1$ terms of the form T_r, no more that $2r-1$ summands in total. We can conclude that each derivative $\partial/\partial z_i \, D_m \Phi_2$ is a sum of at most $(2m+1)!!$ terms, each of these being bounded by 1 or by $t_j \widetilde{\mathbf{x}}_j^2$ for some $j = 1, 2, \ldots, k$, depending on Φ_2. But $\mathbf{E}(t_j \widetilde{\mathbf{x}}^2)^2 \leq \tau^2 y^2$, where $j = 1, 2, \ldots, k$, and $y = n/N$. It follows that

$$\mathbf{E} \left| \frac{\partial}{\partial z_i} \, D_m \Phi \right|^2 \leq (1 + \tau^2 y^2)[(2m + 3)!!]^2$$

for any $i = 1, 2, \ldots, k$. We introduce a displacement $\delta > 0$ of $z_i = \ln t_i$ being the same for all $i = 1, 2, \ldots, k$ and replace the derivatives by finite-differences. Let Δ_m be a finite-difference operator corresponding to D_m. We obtain

$$\Delta_m \Phi_2 = D_m \Phi_2 + \delta \sum_{i=1}^{k} \frac{\partial}{\partial z_i} \, D_m \Phi_2 \Big|_{z=\xi,}$$

where ξ are some intermediate values of z_i, $i = 1, 2, \ldots, k$. By Theorem 3.12, the function $\Delta_m \Phi_2$ is ε-normal evaluable with $\varepsilon = \varepsilon' = 2\varepsilon_2 2^{m+1}/\delta^m$. The quadratic difference

$$\mathbf{E}(\Delta_m \Phi_2 - D_m \Phi_2)^2 \leq \varepsilon'' = \delta^2 k^2 (1 + \tau^2 y^2)[(2m + 3)!!]^2.$$

We conclude that $D_m \Phi_2$ is ε-normal evaluable with $\varepsilon = \sigma \overset{def}{=} \varepsilon' + \varepsilon''$. Choosing $\delta = 2\varepsilon_2^{1/(2+m)}(1 + \tau^2 y^2)^{1/(2+m)}$, we obtain that $\sigma < a\varepsilon_2^{1/(1+m)}$, where the numerical coefficient a depends on m and k. We have proved Theorem 3.13 for the functionals $D_m \Phi_2$.

Now consider functionals of the form $\partial/\partial t_0 \, D_m \Phi_2$. Let us replace the derivative by a finite difference with the displacement

δ of the argument. An additional differentiation with respect to t_0 and the transition to finite differences transform each term T_r into $2r$ summands, where $r \leq m + 1$. Each summand can be increased by a factor of no more than $(1 + \|A\|)$. Choose $\delta = \sqrt{\sigma}$. Reasoning similarly, we conclude that $\partial/\partial t_0 \, D_m \Phi_2$ is ε-normal evaluable with $\varepsilon = 2(m + 1)^2(1 + \|A\|)^2\sqrt{\sigma}$. This completes the proof of Theorem 3.13. $\square$

The next two statements follow immediately.

COROLLARY 1. *The functionals* $\Phi_4 \in \mathfrak{L}_4$ *are* ε-*normal evaluable with* $\varepsilon = \varepsilon_4 = \varepsilon_3$.

COROLLARY 2. *The functionals* $\Phi_5 \in \mathfrak{L}_5$ *are* ε-*normal evaluable in* $\mathfrak{K}_4(M)$ *with* $\varepsilon = \varepsilon_5 \leq a_5^2 m^2 \varepsilon_4$.

Conclusions

Thus, on the basis of spectral theory of large sample covariance matrices that was developed in Chapter 3, it proves to be possible to suggest a method for reliable estimating of a number of sample-dependent functionals, including most popular quality functions for regularized procedures. Under conditions of the multiparametric method applicability, standard quality functions of statistical procedures display specific properties that may be used for the improvement of statistical problem solutions.

The first of these is a decrease in variance produced by an accumulation of random contributions from a large number of boundedly dependent estimators. Under the Kolmogorov asymptotics, in spite of the increasing number of parameters, statistical functionals that uniformly depend on variables have the variance of the order of magnitude N^{-1}, where N is sample size. This means that standard quality functions of multiparametric procedures approach nonrandom limits calculated in the asymptotical theory. In this case, we may leave the problem of quality function estimation and say rather on the *evaluation* of quality functions than on estimation.

The second specific fact established in this chapter is that the principal parts of quality functions prove to be dependent on only two moments of variables and are insensitive to higher ones. If the fourth moment (1) is bounded, the remainder terms prove to be of the order of magnitude of $\sqrt{\gamma + 1/N}$, where N is sample size, and of the measure of quadric variance $\gamma = O(n^{-1})$ for boundedly dependent variables. This fact leads to a remarkable specifically multiparametric phenomenon: the standard quality functions of regularized statistical procedures prove to be approximately independent of distributions. Thus, a number of traditional methods of multivariate analysis developed previously, especially for normal populations, prove to have much wider range of applicability. Solutions of extremum problems under normality assumption retain their extremal properties for a wide class of distributions. For the first time in statistics, we obtain the possibility to construct systematically distribution-free improved and unimprovable procedures. The necessity of a regularization seems not much restrictive since we have the possibility to estimate the effect of regularization and choose best solutions within classes of regularized procedures. Actually, a worthy theory should recommend for practical usage of only obviously stable regularized procedures.

It follows that the multiparametric approach opens a new branch of investigations in mathematical statistics, providing a variety of improved population-free solutions that should replace the traditional restrictively consistent not-always-stable methods of multivariate analysis.

Also, it follows that, for problems of multiparametric statistics, we may propose the *Normal Evaluation Principle* to prove theorems first for normal distributions and then to estimate corrections resulting from non-normality using inequalities obtained above.

CHAPTER 4

ASYMPTOTICALLY UNIMPROVABLE
SOLUTION OF MULTIVARIATE PROBLEMS

The characteristic feature of multiparametric situation is that the variance of sample functionals proves to be small. This effect stabilizes quality functions of estimators and makes them independent on sampling. The observer gets the possibility not to estimate the quality of his methods, but to *evaluate* it, and thus to choose better procedures.

Multiparametric theory suggests the special technique for systematic construction of improved solutions. It can be described as follows.

1. Dispersion equations are derived which connect nonrandom leading parts of functionals with functions, depending on estimators. These equations supply an additional information on the distribution of unknown parameters and spectral properties of true covariance matrices.

2. A class of generalized multivariate statistical procedures is considered which depend on *a priori* parameters and functions. To improve a statistical procedure, it suffices to choose better these parameters or function.

3. Using dispersion equations, the leading parts of quality functions are singled out and expressed, on one hand, in terms of parameters of populations (which is of theoretical interest) and, on the other hand, in terms of functions of statistics only (for applications).

4. The extremum problem is solved for nonrandom leading parts of quality functions or for their limit expressions, and extremum conditions are derived which determine approximately unimprovable or *best-in-the-limit* procedures.

127

5. The accuracy of estimators of the best (unknown) extremum solutions is studied, and their dominating property is established.

In this chapter, this technique is applied to some most usable statistical procedures.

4.1.   ESTIMATORS OF LARGE INVERSE COVARIANCE MATRICES

It is well known that standard linear statistical methods of multivariate analysis do not provide best solutions for finite samples and, moreover, are often not applicable to real data. Most popular linear procedures using the inverse covariance matrix are constructed by the "plugin" method in which the true unknown covariance matrix is replaced by standard sample covariance matrix. However, sample covariance matrices may be degenerate already for the observation dimension $n = 2$. The inversion becomes unstable for large dimension; for $n > N$, where N is sample size, the inverse sample covariance matrix does not exist. In these cases, the usual practice is to reduce artificially the dimension [3] or to apply some regularization by adding positive "ridge" parameters to the diagonal of sample covariance matrix before its inversion. Until [71], only heuristically regularized estimators of the inverse covariance matrices were known, and the problem of optimal regularization had no accurate solution. In this section we develop the successive asymptotical theory of constructing regularized estimators $\widehat{\Sigma}^{-1}$ of the inverse covariance matrices Σ^{-1} approximately unimprovable in the meaning of minimum quadratic losses $n^{-1}\mathrm{tr}(\Sigma^{-1} - \widehat{\Sigma}^{-1})^2$.

In previous chapters, the systematic technique was described for constructing optimal statistical procedures under the Kolmogorov "increasing dimension asymptotics." This technique is based on the progress of spectral theory of increasing random matrices and spectral theory of sample covariance matrices of increasing dimension developed in the previous chapter. The main success of the spectral theory of large random matrices is the derivation of *dispersion equations* (see Section 3.1) connecting limit spectral functions of sample covariance matrices with limit spectral functions of real unknown covariance matrices. They provide an additional information on unknown covariance matrices that will be used for the construction of improved estimators.

In this chapter, first we choose a parametric family of estimators (depending on a scalar or a weighting function of finite variation) and apply the Kolmogorov asymptotics for isolating nonrandom principal part of the quadratic losses. Then, we solve the extremum problem and derive equations that determine the best in the limit parameters of this family and the best weighting function. Further, we approximate these parameters (or function) by statistics and construct the corresponding estimators of the inverse covariance matrix. Then, we prove that this estimator asymptotically dominates the chosen family.

First, we present results and their discussion and then add proofs.

Problem Setting

Let $\mathbf{x}$ be an observation vector from an n-dimensional population $\mathfrak{S}$ with expectation $\mathbf{Ex} = 0$, with fourth moments of all components and a nondegenerate covariance matrix $\Sigma = \mathrm{cov}(\mathbf{x}, \mathbf{x})$. A sample $\mathfrak{X} = \{\mathbf{x}_m\}$ of size N is used to calculate the mean vector $\bar{\mathbf{x}}$ and sample covariance matrix

$$C = N^{-1} \sum_{m=1}^{N} (\mathbf{x}_m - \bar{\mathbf{x}})(\mathbf{x}_m - \bar{\mathbf{x}})^T.$$

We use the following asymptotical setting. Consider a hypothetical sequence of estimation problems

$$\mathfrak{P} = \{(\mathfrak{S}, \ \Sigma, \ N, \ \mathfrak{X}, \ C, \ \widehat{\Sigma}^{-1})_n\}, \quad n = 1, 2, \ldots,$$

where $\mathfrak{S}$ is a population with the covariance matrix $\Sigma = \mathrm{cov}(\mathbf{x}, \mathbf{x})$, $\mathfrak{X}$ is a sample of size N from $\mathfrak{S}$, $\widehat{\Sigma}^{-1}$ is an estimator Σ^{-1} calculated as function of the matrix C (we do not write the indexes n for arguments of $\mathfrak{P}$). Our problem is to construct the best statistics $\widehat{\Sigma}^{-1}$.

We begin by consideration of more simple problem of improving estimators of Σ^{-1} by the introduction of a scalar multiple of C^{-1} (shrinkage estimation) for normal populations. Then, we consider a wide class of estimators for a wide class of populations.

Shrinkage for Inverse Covariance Matrices

Let $\mathfrak{K}^{(1)}$ be parametrically defined family of estimators of the form $\widehat{\Sigma}^{-1} = \alpha C^{-1}$, where α is a nonrandom scalar.

Suppose that the sequence of problems $\mathfrak{P}$ is restricted by following conditions.

1. For each n, the observation vectors $\mathbf{x} \sim \mathbf{N}(0, \Sigma)$, and all eigenvalues of Σ, are located on the segment $[c_1, c_2]$, where $c_1 > 0$ and c_2 do not depend on n.

2. The convergence holds $n^{-1}\text{tr } \Sigma^{-\nu} \to \Lambda_\nu$, $\nu = 1, 2$, in $\mathfrak{P}$.

3. For each n in $\mathfrak{P}$, the inequality holds $N = N(n) > n + 2$, and the ratio $n/N \to y < 1$ as $n \to \infty$.

Remark 1. Under Assumptions 1–3, the limits exist

$$M_{-1} = \underset{n \to \infty}{\text{l.i.m.}} \ n^{-1}\text{tr } C^{-1} = (1 - y)^{-1}\Lambda_{-1},$$

$$M_{-2} = \underset{n \to \infty}{\text{l.i.m.}} \ n^{-1}\text{tr } C^{-2} = (1 - y)^{-2}\Lambda_{-2} + y (1 - y)^{-3}\Lambda_{-1}^2$$

(here and in the following, l.i.m. denotes the limit in the square mean).

Remark 2. Under Assumptions 1–3, the limits exist

$$\underset{n \to \infty}{\text{l.i.m.}} \ n^{-1}\text{tr}(\Sigma^{-1} - \alpha C^{-1})^2 = R(\alpha) = \Lambda_{-2} - 2\alpha\Lambda_{-1}M_{-1} + \alpha^2 M_{-2}.$$

For the standard estimator, $\alpha = 1$ and

$$R = R(1) = y^2(1 - y)^{-2}\Lambda_{-2} + y (1 - y)^{-3}\Lambda_{-1}^2.$$

Remark 3. Under Assumptions 1–3, the value $R(\alpha)$ reaches the minimum for $\alpha = \alpha_{\text{opt}} = (1 - y)^{-1}\Lambda_{-2}/M_{-2} = 1 - y - yM_{-1}^2/M_{-2}$ and

$$R(\alpha_{\text{opt}}) = \Lambda_{-2} - \frac{M_{-1}^2}{M_{-2}} =$$

$$= \left(\frac{(1 - y)^2\Lambda_{-2}}{\Lambda_{-2} + y (1 - y)^{-1}\Lambda_{-1}^2}\right)\left(\frac{\Lambda_{-1}^2}{\Lambda_{-1}^2 + y (1 - y)\Lambda_{-2}}\right)R(1).$$

However, the parameter α_{opt} is unknown to the observer.

Consider a class $\mathfrak{R}^{(2)}$ of estimators of the form $\widehat{\Sigma}^{-1} = \widehat{\alpha}_n C^{-1}$, where the statistics $\widehat{\alpha}_n$ as $n \to \infty$ tends in the square mean to a constant $\alpha \geq 0$ as $n \to \infty$.

Remark 4. Under Assumptions 1–3, the convergence in the square mean holds

$$n^{-1}\text{tr}(\Sigma^{-1} - \widehat{\alpha}_n C^{-1})^2 \to R(\alpha).$$

To estimate the best limit parameter α_{opt}, we construct the statistics

$$\widehat{\alpha}_{\text{opt}} = \widehat{\alpha}_{\text{opt}}(C) = \max\left(0, 1 - \frac{n}{N} - \frac{1}{N}\frac{\text{tr }^2 C^{-1}}{\text{tr }C^{-2}}\right)$$

and consider an estimator $\widehat{\Sigma}_{\text{opt}}^{-1} = \widehat{\alpha}_{\text{opt}}C^{-1}$.

Remark 5. Under Assumptions 1–3, $\underset{n \to \infty}{\text{l.i.m.}} \widehat{\alpha}_{\text{opt}} = \alpha_{\text{opt}}$ and

$$\underset{n \to \infty}{\text{l.i.m.}} n^{-1}\text{tr}(\Sigma^{-1} - \widehat{\alpha}_{\text{opt}}\ C^{-1})^2 =$$
$$= R(\alpha_{\text{opt}}) = \underset{\mathfrak{R}^{(2)}}{\inf}\ \underset{n \to \infty}{\text{l.i.m.}}\ n^{-1}\text{tr}(\Sigma^{-1} - \widehat{\alpha}_n C^{-1})^2.$$

In this meaning, the estimators $\widehat{\Sigma}_{\text{opt}}^{-1} = \widehat{\alpha}_{\text{opt}}C^{-1}$ of matrices Σ^{-1} asymptotically dominate the class of estimators $\mathfrak{R}^{(2)}$.

In the case when $y \to 1$, the standard consistent estimator of the matrix Σ^{-1} becomes degenerate and its quadratic risk increases infinitely, whereas the optimum shrinkage factor $\alpha^{\text{opt}} \to 0$, and the limit quadratic risk of the shrinkage estimator $\widehat{\Sigma}_{\text{opt}}^{-1}$ remains bounded tending to Λ_{-1}^2. Thus, the optimum shrinkage proves to be sufficient to suppress the increasing scatter of entries of ill-conditioned matrices C^{-1}. This provides a reason to recommend the estimator $\widehat{\Sigma}_{\text{opt}}^{-1}$ for the improvement of linear regression analysis, discriminant analysis, and other linear multivariate procedures.

Generalized Ridge Estimators

Regularized "ridge" estimators of the inverse covariance matrices are often used in applied statistics. The corresponding algorithms are included in many packages of applied statistical programs. In the subsequent sections, we develop a theoretical approach allowing

1. to find the dependence of the quadratic risk on the ridge parameter,

2. to study the effect of combined shrinkage-ridge estimators,

3. to calculate the quadratic risk of linear combinations of shrinkage-ridge estimators of the inverse covariance matrices,

4. to offer an asymptotically ε-unimprovable estimator.

Consider a family $\mathfrak{K}^{(3)}$ of nondegenerating estimators Σ^{-1} of the form $\widehat{\Sigma}^{-1} = \Gamma(C)$, where

$$\Gamma(C) = \int_{t \geq 0} H(t) \, d\eta(t), \qquad H(t) = (I + tC)^{-1},$$

and $\eta(t)$ is an arbitrary function of t with bounded variation on $[0, \infty)$.

We search estimators of the matrix Σ^{-1} that asymptotically dominate the class $\mathfrak{K}^{(3)}$ with respect to square losses.

Define the maximum fourth moment of the projection $\mathbf{x}$ onto nonrandom axes

$$M = \sup_{|\mathbf{e}|=1} \mathbf{E}(\mathbf{e}^T \mathbf{x})^4 > 0, \tag{1}$$

where $\mathbf{e}$ are nonrandom unit vectors, and the special measure of the quadratic forms variance

$$\nu = \sup_{\|\Omega\|=1} \text{var}(\mathbf{x}^T \Omega \mathbf{x}/n), \quad \text{and} \quad \gamma = \nu/M, \tag{2}$$

where Ω are nonrandom, symmetric, positively semidefinite matrices of unit spectral norms.

Let us restrict the sequence $\mathfrak{P}$ with the following requirements.

A. For each n, the parameters $M < c_0$ and all eigenvalues of Σ lay on a segment $[c_1, c_2]$, where c_0, $c_1 > 0$ and $c_2 < \sqrt{c_0}$ do not depend on n.

B. The parameters γ vanish as $n \to \infty$.

C. The ratio $n/N \to y$, where $0 < y < 1$.

D. For $u \geq 0$, the convergence holds

$$F_{0\Sigma}(u) \overset{def}{=} n^{-1} \sum_{i=1}^{n} \text{ind}(\lambda_i \leq u) \to F_\Sigma(u),$$

where λ_i are eigenvalues of Σ, $i = 1, \ldots, n$. Under Assumption D for any $k > 0$, the limits exist $\Lambda_k = \lim_{n \to \infty} n^{-1} \text{tr} \Sigma^k$.

The Assumptions A–D provide the validity of spectral theory of sample covariance matrices developed in Section 3.1. Let us gather some inferences from theorems proved in Section 3.1 in the form of a lemma.

LEMMA 4.1. *Under Assumptions A–D the following is true.*

1. *For all (real or complex) z except $z > 0$, the limit function exists*

$$h(z) = \underset{n \to \infty}{\text{l.i.m.}} \, n^{-1} \text{tr}(I - zC)^{-1} = \int (1 - zs(z)u)^{-1} dF_\Sigma(t); \quad (3)$$

this function satisfies the Hölder inequality

$$|h(z) - h(z')| \leq |z - z'|^\varsigma, \qquad 0 < \varsigma < 1, \qquad \text{Im } z \cdot \text{Im } z' \neq 0$$

and as $|z| \to \infty$, the function $zh(z) = -(1 - y)^{-1} \Lambda_{-1} + O(|z|^{-1})$.

2. *For each (complex) z except $z > 0$, we have*

$$\mathbf{E}(I - zC)^{-1} = (I - zs(z)\Sigma)^{-1} + \Omega_n(z), \qquad (4)$$

where $s(z) = 1 - y + yh(z)$ and as $n \to \infty$ $\|\Omega_n(z)\| \to 0$.

3. *For $u \geq 0$ as $n \to \infty$, we have the weak convergence in the square mean*

$$F_n(u) = n^{-1} \sum_{j=1}^{n} \mathrm{ind}(\lambda_j \leq u) \to F(u), \tag{5}$$

where λ_j are eigenvalues of the matrix C; the equation holds

$$h(z) = \int (1 - zu)^{-1} \, dF(u).$$

4. *If $c_1 > 0$ and $y > 0$ for each $u > 0$, the continuous derivative $F'(u) \leq (c_1 y u)^{-1/2}$ exists that vanishes for $u < u_1$ and for $u > u_2$, where $u_1 = c_1(1 - \sqrt{y})^2$ and $u_2 = c_2(1 + \sqrt{y})^2$.*

First we prove the convergence of the quadratic risk for the estimator $\widehat{\Sigma}^{-1} = \Gamma(C)$

$$R_n = R_n(\Gamma) = \mathbf{E} n^{-1} \mathrm{tr}(\Sigma^{-1} - \Gamma(C))^2. \tag{6}$$

We also consider a scalar function $\Gamma(u)$ corresponding to the matrix $\Gamma(C)$ (the matrix $\Gamma(C)$ is diagonalized along with C with eigenvalues $\Gamma(\lambda)$ that correspond to the eigenvalues λ of matrix C),

$$\Gamma(u) = \int (1 + ut)^{-1} \, d\eta(t).$$

THEOREM 4.1. *Let conditions A–D be satisfied. Then,*

$$R(\Gamma) \overset{\text{def}}{=} \lim_{n \to \infty} R_n(\Gamma) = \Lambda_{-2} - 2\int (\Lambda_{-1} + th(-t)s(-t)) \, d\eta(t) +$$

$$+ \int \int \frac{th(-t) - t'h(-t')}{t - t'} \, d\eta(t) \, d\eta(t') =$$

$$= \Lambda_{-2} - 2(1 - y) \int u^{-1} \Gamma(u) \, dF(u) +$$

$$+ 2y \int \int \frac{\Gamma(u) - \Gamma(u')}{u' - u} \, dF(u) \, dF(u') + \int \Gamma^2(u) \, dF(u). \tag{7}$$

Notice that $R(\Gamma)$ is quadratic with respect to $\Gamma(\cdot)$, and we can transform (7) to the form convenient for minimization. Consider the statistics

$$\hat{g}_n(w) = \int (w - u)^{-1} dF_n(u).$$

If $c_1 > 0$, $y > 0$, and Im $w \neq 0$, we have the convergence in the square mean $\hat{g}_n(w) \to g(w) = w^{-1}h(w^{-1})$. The (complex) function $g(w)$ satisfies the Hölder condition and can be continuously extended to the half-axis $u > 0$, and for $u > 0$

$$\text{Re } g(u) = \int_P (u - u')^{-1} \, dF(u')$$

in the principal value meaning. Define

$$\Gamma^{\text{opt}}(u) = (1 - y)u^{-1} + 2y \text{ Re } g(u), \qquad u > 0. \qquad (8)$$

Remark 6. The relation (7) can be rewritten as follows:

$$R(\Gamma) = R^{\text{opt}} + \int (\Gamma(u) - \Gamma^{\text{opt}}(u))^2 \, dF(u),$$

$$\text{where} \quad R^{\text{opt}} = \Lambda_{-2} - \int (\Gamma^{\text{opt}}(u))^2 \, dF(u).$$

Example 1. Equation (4) allows an explicit solution for a two-parametric "ρ-model" of limit spectral distribution $F_0(u) = F_0(u, \sigma, \rho)$ of eigenvalues of the matrix Σ that is defined by the density function

$$F_0'(u) = \begin{cases} (2\pi\rho)^{-1}(1 - \rho)u^{-2}\sqrt{(c_2 - u)(u - c_1)}, & \text{for } u_1 \leq u \leq u_2, \\ 0 & \text{otherwise,} \end{cases}$$

where $u > 0$, $0 \leq \rho < 1$, $c_1 = \sigma^2(1 - \sqrt{\rho})^2$, $c_2 = \sigma^2(1 + \sqrt{\rho})^2$, $\sigma > 0$. For this model (see in Section 3.1), $\Lambda_{-1} = 1/\kappa$, $\Lambda_{-2} = (1 + \rho)/\kappa^2$, where $\kappa = \sigma^2(1 - \rho)^2$, and the function $h(z)$ satisfies the equations

$$(1 - h(z))(1 - \rho h(z)) = \kappa z h(z) s(z),$$

$$h(z) = \frac{1 + \rho + \kappa(1 - y)z - \sqrt{(1 + \rho + \kappa(1 - y)z)^2 - 4(\rho - \kappa z y)}}{2(\rho - \kappa y z)},$$

where $s(z) = 1 - y + y h(z), \ 0 < y < 1$.

For this special case, the extremal function (8) is

$$\Gamma^{\mathrm{opt}}(u) = (1 - y + 2y \ \mathrm{Re} \ h(u^{-1})) u^{-1} = (\rho + y) \ (\rho u + \kappa y)^{-1}.$$

The equation

$$\Gamma^{\mathrm{opt}}(u) = \int_{t \geq 0} (1 + ut)^{-1} \, d\eta^{\mathrm{opt}}(t), \quad u \geq 0$$

has a solution

$$\eta^{\mathrm{opt}}(t) = \begin{cases} 0 & \text{for } t < \rho \kappa^{-1} y^{-1}, \\ (\rho + y)\kappa^{-1} y^{-1} & \text{for } t \geq \rho \kappa^{-1} y^{-1}, \quad \rho > 0. \end{cases}$$

Calculating $R^{\mathrm{opt}} = R^{\mathrm{opt}}(\Gamma)$ we find that $R^{\mathrm{opt}} = \rho y (\rho + y)^{-1} \kappa^{-2}$. As $\rho \to 0$ (passing to unit covariance matrices) or $y \to 0$ (passing to the traditional asymptotics), we obtain $R^{\mathrm{opt}} \to 0$.

The optimum estimator of Σ^{-1} in this case is $(\rho + y)(\rho C + \kappa y I)^{-1}$.

THEOREM 4.2. *Under Assumptions A–D, for the statistics*

$$\widehat{R}_{n\varepsilon}(\Gamma) \overset{def}{=} \widehat{\Lambda}_{-2} - 2(1 - nN^{-1}) \int_{u \geq \varepsilon} u^{-1} \Gamma(u) \, dF_n(u) \ +$$

$$+ \ 2nN^{-1} \iint_{u,u' \geq \varepsilon} \frac{\Gamma(u) - \Gamma(u')}{u' - u} \, dF_n(u) \, dF_n(u') + \int \Gamma^2(u) \, dF_n(u),$$

(9)

we have $\lim\limits_{\varepsilon \to +0} \ \mathrm{l.i.m.}\limits_{n \to \infty} \widehat{R}_n(\Gamma) = R(\Gamma)$.

Example 2. Consider shrinkage-ridge estimators of the form $\widehat{\Sigma}^{-1} = \Gamma(C) = \alpha(I + tC)^{-1}$, where $\alpha > 0$ is the shrinkage coefficient and $1/t$ is the regularizing ridge parameter. In this case, the limit quadratic risk (7) equals

$$R = R(\Gamma) = \Lambda_{-2} - 2\alpha\left(\Lambda_{-1} + zh(z)s(z)\right) + \alpha^2\frac{d}{dz}(zh(z)), \quad z = -t.$$

As a statistics approximating this quantity, we can offer

$$\widehat{R}_{n\varepsilon} = \widetilde{\Lambda}_{-2} - 2\alpha\left(\widetilde{\Lambda}_{-1} + zh_n(z)s_n(z)\right) + \alpha^2\frac{d}{dt}(zh_n(z)),$$

where $z = -t$, $h_n(z) = \operatorname{tr}(I + tC)^{-1}$, $s_n(z) = 1 - nN^{-1}$ $+ nN^{-1}h_n(z)$. By virtue of Theorem 4.2, $\lim\limits_{\varepsilon \to +0} \operatorname{l.i.m.}\limits_{n \to \infty} \widehat{R}_n = R$. The best shrinkage coefficient can be easily found in an explicit form. The minimization in t can be performed numerically.

Asymptotically Unimprovable Estimator

Now we construct an estimator of (8) and show its dominating property.

In the general case, the function $\widehat{R}_n(\Gamma)$ of the form (9) reaches no minima for any smooth functions $\Gamma(u)$. The estimator $\widehat{g}_n(u)$ of the continuous function $g(u)$ is singular for $u > 0$. To suggest a regular estimator, we perform a smoothing of $g_n(u)$. We construct a smoothed function (8) of the form

$$\Gamma_\varepsilon^{\mathrm{opt}}(u) = \operatorname{Re}[(1 - y)w^{-1} + 2y\operatorname{Re}\, g(w)],$$

where $w = u - i\varepsilon$, $\varepsilon > 0$.

For $\Gamma_\varepsilon^{\mathrm{opt}}(u)$, one can offer a "natural estimator"

$$\widehat{\Gamma}_{n\varepsilon}^{\mathrm{opt}}(u) = \operatorname{Re}[(1 - nN^{-1})w^{-1} + 2nN^{-1}\operatorname{Re}\, g_n(w)], \qquad (10)$$

where $w = u - i\varepsilon$, $\varepsilon > 0$.

LEMMA 4.2. *Let $u \geq c_1 > 0$. For $\varepsilon \to +0$, we have*

$$|\Gamma^{\mathrm{opt}}(u) - \Gamma_\varepsilon^{\mathrm{opt}}(u)| = O(\varepsilon^\varsigma), \quad \varsigma > 0.$$

For any $\varepsilon > 0$ and $n \to \infty$,

$$\mathbf{E}|\widehat{\Gamma}_{n\varepsilon}^{\mathrm{opt}}(u) - \Gamma_\varepsilon^{\mathrm{opt}}(u)|^2 \to 0.$$

This assertion follows immediately from Assumption C, (5), and (10).

Given estimator $\widehat{\Sigma}^{-1} = G(C)$, define the quadratic loss function

$$L_n = L_n(G) = n^{-1}\mathrm{tr}(\Sigma^{-1} - G(C))^2.$$

We offer an estimator $\widehat{\Sigma}^{-1} = \widehat{\Gamma}_{n\varepsilon}^{\mathrm{opt}}$ that presents a matrix diagonalized along with the matrix C with the eigenvalues

$$\widehat{\Gamma}_{n\varepsilon}^{\mathrm{opt}}(\lambda_i) = \left(1 - \frac{n}{N}\right)\frac{\lambda_i}{\lambda_i^2 + \varepsilon^2} + \frac{2}{N}\sum_{j=1}^{n}\frac{\lambda_i - \lambda_j}{(\lambda_i - \lambda_j)^2 + \varepsilon^2}, \quad (11)$$

where λ_i are the corresponding eigenvalues of C, $i = 1, \ldots, n$.

THEOREM 4.3. *Under Assumptions A–D*

$$\lim_{\varepsilon \to +0} \;\mathrm{l.i.m.}_{n \to \infty} \; \mathbf{E}|L_n(\widehat{\Gamma}_{n\varepsilon}^{\mathrm{opt}}) - R(\Gamma^{\mathrm{opt}})|^2 = 0.$$

Corollary. The statistics $\widehat{\Gamma}_{n\varepsilon}^{\mathrm{opt}}(C)$ is an estimator of Σ^{-1} that asymptotically ε-dominates the class $\mathfrak{K}^{(2)}$ in the square mean with respect to the square losses $L_n(\Gamma)$.

For applications, it is of importance to solve the problem of an optimal choice of smoothing factor $\varepsilon > 0$. The necessity of smoothing is due to the essence of the formulated extremum problems. The improving effect of (11) is achieved by an averaging within groups of boundedly dependent variables (eigenvalues) that makes it possible to use thus produced nonrandom regularities. These groups must be sufficiently small to produce more regularities and sufficiently large for these regularities to be stable. For the problem of estimating expectation vectors considered in Section 2.4 in the increasing dimension asymptotics $n \to \infty$, $N \to \infty$, $n/N \to y > 0$, the optimal dependence $\varepsilon = \varepsilon(n)$ was found explicitly. This dependence is of the form $n = n^\alpha$, where $\alpha > 0$ is not large positive magnitude. In application to real problems, this fact seems to impose rather restrictive requirements to number of dimensions. However, numerical experiments (see Appendix) indicate that the theoretical upper estimates of the remainder terms are strongly overstated, and the quadratic risk functions are well described by the principal terms of the increasing dimension asymptotics even for small n and N.

Remark 7. Consider the same "ρ-model" of limit spectra of sample covariance matrices as in Example 1. Let *it be known a priori* that the populations have the distribution functions $F_{0n}(u)$ of eigenvalues of Σ tending to $F_0(u) = F_0(u, \sigma, \rho)$. Then, it suffices to construct consistent estimators of the parameters σ^2 and ρ.

Consider the statistics

$$\widehat{M}_\nu = n^{-1}\mathrm{tr}\ C^\nu, \quad \nu = 1, 2, \quad \widehat{\sigma}^2 = \widehat{M}_2/\widehat{M}_1, \quad \widehat{\rho} = 1 - \widehat{M}_1^2/\widehat{M}_2.$$

For these, under Assumptions A–D, in [71] the limits were found

$$\mathop{\mathrm{plim}}_{n\to\infty}\ \widehat{M}_1 = \Lambda_1 > 0, \qquad \mathop{\mathrm{plim}}_{n\to\infty}\ \widehat{M}_2 = \Lambda_2 + y\Lambda_1^2.$$

It follows that the limits exist $\mathop{\mathrm{plim}}\limits_{n\to\infty}\ \widehat{\sigma}^2 = \sigma^2 > 0$, and $\mathop{\mathrm{plim}}\limits_{n\to\infty}\ \widehat{\rho} = \rho$. Let us construct the estimator $\widetilde{\Sigma}^{-1}$ of the matrix Σ^{-1}

$$\widetilde{\Sigma}^{-1} = \widehat{\sigma}^2(\widehat{\rho} + nN^{-1})(\widehat{\rho}\ \widehat{\sigma}^2 C + n^{-1}N^{-1}\mathrm{tr}^2 C \cdot I)^{-1}.$$

The matrices $\widetilde{\Sigma}^{-1}$ have only uniformly bounded eigenvalues with probabilities $p_n \to 1$ as $n \to \infty$. We conclude that $\|\widetilde{\Sigma}^{-1} - \Gamma^{\mathrm{opt}}(C)\| \to 0$ in probability. By Theorem 4.1 and Remark 5, we have

$$\mathop{\mathrm{plim}}_{n\to\infty}\ n^{-1}\mathrm{tr}(\Sigma^{-1} - \widetilde{\Sigma}^{-1})^2 = R^{\mathrm{opt}}.$$

Thus, for the populations with $F_{0n}(u) \to F_0(u, \sigma, \rho)$, the family of estimators $\{\widetilde{\Sigma}^{-1}\}$ has the quadratic losses almost surely not greater than the quadratic losses of any estimator from the class $\mathfrak{R}^{(2)}$.

Proofs for Section 4.1.

First, we prove that, under Assumptions A–D, the ε-regularized statistics $n^{-1}\mathrm{tr}C^{-\nu}$, $\nu \to 1, 2$, for all distributions tend to the same limits as for normal ones. Denote $\widetilde{C} = C - i\varepsilon I$.

LEMMA 4.3. *Under Assumptions A–D, the limits exist*

$$M_{-1} = \lim_{\varepsilon \to +0} \text{ l.i.m.} _{n \to \infty} n^{-1} \text{tr} \widetilde{C}^{-1} = (1-y)^{-1} \Lambda_{-1},$$

$$M_{-2} = \lim_{\varepsilon \to +0} \text{ l.i.m.} _{n \to \infty} n^{-1} \text{tr} \widetilde{C}^{-2} = (1-y)^{-1} \Lambda_{-2} + y(1-y)^{-3} \Lambda_{-1}^2.$$

Proof. We start from the last expression in (3). For fixed $\varepsilon > 0$ and $z = i(1/t + \varepsilon)^{-1}$, we have for increasing $t > 0$:

$$zh(z) = -(1-y)^{-1}\Lambda_{-1} - ((1-y)^{-2}\Lambda_{-2} + y(1-y)^{-3}\Lambda_{-1}^2)z^{-1} + \xi(z),$$

where $|\xi(z)| < c^{-3}t^{-2}$. At the same time, the function $zh(z)$ is the square mean limit of the function $zn^{-1}\text{tr}(I - zC)^{-1}$, which can be expanded as $t \to \infty$ in the series

$$zn^{-1}\text{tr}(I - zC)^{-1} = -n^{-1}\text{tr}\widetilde{C}^{-1} + n^{-1}\text{tr}\widetilde{C}^{-2}t^{-1} + \zeta(z),$$

where $\mathbf{E}|\zeta(z)|^2 = O(\varepsilon^{-3}t^{-2})$. Comparing these expressions as $t \to \infty$, we obtain the limits in the lemma formulation. $\square$

It follows that $\lim_{\varepsilon to +0} \text{ l.i.m.} _{n \to \infty} \widetilde{\Lambda}_{-2} = \Lambda_{-2}$.

THEOREM 4.1 (proof).
The first theorem statement follows from 3. By definition (6),

$$R_n(\Gamma) = n^{-1}\text{tr}\Sigma^{-2} - 2\mathbf{E}n^{-1}\text{tr}\Sigma^{-1}\Gamma(C) + \mathbf{E}n^{-1}\text{tr } \Gamma^2(C). \quad (12)$$

The first addend in the right-hand side of (12) tends to Λ_{-2}. Let T be a large positive number. In view of Lemma 4.1, the expectation of the second addend in the right-hand side of (12) equals

$$-2n^{-1}\text{tr}[\Sigma^{-1} \int_{t<T} \mathbf{E}H \; d\eta(t)] + o_T =$$

$$= -2n^{-1}\text{tr}[\Sigma^{-1} \int_{t<T} (I + ts\Sigma^{-1}) \; d\eta(t)] + o_n(T) + o_T,$$

where $H = (I + tC)^{-1}$, $s = s(-t)$, $o_n(T) \to 0$ as $n \to \infty$ for any fixed $T > 0$. The value o_T is a contribution of large $t \geq T$, which uniformly (with respect to n) decreases as $t \to \infty$. We use once more the expression for $\mathbf{E}H$ in Lemma 4.1 and find that the second term of the right-hand side of (12) can be rewritten as

$$-2 \int_{t<T} n^{-1} \mathrm{tr}(\Sigma^{-1} - ts\mathbf{E}H) \, d\eta(t) + o_n'(T) + o_T =$$

$$= -2 \int_{t<T} (\Lambda_{-1} - tsh) \, d\eta(t) + o_n''(T) + o_T,$$

where $h = h(-t)$, $s = s(-t)$, $o_n'(T) \to 0$, and $o_n''(T) \to 0$ as $n \to \infty$ for any fixed $T > 0$, and $o_T \to 0$ as $T \to \infty$. By the first statement of Lemma 4.1, the value $|tsh|$ is bounded as $t \to \infty$. Tending $n \to \infty$ and $T \to \infty$, we obtain the second term of the right-hand side of (7).

Further, let $T > 0$. Consider the expression

$$\mathbf{E}n^{-1}\mathrm{tr}\,\Gamma^2(C) = \mathbf{E} \iint_{|t-t'|>\varepsilon} n^{-1}\mathrm{tr}\, \frac{tH(-t) - t'H(-t')}{t' - t} \, d\eta(t)d\eta(t'),$$

$$(13)$$

where $H(-t) = (I+tC)^{-1}$ on the square $0 \leq t, t' \leq T$, $\varepsilon > 0$. In the region where $|t - t'| > \varepsilon$, we have $h_n(-t) \overset{def}{=} n^{-1}\mathrm{tr}\, H(-t) \overset{2}{\to} h(-t)$ as $n \to \infty$ for any fixed $\varepsilon > 0$ and $t < T$ uniformly with respect to t (here and in the following, the sign $\overset{2}{\to}$ denotes convergence in the square mean). Therefore, in this region, the integrand of (13) uniformly converges to $(th(-t) - t'h(-t'))/(t - t')$ in the square mean, and we obtain the principal part of the last term in (7).

Let us prove now that the contribution of the region $|t - t'| < \varepsilon$ is small. Let us expand the integrand in (13) with respect to $x = |t - t'|$ near the point $t > 0$. Note that

$$\left|\frac{d}{dt}(th_n(-t))\right| \leq 1, \quad \mathbf{E}\left|\frac{d^2}{dt^2}th_n(-t)\right| \leq 2\mathbf{E}n^{-1}\mathrm{tr}\, C^{-1} < c,$$

where the quantity c does not depend on n and t. It follows that the region $|t - t'| < \varepsilon$ contributes $O(\varepsilon)$ to (13). We conclude that $\mathbf{E}n^{-1}\mathrm{tr}\,\Gamma^2(C)$ converges to the third term in (7). Theorem 4.1 is proved. $\square$

THEOREM 4.2 (proof).

We start from Theorem 4.1. By Lemma 4.3, $\Lambda_{-1} = (1-y) \int u^{-1} \, dF(u)$. We have the identities

$$\int (\Lambda_{-1} - (1-y)th(-t)) \, d\eta(t) = (1-y) \int u^{-1} \Gamma(u) \, dF(u),$$

$$\int th^2(-t) \, d\eta(t) = \iint \frac{\Gamma(u) - \Gamma(u')}{u - u'} \, dF(u) \, dF(u'),$$

where the integrand is extended by continuity to $u = u'$. The weak convergence in the square mean $F_n(u) \to F(u)$ implies the convergence in the square mean

$$n^{-1}\mathrm{tr}\ \Gamma^2(C) = \int \Gamma^2(u) \, dF_n(u) \to \int \Gamma^2(u) \, dF(u).$$

In view of the convergence in the square mean $F_n(u) \to F(u)$, for each $\varepsilon > 0$, the integrals in (8) converge as $n \to \infty$ uniformly to the integrals above. For $\varepsilon < u_1$, by Lemma 4.1, $F_n(\varepsilon) \to 0$ in the square mean. As $\varepsilon \to +0$, we obtain the assertion of our theorem. $\square$

Now, we establish the convergence of the square losses of the optimal theoretical estimator $\Gamma_\varepsilon^{\mathrm{opt}}(C)$.

LEMMA 4.4. *Under Assumptions A–D for $\varepsilon > 0$,*

$$L_n(\Gamma_\varepsilon^{\mathrm{opt}}) = R(\Gamma_\varepsilon^{\mathrm{opt}}) + \xi_n(\varepsilon) + O(\varepsilon),$$

where $\mathbf{E}|\xi_n(\varepsilon)|^2 \to 0$ *as* $n \to \infty$ *for fixed* $\varepsilon > 0$, *and the upper estimate* $O(\varepsilon)$ *is uniform in* n.

Proof. For fixed $\varepsilon > 0$, the function $\Gamma_\varepsilon^{\mathrm{opt}}(u)$ is continuous and has bounded variation for $u > 0$. The quantity

$$L_n(\Gamma_\varepsilon^{\mathrm{opt}}) = n^{-1}\mathrm{tr}\ \Sigma^{-2} - 2n^{-1}\mathrm{tr}[\Sigma^{-1}\Gamma_\varepsilon^{\mathrm{opt}}(C)] + n^{-1}\mathrm{tr}[\Gamma_\varepsilon^{\mathrm{opt}}(C)]^2,$$

where $\Gamma_\varepsilon(C)$ ia a matrix corresponding to the scalar function $\Gamma_\varepsilon(u)$. Note that $n^{-1}\mathrm{tr}\ \Sigma^{-2} \to \Lambda_{-2}$ in the right-hand side and

$$n^{-1}\mathrm{tr}[\Gamma_\varepsilon^{\mathrm{opt}}(C)]^2 = \int [\Gamma_\varepsilon^{\mathrm{opt}}(u)]^2 \, dF_n(u) \xrightarrow{2} \int [\Gamma_\varepsilon^{\mathrm{opt}}(u)]^2 \, dF(u).$$

Let us compare these expressions with (7). We notice that it suffices to prove the convergence in the square mean

$$n^{-1}\mathrm{tr}\Sigma^{-1}\varphi(C) \xrightarrow{2} \int (1-y)^{-1}u^{-1}\varphi(u)\,dF(u)$$

$$-y\iint \frac{\varphi(u)-\varphi(u')}{u'-u}\,dF(u)dF(u') \tag{14}$$

for $\varphi(u) = \Gamma_\varepsilon^{\mathrm{opt}}(u)$, where the latter expression in the integrand is extended by continuity to $u = u'$. This relation is linear with respect to $\varphi(u)$, and it is sufficient to prove (14) for $\varphi(u) = w^{-1}$ and $\varphi(u) = \mathrm{Re}\, g_n(w)$, $w = u - i\varepsilon$, $\varepsilon > 0$. As $n \to \infty$, we have the convergence in the square mean

$$n^{-1}\mathrm{tr}\big(\Sigma^{-1}(C - i\varepsilon I)^{-1}\big) \xrightarrow{2} (1-y)^{-1}\Lambda_{-2} + O(\varepsilon).$$

Consequently, we have

$$(1-y)\int u^{-1}w^{-1}\,dF(u) - y\Big[\int w^{-1}\,dF(u)\Big]^2 =$$

$$= (1-y)M_{-2} - yM_{-1}^2 + O(\varepsilon) = (1-y)^{-1}\Lambda_{-2} + O(\varepsilon).$$

Now let $\varphi(u) = g(w)$, $\quad w = u - i\varepsilon$, $\quad \varepsilon > 0$. We find that

$$n^{-1}\mathrm{tr}\big[\Sigma^{-1}g\,((C - i\varepsilon I)^{-1})\big] = \int n^{-1}\mathrm{tr}\big[\Sigma^{-1}(C - z^{-1}I)^{-1}\big]\,dF(u)$$

$$= -\int zn^{-1}\mathrm{tr}\big[\Sigma^{-1}(I - zC)^{-1}\big]\,dF(u), \tag{15}$$

where $z = (u - i\varepsilon)^{-1}$. Here the matrix $g(A)$ is a matrix diagonalized along with A with eigenvalues $g(\lambda_i)$ on the diagonal, where λ_i are corresponding eigenvalues of A. By Lemma 4.1 for fixed $\varepsilon > 0$ and $u < u_2 = c_2(1 + \sqrt{y})^2$, we have

$$\mathbf{E}n^{-1}\mathrm{tr}\big(\Sigma^{-1}(I - zC)^{-1}\big) =$$

$$= n^{-1}\mathrm{tr}\Sigma^{-1} + zs(z)\mathbf{E}n^{-1}\mathrm{tr}(I - zC)^{-1} + o'_n(\varepsilon) =$$

$$= \Lambda_{-1} + zs(z)h(z) + o''_n(\varepsilon),$$

where $z = (u - i\varepsilon)^{-1}$, and uniformly in u $o'_n(\varepsilon) \to 0$ and $o''_n(\varepsilon) \to 0$ as $n \to \infty$. In view of Lemma 4.1 as $\varepsilon \neq 0$ and $n \to \infty$, we find that $\mathrm{var}[n^{-1}\mathrm{tr}\,(\Sigma^{-1}(I - zC)^{-1})] \to 0$ uniformly with respect to u, and the right-hand sides of (15) converge to $\Lambda_1 + zsh$ in the square mean. Substituting $g(z^{-1}) = zh(z)$ and $s(z) = 1 - y + yz^{-1}g(z^{-1})$, we obtain

$$n^{-1}\mathrm{tr}(C - i\varepsilon I)^{-1} = \int (u - i\varepsilon)^{-1}\,dF_n(u) \to \int (u - i\varepsilon)^{-1}\,dF(u),$$

where the right-hand side equals $-zh(z)$ and $z = i/\varepsilon$. As $\varepsilon \to +0$, this quantity tends to M_{-1}. Thus, the right-hand side of (15) equals

$$-\Lambda_1 M_1 - (1-y)\int zg(z^{-1})\,dF(u) - y\int g^2(z^{-1})\,dF(u) + \xi_n(\varepsilon) + o_\varepsilon,$$
$$(16)$$

where $z = (u - i\varepsilon)^{-1}$, $o_\varepsilon \to 0$ as $\varepsilon \to +0$; for fixed $\varepsilon > 0$, we have the convergence $\mathbf{E}|\xi_n(\varepsilon)|^2 \to 0$ as $n \to \infty$. Recall that the arguments u are bounded in the integration region. By virtue of the Hölder inequality for $g(z)$, the contribution of the difference between z and u^{-1} in (16) is of the order of magnitude ε^ζ as $\varepsilon \to +0$, where $\zeta > 0$. We have the identity

$$2\int n^{-1}\mathrm{Re}\,g(u)\,dF(u) = 2\int \left[\int_p (u - u')^{-1}\,dF(u')\right] u^{-1}\,dF(u) =$$
$$= -\left[\int u^{-1}\,dF(u)\right]^2 = -M_{-1}^2,$$

where $M_{-1} = (1 - y)^{-1}\Lambda_{-1}$. Consequently, (16) equals

$$(1-y)\int u^{-1}g^*(u)\,dF(u) - y\int g^2(u)\,dF(u) + o_\varepsilon + \xi_n(\varepsilon), \quad (17)$$

where the asterisk denotes complex conjugation, and $o_\varepsilon \to 0$ as $\varepsilon \to +0$.

On the other hand, substitute $\varphi(u) = g(w)$, where $w = u - i\varepsilon$, to the right-hand side of (14). Using the identity

$$\iint \frac{g(u - i\varepsilon) - g(u' - i\varepsilon)}{u' - u}\,dF(u)\,dF(u') = \int g^2(u - i\varepsilon)\,dF(u),$$

we obtain that for $\varphi(u) = g(w)$ and $w = u - i\varepsilon$, the right-hand side of (14) equals

$$(1-y) \int u^{-1} g(w) \, dF(u) - y \int g^2(w) \, dF(u). \qquad (18)$$

In view of the Hölder condition, the real parts of (17) and (18) differ by $o_\varepsilon + \xi_n(\varepsilon)$. Thus, the convergence (14) is proved for $\varphi(u) = \operatorname{Re} g(w)$, and for $\varphi(u) = \Gamma_\varepsilon^{\mathrm{opt}}(u)$. Lemma 4.3 statement follows. $\square$

THEOREM 4.3 (proof).

First note that for $\varepsilon > 0$ and $u > 0$, the scalar function

$$\Gamma_{n\varepsilon}^{\mathrm{opt}}(u) = \Gamma_\varepsilon^{\mathrm{opt}}(u) + u^{-1} o_n + \xi_n(u, \varepsilon). \qquad (19)$$

Here $o_n \to 0$ and $\mathbf{E}\xi_n^2(u, \varepsilon) \to 0$ as $n \to \infty$. For $u \geq c_1 > 0$, the function $\Gamma_\varepsilon^{\mathrm{opt}}(u) = \Gamma^{\mathrm{opt}}(u) + r(\varepsilon)$, where $r(\varepsilon) = O(\varepsilon^\zeta)$ as $\varepsilon \to +0$, $\zeta > 0$.

Consider the difference

$$L_n(\Gamma_{n\varepsilon}^{\mathrm{opt}}) - L_n(\Gamma_\varepsilon^{\mathrm{opt}}) = n^{-1} \operatorname{tr}\left[Q(\Gamma_{n\varepsilon}^{\mathrm{opt}}(C) - \Gamma_\varepsilon^{\mathrm{opt}}(C))\right], \qquad (20)$$

where $Q = -2\Sigma^{-1} + \Gamma_{n\varepsilon}^{\mathrm{opt}}(C) + \Gamma_\varepsilon^{\mathrm{opt}}(C)$. Eigenvalues of Q are bounded for $\varepsilon > 0$, and in view of (19), the right-hand side of (20) is not greater than $o(n^{-1}) \operatorname{tr} C^{-1} + |\xi_n(u, \varepsilon)|$. Therefore, $L_n(\Gamma_{n\varepsilon}^{\mathrm{opt}})$ tends to $L_n(\Gamma_\varepsilon^{\mathrm{opt}})$ in the square mean as $n \to \infty$. Lemma 4.4 implies that $\mathbf{E}|L_n(\Gamma_\varepsilon^{\mathrm{opt}}) - R(\Gamma_\varepsilon^{\mathrm{opt}})|^2 \to 0$.

We apply Lemma 4.2 and Remark 6 and conclude that the assertion of Theorem 4.3 is true. $\square$

4.2.   MATRIX SHRINKAGE ESTIMATORS OF EXPECTATION VECTORS

Until recently, efforts to improve estimators of the expectation value vector by shrinkage were restricted to a special case of shrinkage estimators in the form of a scalar multiple of the sample mean vector depending only on length of sample mean vector (see Introduction). Such approach is natural for independent components of the observation vector. In case of dependent variables, it is more natural to shrink the observation vector components by weighting in the system of coordinates where the covariance matrix is diagonal, or where the sample covariance matrix is diagonal. This is equivalent to multiplying the observation vector by matrix function depending on sample covariance matrix. The effect of such matrix shrinkage was investigated in [75].

As it is clear from Introduction, the gain of shrinkage can be measured approximately by the ratio of the dimension to the sample size. This fact suggests that the effect of shrinkage can be adequately investigated under the Kolmogorov asymptotic approach when sample size increase along with the dimensionality so that their ratio tends to a constant. In Section 2.4, this asymptotics was applied to seek best component-wise shrinking coefficients under the assumption that variables are independent and normal.

In this section, we develop the *distribution-free* shrinking techniques for the case of a large number of *dependent variables* . This is achieved by multiplying sample mean vector by matrix $\Gamma(C)$ that is diagonalized together with sample covariance matrix C, and finding asymptotically optimum scalar weighting function $\Gamma(\lambda)$ of eigenvalues λ of matrices C.

Let $\bar{\mathbf{x}}$ be an n-dimensional sample mean vector calculated over a sample $\mathfrak{X} = \{\mathbf{x}_m\}$ of size N and sample covariance matrix be

$$C = N^{-1} \sum_{m=1}^{N} (\mathbf{x}_m - \bar{\mathbf{x}})(\mathbf{x}_m - \bar{\mathbf{x}})^T.$$

Consider a class $\mathfrak{K}$ of estimators of expectation value vectors $\mathbf{Ex} = \vec{\mu} = (\mu_1, \ldots, \mu_n)$ of the form

$$\widehat{\mu} = \Gamma(C)\bar{\mathbf{x}}, \tag{1}$$

where the matrix function $\Gamma(C)$ can be diagonalized together with C with $\Gamma(\lambda)$ as eigenvalues, where λ are corresponding eigenvalues of C; the scalar function $\Gamma : \mathbb{R}^1 \to \mathbb{R}^1$ (denoted by the same letter)

$$\Gamma(u) = \int_{t \geq 0} (1 + ut)^{-1} \, d\eta(t), \tag{2}$$

has finite variation on $[0, \infty)$, is continuous except, maybe, of a finite number of points, and has sufficiently many moments $\int u^k |\, d\eta(u)|$, $k = 1, 2 \ldots$

Our problem is to find a function $\Gamma(u)$ minimizing the square losses

$$L_n = L_n(\eta) = (\vec{\mu} - \widehat{\mu})^2. \tag{3}$$

Limit Quadratic Risk for Estimators of Vectors

We use the increasing dimension approach in the limit form as follows. Consider a sequence of problems

$$\mathfrak{P} = \{(\mathfrak{S}, \vec{\mu}, N, \mathfrak{X}, \bar{\mathbf{x}}, C, \widehat{\mu})_n\}, \quad n = 1, 2, \ldots$$

(we do not write out the subscripts for the arguments of $\mathfrak{P}$), in which the expectation value vectors $\vec{\mu} = \mathbf{Ex}$ are estimated by samples $\mathfrak{X}$ of size N from populations $\mathfrak{S}$ with sample means $\bar{\mathbf{x}}$ and sample covariance matrices C, and estimators of $\vec{\mu}$ are constructed using an a priori chosen function $\Gamma(u)$.

We restrict the populations with the only requirement that all eight moments of all variables exist. Define

$$M_8 = \max\left[(\vec{\mu}^2)^4, \ \mathbf{E} \sup_{|e|=1} (e^T \overset{\mathrm{o}}{\mathbf{x}})^8 \right] > 0,$$

$$\gamma = \sup_{\|\Omega\| < 1} \mathrm{var}(\overset{\mathrm{o}}{\mathbf{x}}{}^T \Omega \overset{\mathrm{o}}{\mathbf{x}}) / \sqrt{M_8},$$

where $\overset{\mathrm{o}}{\mathbf{x}} = \mathbf{x} - \vec{\mu}$ is the centered observation vector, e is a non random unity vector, and Ω are nonrandom symmetric matrices of spectral norm not greater than 1.

Define empirical distribution functions

$$F_{0n}(u) = n^{-1} \sum_{i=1}^{n} \mathrm{ind}(\lambda_i^0 \leq u), \quad G_n(u) = \sum_{i=1}^{n} \mu_i^2 \, \mathrm{ind}(\lambda_i^0 \leq u),$$

$$F_n(u) = n^{-1} \sum_{i=1}^{n} \mathrm{ind}(\lambda_i \leq u),$$

where λ_i^0 are eigenvalues of Σ, λ_i are eigenvalues of C, and μ_i are components of μ in the system of coordinates, in which the matrix Σ is diagonal, $i = 1, \ldots, n$.

To derive limit relations, we restrict $\mathfrak{P}$ with the following conditions.

A. The parameters $0 < M_8 < c_0$ and $\gamma \to 0$ in $\mathfrak{P}$, where c_0 does not depend on n.

B. For each n, all eigenvalues of Σ are located on a segment $[c_1, c_2]$, where $c_1 > 0$ and c_2 do not depend on n.

C. The ratios $n/N \to y$.

D. For $u \geq 0$, the weak convergence holds $F_{0n}(u) \to F_0(u)$.

E. For $u \geq 0$ almost everywhere, the convergence holds $G_n(u) \to G(u)$.

Under these conditions, the limit exists

$$B \stackrel{def}{=} \lim_{n \to \infty} \vec{\mu}^2 = G(c_2).$$

We start from the results of spectral theory of large-dimensional covariance matrices developed in Chapter 3. We consider the resolvent $H = H(z) = (I - zC)^{-1}$ of sample covariance matrices C and use the following corollary of theorems from Section 3.1.

Let $\mathfrak{G} = \mathfrak{G}(\varepsilon)$ be a region of the complex plane outside some ε-neighborhood of the axis $z > 0$. We formulate the following corollary of theorems proved in Section 3.1.

LEMMA 4.5. *Under Assumptions A–E,*

1. *The limits exist*

$$h(z) = \mathop{\mathrm{l.i.m.}}_{n \to \infty} n^{-1} \mathrm{tr}(I - zC)^{-1} = \lim_{n \to \infty} n^{-1} \mathrm{tr}(I - zs(z)\Sigma)^{-1},$$

where the convergence is uniform in $\mathfrak{G}$;

2. *the limits exist*

$$b(z) = \operatorname*{l.i.m.}_{n \to \infty} \vec{\mu}^T (I - zC)^{-1} \vec{\mu}, \quad k(z) = \operatorname*{l.i.m.}_{n \to \infty} \bar{\mathbf{x}}^T (I - zC)^{-1} \bar{\mathbf{x}},$$

where the convergence is uniform in $\mathfrak{G}$, and

$$k(z) = \begin{cases} b(z) + y(h(z) - 1)/s(z) & \text{if } z \neq 0, \\ B + y\Lambda_1 & \text{if } z = 0, \end{cases}$$

and $\Lambda_1 = \lim\limits_{n \to \infty} n^{-1} \operatorname{tr} \Sigma$;

3. *for $u \geq 0$ almost everywhere, the limit exists $F(u) = \operatorname*{l.i.m.}_{n \to \infty} F_n(u)$, and $F(u_2) = 1$, where $u_2 = c_2(1 + \sqrt{y})^2$;*

4. *the equations hold*

$$h(z) = \int (1 - zu)^{-1} \, dF(u) = \int (1 - zs(z))^{-1} \, dF_0(u),$$

$$b(z) = \int (1 - zs(z)u)^{-1} \, dG(u);$$

5. *the inequality holds $|h(z) - h(z')| < c|z - z'|^\varsigma$, where $c, \varsigma > 0$;*

6. *if $y < 1$ and $|z| \to \infty$, then*

$$h(z) \to 0, \ b(z) \to 0, \ and \ k(z) \to 0 \ so \ that$$

$$zh(z) \approx -\Lambda_{-1}(1 - y)^{-1}, \quad zb(z) \approx -(1 - y)^{-1} \int u^{-1} dG_0(u);$$

if $y > 1$ and $|z| \to \infty$, then

$$b(z) \to b(\infty) \ and \ k(z) \to k(\infty)$$

so that with the accuracy up to $O(|z|^{-2})$ we have

$$h(z) \approx 1 - y^{-1} - \lambda_0 y^{-1} z^{-1},$$

$$b(z) \approx b(\infty) - \beta \int \frac{u}{(1 + \lambda_0 u)^2} \, dG(u) \cdot z^{-1},$$

$$k(z) - b(z) \approx k(\infty) - b(\infty) - \beta \lambda_0^{-2} z^{-1},$$

where λ_0 and β are roots of the equations

$$\int (1 + \lambda_0 t)^{-1} \, dF_0(u) = 1 - y^{-1}, \quad \beta = \frac{\lambda_0}{y} \int \frac{u}{(1 + \lambda_0 u)^2} \, dF_0(u),$$

$$b(\infty) = \int (1 + \lambda_0 u)^{-1} \, dG(u), \quad k(\infty) = b(\infty) + \frac{1}{\lambda_0}.$$

First, we establish the convergence of expressions including two resolvents $H(z) = (I - zC)^{-1}$ with different arguments.

Denote the limits in the square mean by l.i.m. and the convergence in the square mean by the sign $\overset{2}{\rightarrow}$.

LEMMA 4.6. *Under Assumptions A–E uniformly in $z, z' \in \mathfrak{G}$ as $n \to \infty$, the convergence holds*

$$\vec{\mu}^T H(z) H(z') \vec{\mu} \overset{2}{\rightarrow} \frac{zb(z) - z'b(z')}{z - z'},$$

$$\bar{\mathbf{x}}^T H(z) H(z') \bar{\mathbf{x}} \overset{2}{\rightarrow} \frac{zk(z) - z'k(z')}{z - z'}. \tag{4}$$

Proof. Let $\varepsilon > 0$ be arbitrarily small. Denote $b_n(z) = \vec{\mu}^T H(z) \vec{\mu}$, $k_n(z) = \bar{\mathbf{x}}^T H(z) \bar{\mathbf{x}}$. We have $H(z) H(z') = (zH(z) - z'H(z'))/(z - z')$. By Lemma 4.5, for $z, z' \in \mathfrak{G}$ and $|z - z'| > \varepsilon > 0$ uniformly, the convergence holds

$$X \overset{def}{=} \vec{\mu}^T H(z) H(z') \vec{\mu} = \frac{zb_n(z) - z'b_n(z')}{z - z'} \overset{2}{\rightarrow} \frac{zb(z) - z'b(z')}{z - z'},$$

$$Y \overset{def}{=} \bar{\mathbf{x}}^T H(z) H(z') \bar{\mathbf{x}} = \frac{zk_n(z) - z'k_n(z')}{z - z'} \overset{2}{\rightarrow} \frac{zk(z) - z'k(z')}{z - z'}.$$

Suppose $|z - z'| < \varepsilon$. It suffices to prove that X and Y can be written in the form

$$X = \frac{d}{dz}(zb(z)) + \xi_n + \eta(\varepsilon), \quad Y = \frac{d}{dz}(zk(z)) + \xi_n + \eta(\varepsilon),$$

where $\xi_n \overset{2}{\to} 0$ as $n \to \infty$ uniformly with respect to z and ε, and $\mathbf{E}|\eta(\varepsilon)|^2 \to 0$ as $\varepsilon \to +0$ uniformly in z and n. Using the identity

$$H(z)H(z') = \frac{d}{dz}(zH(z)) + (z' - z)CH^2(z)H(z'),$$

we obtain

$$X = \frac{d}{dz}(zb_n(z)) + \xi_n + \eta(\varepsilon),$$

where $\eta(\varepsilon) = (z' - z)\vec{\mu}^T CH^2(z)H(z')\vec{\mu}$. Here $\xi_n \overset{2}{\to} 0$ since the second derivatives $b_n''(z)$ and $b''(z)$ exist and are uniformly bounded, and

$$\mathbf{E}\left|\frac{d^2}{dz^2}zb_n(z)\right|^2 = 2\mathbf{E}\left|\vec{\mu}^T CH^3(z)\mu\right|^2 = O(1)\mathbf{E}\left|\vec{\mu}^T C\mu\right|^2 = O(1).$$

As $\varepsilon \to +0$, we have

$$\mathbf{E}|\eta(\varepsilon)|^2 = O(\varepsilon^2)\mathbf{E}\left|\vec{\mu}^T C\vec{\mu}\right|^2 = O(\varepsilon^2)$$

uniformly in n. This proves the first statement of our lemma.

Analogously, we rewrite the expression for Y in the form

$$Y = \frac{d}{dz}(zk_n(z)) + \xi_n + \eta(\varepsilon),$$

where $\eta(\varepsilon) = (z' - z)\bar{\mathbf{x}}^T CH(z)H(z')\bar{\mathbf{x}}$. Here $\xi_n \overset{2}{\to} 0$ in view of the convergence $k_n(z) \overset{2}{\to} k(z)$ and the existence and uniform boundedness of the moments

$$\mathbf{E}\left|\frac{d^2}{dz^2}zk_n(z)\right|^2 = O(1)\mathbf{E}|\bar{\mathbf{x}}^T C\bar{\mathbf{x}}|^2 \le O(1)\mathbf{E}|\bar{\mathbf{x}}^T S\bar{\mathbf{x}}|^2.$$

Indeed, $\mathbf{E}|\bar{\mathbf{x}}^T S\bar{\mathbf{x}}|^2 \le 2(\mathbf{E}|\vec{\mu}^T S\vec{\mu}|^2 + \mathbf{E}|\overset{\circ}{\bar{\mathbf{x}}}{}^T S\overset{\circ}{\bar{\mathbf{x}}}|^2)$, $\overset{\circ}{\bar{\mathbf{x}}} = \bar{\mathbf{x}} - \vec{\mu}$, and

$$\mathbf{E}|\overset{\circ}{\bar{\mathbf{x}}}{}^T S\overset{\circ}{\bar{\mathbf{x}}}|^2 \le N^{-2}\mathbf{E}(\mathrm{tr}^2\ S + 2\ \mathrm{tr}\ S^4) = O(1).$$

Therefore,

$$\mathbf{E}|\eta_n(z)|^2 \le O(\varepsilon^2)\mathbf{E}|\bar{\mathbf{x}}^T C\bar{\mathbf{x}}|^2 = O(\varepsilon^2).$$

This completes the proof of Lemma 4.6. $\square$

THEOREM 4.4. *Under Assumptions A–E, we have*

$$R = R(\eta) \overset{def}{=} \underset{n \to \infty}{\text{l.i.m.}} L_n(\eta) =$$

$$= B - 2 \int b(-t) \, d\eta(t) + \iint \frac{tk(-t) - t'k(-t')}{t - t'} \, d\eta(t) \, d\eta(t'),$$

where the expression in the integrand is extended by continuity to $t = t'$.

Proof. We have

$$L_n(\eta) = \bar{\mu}^2 - 2 \int \bar{\mu}^T (I + tC)^{-1} \bar{\mathbf{x}} \, d\eta(t) +$$

$$+ \iint \bar{\mathbf{x}}^T (I + tC)^{-1}(I + t'C)^{-1} \bar{\mathbf{x}} \, d\eta(t) \, d\eta(t'). \qquad (5)$$

We have $\bar{\mu}^2 \to B$ as $n \to \infty$. Denote $H = (I + tC)^{-1}$, $H' = (I + t'C)^{-1}$. The product $HH' = (tH - t'H')/(t - t')$. In the right-hand side of (5), we obtain random values converging to the limits

$$\bar{\mu}^T H \bar{\mathbf{x}} \overset{2}{\to} b(-t), \quad \bar{\mathbf{x}}^T H H' \bar{\mathbf{x}} \overset{2}{\to} (tk(-t) - t'k(-t'))/(t - t')$$

as $n \to \infty$ uniformly with respect to t, t' by Lemma 4.5 and Lemma 4.6. We conclude that we can perform the limit transition in the integrands in (5). This proves Theorem 4.4. $\square$

Example. Let the matrices $\Gamma(C)$ have a "ridge" form: the function $\eta(v) = 0$ for $v < t$ and $\eta(v) = \alpha > 0$ for $v \geq t$, corresponding to the estimator $\hat{\mu} = \alpha(I + tC)^{-1}$. In this case,

$$R = B - 2\alpha b(-t) + \alpha^2 \frac{d}{dt}(t \cdot k(-t)).$$

Let $t = 0$, $\alpha = 1$, $\hat{\mu} = \bar{\mathbf{x}}$ (the standard estimator). Then, the quadratic risk $R = R^{\text{st}} \overset{def}{=} y\Lambda_1$.

Let $t = 0$, $\hat{\mu} = \alpha \bar{\mathbf{x}}$, where α is a constant. Then, we obtain that $R = B(1 - \alpha)^2 + \alpha^2 y\Lambda_1$, and the minimum of R equal to $R^{\text{st}} B/(B + y\Lambda_1)$ is attained for $\alpha = B/(B + y\Lambda_1)$.

Let $t \neq 0$. The minimum of R equal to $R^{\mathrm{opt}} = B - \alpha^0 b(-t)$ is attained for $\alpha = \alpha^0 = b(-t)/(t \, k(-t))$.

In a special case of the "ρ-model" (see Section 3.1) of limit spectra of the matrices Σ with a special choice of identical $\mu_i^2 = \mu^2/n$ for $i = 1, \ldots, n$, we can express the values α^0 and R^{opt} in the form of rational functions of $h(-t)$. Then, $b(z) = Bh(z)$. For $\rho = 0$, we obtain $R = R^{\mathrm{st}} B(B + y\Lambda_1)^{-1}(1 - y(1 - h)^2)$, and the minimum is attained for $h = 1$ with $t = 0$, where $h = h(-t)$.

Corollary. Under assumptions of Theorem 4.4, the equality

$$\int \frac{tk(-t) - t'k(-t')}{t - t'} \, d\eta^0(t') = b(-t), \quad t \geq 0$$

is sufficient for $R(\eta)$ to have a minimum for $\eta(t) = \eta^0(t)$.

Minimization of the Limit Quadratic Risk

To find a solution of this equation, we use the analytic proper-ties of functions $h(z)$, $s(z)$, $b(z)$, and $k(z)$. Define

$$\widetilde{h}(z) = h(z) - h(\infty), \quad \widetilde{b}(z) = b(z) - b(\infty), \quad \widetilde{k}(z) = k(z) - k(\infty).$$

LEMMA 4.7. *Suppose conditions A–E hold and $y \neq 1$. Then, for any small $\sigma > 0$, we have*

$$\widetilde{R}(\Gamma) \stackrel{\mathrm{def}}{=} B - \frac{1}{\pi i} \int_{\mathfrak{L}} \frac{\widetilde{b}(z)}{z} \Gamma\left(\frac{1}{z}\right) dz + \frac{1}{2\pi i} \int_{\mathfrak{L}} \frac{\widetilde{k}(z)}{z} \Gamma^2\left(\frac{1}{z}\right) dz =$$

$$= \begin{cases} R(\Gamma) & \text{if } y < 1, \\ R(\Gamma) + 2\Gamma(0)b(\infty) - \Gamma^2(0)k(\infty) & \text{if } y > 1, \end{cases} \qquad (6)$$

where the contour of the integration $\mathfrak{L} = (\sigma - i\infty, \, \sigma + i\infty)$.

Proof. The functions $h(z)$, $b(z)$, and $k(z)$ are analytical and have no singularities for Re $z < \sigma$, where $0 < \sigma < u_2^{-1}$. In the half-plane to right of $\mathfrak{L}$, the functions $\widetilde{b}(z)$ and $\widetilde{k}(z)$ decrease as

$O(|z|^{-1})$ for $|z| \to \infty$ and the function $\Gamma(z^{-1})z^{-1} = O(|z|^{-1})$. Therefore, the integration contour $\mathfrak{L}$ may be closed by a semicircle of radius $r = |z - \sigma| \to \infty$. We change the order of integration and use the residue theorem. It follows

$$\frac{1}{2\pi i} \int_{\mathfrak{L}} \frac{\widetilde{b}(z)}{z} \Gamma\left(\frac{1}{z}\right) dz =$$

$$= \int \frac{1}{2\pi i} \oint \widetilde{b}(z)(z+t)^{-1} dz \, d\eta(t) = \int \widetilde{b}(-t) \, d\eta(t).$$

The latter integral in the right-hand side of (6) can be rewritten as

$$\frac{1}{2\pi i} \int_{\mathfrak{L}} \frac{\widetilde{k}(z)}{z} \Gamma^2\left(\frac{1}{z}\right) dz = \iint \frac{1}{2\pi i} \oint \frac{z\widetilde{k}(z)}{(z+t)(z+t')} \, dz \, d\eta(t) \, d\eta(t').$$

For $t \neq t'$, we calculate two residues at the points $z = -t$ and $z = -t'$ and obtain in the integrand

$$\frac{t\widetilde{k}(-t) - t'\widetilde{k}(-t')}{t - t'} = \frac{tk(-t) - t'k(-t')}{t - t'} - k(\infty).$$

For $t = t'$, the residue of the second order yields $\dfrac{d}{dt}(tk(-t))$. We gather summands and obtain the right-hand side of (6). Lemma 4.7 is proved. □

Now we consider a special case when the above integrals can be reduced to integrals over a segment. By Lemma 4.6, for any $v > 0$, we have $h(z) \to h(v)$, where $z = v + i\varepsilon$, $\varepsilon \to +0$, and $\text{Im } h(v) > 0$ if and only if $dF(v^{-1}) > 0$. For $b(z)$ to be continuous, an additional assumption is required. We will need the Hölder condition

$$|b(z) - b(z')| < c|z - z'|^{\varsigma}, \quad c, \varsigma > 0. \tag{7}$$

Remark 1. Suppose Assumptions A–E are valid and $0 < y < 1$. Then, the inequality (7) follows from the condition

$$\sup_{u \geq 0} \left| \frac{dG(u)}{dF_0(u)} \right| < a_0, \tag{8}$$

where a_0 is a constant.

Indeed, suppose (8) holds. Then, the limits exist $b(v) = \lim b(v + i\varepsilon)$ and $k(v) = y(h(v) - 1)/(vs(v)) = \lim k(z)$ as $z = v + i\varepsilon \to v$.

Define the region $\mathfrak{V} = \{v > 0 : \operatorname{Im} k(v) > 0\}$, and the function

$$\Gamma^0(u) = \operatorname{Im} b(u^{-1})/\operatorname{Im} k(u^{-1}), \quad u^{-1} \in \mathfrak{V}.$$

Let $\Gamma^0(0) = 0$ for $y \leq 1$ and $\Gamma^0(0) = b(\infty)/k(\infty)$ for $y > 1$. Note that $0 \leq \Gamma^0(u) \leq 1$.

THEOREM 4.5. *Suppose conditions A–E are fulfilled, (7) holds, and $y \neq 1$.*
Then, $R = R(\eta) = R(\Gamma)$ is such that

$$R = R^{\text{opt}} + \frac{1}{\pi} \int_{\mathfrak{V}} \operatorname{Im} k(v) \left(\Gamma\left(\frac{1}{v}\right) - \Gamma^0\left(\frac{1}{v}\right) \right)^2 v^{-1} dv +$$

$$+ k(\infty)(\Gamma(0) - \Gamma^0(0))^2, \tag{9}$$

where

$$R^{\text{opt}} = B - \frac{1}{\pi} \int_{\mathfrak{V}} \frac{(\operatorname{Im} b(v))^2}{\operatorname{Im} k(v)} v^{-1} dv - \Gamma^0(0)b(\infty).$$

Proof. Given $\operatorname{Im} z > 0$, we contract the contour L to the beam $z \geq \sigma > 0$ using analytical properties of functions in (6). The function

$$\alpha(z) = \frac{1}{z} \Gamma\left(\frac{1}{z}\right) = \int_{t \geq 0} (z + t)^{-1} d\eta(t)$$

has a bounded derivative for $\mathrm{Re}\, z > \sigma^{-1}$ and the expressions in the integrands in (6) satisfy the Hölder inequality. The contribution of large $|z|$ can be made arbitrarily small. Under the contraction of L to real points outside of $\mathfrak{V}$, contributions of integrals of $b(z)$ and $k(z)$ along oppositely directed parallel beams on both sides from the axis of abscissae mutually cancel. The contributions of real parts of $b(z)$ and $k(z)$ mutually cancel, while contributions of imaginary parts double. We obtain

$$R(\Gamma) = B - \frac{2}{\pi} \int\limits_{\mathfrak{V}} \mathrm{Im}\ b(v)\Gamma\left(\frac{1}{v}\right) v^{-1}dv +$$

$$+ \int\limits_{\mathfrak{V}} \mathrm{Im}\ k(v)\Gamma^2\left(\frac{1}{v}\right) v^{-1}dv - 2\Gamma(0)b(\infty) + \Gamma^2(0)k(\infty). \qquad (10)$$

We examine that the right-hand side of (9) coincides with the right-hand side of (10). This is the proof of Theorem 4.5. $\square$

Denote $\mathfrak{U}_0 = \{u : u = 0 \text{ or } u^{-1} \in \mathfrak{V}\}$.

Remark 2. Suppose conditions A–E hold, $y \neq 1$, and there exists a function of finite variation $\eta^0(t)$ such that

$$\int\limits_{t \geq 0} (1 + ut)^{-1}\ d\eta^0(t) = \Gamma^0(u)\ \text{ for }\ u \in \mathfrak{U}_0.$$

Then $\displaystyle\lim_{n \to \infty}\mathrm{l.i.m.}\ (\vec{\mu} - \Gamma^0(C)\bar{\mathbf{x}})^2 = R^{\mathrm{opt}}$.

The estimator $\widehat{\mu} = \Gamma^0(C)\bar{\mathbf{x}}$ of the vector $\vec{\mu}$ may be called the "best in the limit."

Example. In a special case, let $\mu_i^2 = \vec{\mu}^2/n$, $i = 1, \ldots, n$, for all n in the system of coordinates where Σ is diagonal. By Lemma 4.5, we have $b(\infty) = 0$ if $y < 1$ and $b(\infty) = B(1 - y^{-1})$ if $y > 0$. The functionals $\widetilde{R}(\Gamma)$ and $R(\Gamma)$ can be expressed in the form of integrals over the limit spectrum of matrices C. We find that

$$R(\Gamma) = B - 2B\int\Gamma(u)\ dF(u) + \int q(u)\Gamma^2\,(u)\ dF(u),$$

where

$$q(u) = \begin{cases} B + yu|s(u^{-1})|^{-2} & \text{if } u^{-1} \in \mathfrak{V}, \\ 0 & \text{if } y < 1 \text{ and } u = 0, \\ B + (1 - y^{-1})^{-1}\lambda_0^{-1} & \text{if } y > 1 \text{ and } u = 0. \end{cases}$$

Assume that the function $F_0(u)$ has a special form defined by the "ρ-model" of limit spectrum of the matrices Σ. In this case, the function $h(-t)$ equals

$$2\left(1 + \rho + \kappa(1 - y)t + \sqrt{(1 + \rho + \kappa(1 - y)t)^2 - 4\rho + 4\kappa yt}\right),$$

where $\rho < 1$ and $\kappa = \sigma^2(1 - \rho^2)$ are the model parameters. We find that $|s(v)|^2 = (\rho + y(1 - \rho))(\rho + \kappa yv)^{-1}$ and the function

$$\Gamma^0(u) = \frac{B}{q(u)} = \frac{\alpha^0}{1 + ut^0},$$

where

$$\alpha^0 = \frac{B[\rho + y(1 - \rho)]}{B[\rho + y(1 - \rho)] + \kappa y^2}, \qquad t^0 = \frac{y\rho}{B[\rho + y(1 - \rho)] + \kappa y^2}.$$

Let $\eta^0(t')$ be a stepwise function with a jump α^0 at the point $t' = t$. We calculate $R(\Gamma^0)$ passing back to the contour L and calculating the residue at the point $z = -t$. We obtain $R(\Gamma^0) = B(1 - \alpha^0 h(-t^0))$. As $y \to 0$, we have $\alpha^0 \to 1$, $t^0 \to 0$, and $R^0 \to 0$, thus showing the advantage of the standard estimator when the dimension increases slower than sample size. If $\Sigma = \sigma^2 I$ for all n, then the values $\rho = 0$, $\alpha^0 = B/(B + y\Lambda_1)$, $\Lambda_1 = \sigma^2$, $t^0 = 0$, $\Gamma^0(C) = \alpha^0 I$, and $R^0 = \alpha R^{\text{st}}$, where $R^{\text{st}} = y\Lambda_1$. The corresponding optimum estimator has the shrinkage form $\hat{\mu} = \alpha^0 \bar{\mathbf{x}}$. As $y \to 1$, the values $\alpha^0 \to B/(B + \kappa)$, $t^0 \to \rho/(B + \kappa)$, $\Gamma^0(C) \to B[(B + \kappa)I + \rho C]^{-1}$, and

$$\frac{R^{\text{opt}}}{R^{\text{st}}} \to \theta\left(1 - \rho + \frac{\rho\theta}{1 + \sqrt{1 + \Lambda_1\rho(1 - \rho)^2/(B + \kappa)}}\right),$$

where $\theta = B/(B+\kappa)$. The maximum effect of the estimator $\widehat{\mu} = \Gamma^0(C)\bar{\mathbf{x}}$ is achieved for $B \ll \kappa$, and this corresponds to the case of small $\vec{\mu}$ when $\vec{\mu}^2 \ll (n^{-1}\mathrm{tr}\,\Sigma^{-1})^{-1}$, when components of $\vec{\mu}$ are much less in absolute value than the standard deviation of components of the sample mean vector, or for a wide spectrum of matrices Σ. If $y \to \infty$, then $R^{\mathrm{st}} \to \infty$, whereas $\alpha^0 \to 0$, and $R^0 \to B < \infty$, whereas $R = y \to \infty$ for the standard estimator.

Statistics to Approximate Limit Risk

Now we pass to the construction of estimators for the limit functions. Denote

$$h_n(z) = n^{-1}\mathrm{tr}(I - zC)^{-1}, \quad b_n(z) = \vec{\mu}^T(I - zC)^{-1}\vec{\mu},$$

$$s_n(z) = 1 + nN^{-1}(h_n(z) - 1), \quad k_n(z) = b_n(z) + nN^{-1}\frac{h_n(z) - 1}{z s_n(z)}.$$

From Lemma 4.5, it follows that the convergence in the square mean holds

$$h_n(z) \xrightarrow{2} h(z), \quad b_n(z) \xrightarrow{2} b(z), \quad s_n(z) \xrightarrow{2} s(z),$$

$$n^{-1}\mathrm{tr}\,C \xrightarrow{2} \Lambda_1, \quad n^{-1}\mathrm{tr}\,C^2 \xrightarrow{2} \Lambda_2 + y\Lambda_1^2,$$

$$\lim_{T\to\infty}\ \underset{n\to\infty}{\mathrm{l.i.m.}}\ Ts_n(-T) = \lambda_0, \quad \lim_{T\to\infty}\ \underset{n\to\infty}{\mathrm{l.i.m.}}\ k_n(-T) = k(\infty).$$

The asymptotically extremum estimator $\widehat{\mu} = \Gamma^0(C)\bar{\mathbf{x}}$ involves the function $\mathrm{Im}\,b(u^{-1})/\mathrm{Im}\,k(u^{-1})$, which should be estimated by observations. But the natural estimator

$$\Gamma_n^0(u) = \mathrm{Im}\,b_n(u^{-1})/\mathrm{Im}\,k_n(u^{-1})$$

is singular for $u > 0$ and may not approach $\Gamma^0(u)$ as $n \to \infty$. We introduce a smoothing by considering $b_n(z)$ and $k_n(z)$ for complex z with $\mathrm{Im}\,(z) > 0$. In applications, the character of smoothing may be essential. To reach a uniform smoothing, we pass to functions of the inverse arguments

$$
\begin{aligned}
a_n(z) &= z^{-1}b_n(z^{-1}) = \vec{\mu}^T(zI - C)\vec{\mu}, \\
g_n(z) &= z^{-1}h_n(z^{-1}) = n^{-1}\mathrm{tr}\ (zI - C)^{-1}, \\
l_n(z) &= z^{-1}k_n(z^{-1}) = \bar{\mathbf{x}}^T(zI - C)^{-1}\bar{\mathbf{x}}.
\end{aligned}
$$

Remark 3. Under Assumptions A–E, the functions $g_n(z)$, $a_n(z)$, and $l_n(z)$ converge in the square mean uniformly with respect to $z \in \mathfrak{G}$ to the limits $g(z)$, $a(z)$, $l(z)$, respectively, such that

$$
g(z) = \int (z - s(z^{-1})u)^{-1}\,dF_0(u), \ \ a(z) = \int (z - s(z^{-1})u)^{-1}dG(u),
$$
$$
l(z) = a(z) + y\ (zg(z) - 1)/s(z^{-1}). \tag{11}
$$

Remark 4. Under Assumptions A–E for $y \neq 1$, the functions $g(z)$, $a(z)$, and $l(z)$ are regular with singularities only at the point $z = 0$ and on the segment $[0, u_2]$. The functions $\widetilde{a}(z) = a(z) - b(\infty)/z$ and $\widetilde{l}(z) = l(z) - k(\infty)/z$ are bounded. As $z \to u > u_2$, we have $\mathrm{Im}\ g(z) \to 0$, $\mathrm{Im}\ a(z) \to 0$, and $\mathrm{Im}\ l(z) \to 0$.

Now we express (6) in terms of $g(z)$, $s(z)$, and $l(z)$.

LEMMA 4.8. *If conditions A–E hold and $y \neq 1$, then, as $\varepsilon \to +0$, the function $\widetilde{R}(\Gamma)$ defined by (6) equals*

$$
B - \frac{2}{\pi}\int_0^\infty \mathrm{Im}\ \widetilde{a}(u - i\varepsilon)\Gamma(u)\,du + \frac{1}{\pi}\int_0^\infty \mathrm{Im}\ \widetilde{l}(u - i\varepsilon)\Gamma^2(u)\,du + O(\varepsilon).
$$
$$
\tag{12}
$$

Proof. Functions in the integrands in (6) are regular and have no singularities for $\mathrm{Re}\ z \geq \sigma > 0$ outside the beam $z \geq \sigma$. As $|z| \to \infty$, there exists a real $T > 0$ such that $\widetilde{b}(z)$ has no singularities also for $|z| > T$. Let us deform the contour $(\sigma - i\infty,$

$\sigma + i\infty)$ in the integrals (6) into a closed contour $\mathcal{L}_1$ surrounding an ε-neighborhood of the segment $[\sigma, T]$. Substitute $w = z^{-1}$. We find that

$$\widetilde{R}(\Gamma) = B - \frac{1}{\pi i} \int\limits_{\mathcal{L}_2} \widetilde{a}(w)\Gamma(w)\, dw + \frac{1}{2\pi i} \int\limits_{\mathcal{L}_2} \widetilde{l}(w)\Gamma^2(w)\, dw,$$

where $\mathcal{L}_2$ is surrounding the segment $[w_0, T]$, where $w_0 = T^{-1}$ and $t = \sigma^{-1}$. If $\operatorname{Re} w \geq 0$, then the analytical function $\Gamma(w)$ is bounded by the inequality $|\Gamma(w)| \leq 1$, and $\widetilde{a}(w)$, $b(z)$, and $\widetilde{l}(w)$ tend to a constant as $w \to u$, where $u = 1/\operatorname{Re} z > 0$. Since the functions $\widetilde{a}(w)$ and $\widetilde{l}(w)$ are analytical, we can deform the contour $\mathcal{L}_2$ into the contour $\widetilde{\mathcal{L}}_2 = (0-i\varepsilon, 0+i\varepsilon, T+i\varepsilon, T-i\varepsilon, 0-i\varepsilon)$, where $T > T_0 > 0$ is sufficiently large. Contributions of integrals along vertical segments of length 2ε are $O(\varepsilon)$ as $\varepsilon \to +0$. Real parts of the integrands on segments $[i\varepsilon, \tau + i\varepsilon]$ and $[\tau - i\varepsilon, -i\varepsilon]$ cancel, while the imaginary ones double. We obtain

$$\widetilde{R}(\Gamma) = B - \frac{2}{\pi} \int\limits_0^T \operatorname{Im}[\widetilde{a}(w)\Gamma(w)]\, du + \frac{1}{\pi} \int\limits_0^T \operatorname{Im}[\widetilde{l}(w)\Gamma^2(w)]\, du + O(\varepsilon),$$

where $w = u - i\varepsilon$. Substitute

$$\Gamma(u - i\varepsilon) = \Gamma(u) + i\varepsilon \int \frac{t}{1 + ut}\, \frac{1}{1 + ut - i\varepsilon t}\, d\eta(t).$$

Comparing with (12), we see that it is left to prove that the difference $\Gamma(u - i\varepsilon) - \Gamma(u)$ gives a contribution $O(\varepsilon)$ into $\widetilde{R}(\Gamma)$. Consider the integral

$$\int\limits_{\mathcal{L}_3} \widetilde{a}(w)\left(\int \frac{1}{1 + wt}\, \frac{t}{1 + (w + i\varepsilon)t}\right) dw, \qquad (13)$$

where the integration contour is $\mathcal{L}_3 = (0 - i\varepsilon, \infty - i\varepsilon)$. If $\operatorname{Im} w < \varepsilon$, then the integrand has no singularities and is $O(|w|^{-2})$ as $|w| \to \infty$. It means that we can replace the contour $\mathcal{L}_3$ by the contour

$\mathcal{L}_4 = (0 - i\varepsilon, 0 - i\infty)$. The function $\widetilde{a}(w)$ is uniformly bounded on $\mathcal{L}_4$, and it follows that the integral (13) is uniformly bounded. Analogously, the integral with $\widetilde{l}(w)$ is also bounded. It follows that we can replace $\Gamma(u - i\varepsilon)$ in $\widetilde{R}(\Gamma)$ by $\Gamma(u)$ with the accuracy to $O(\varepsilon)$. We have proved the statement of Lemma 4.8. $\square$

Statistics to Approximate the Extremum Solution

Let us construct an estimator of the extremal limit function $\Gamma^0(u)$. Let $\varepsilon > 0$. Denote

$$\Gamma^0_\varepsilon(u) = \begin{cases} \operatorname{Im} a(u - i\varepsilon)/\operatorname{Im} l(u - i\varepsilon) \le 1, & \text{if } u \ge 0, \\ 0, & \text{if } u < 0; \end{cases}$$

$$R^{\text{opt}}_\varepsilon = B - \frac{1}{\pi} \int\limits_0^\infty \frac{(\operatorname{Im} a(w))^2}{\operatorname{Im} l(w)} \, du - d,$$

where $w = u - i\varepsilon$, $d = 0$ if $y < 1$, and $d = b^2(\infty)/k(\infty)$ if $y > 1$. From (12), we obtain that

$$R(\Gamma) = R^{\text{opt}}_\varepsilon + \frac{1}{\pi} \int\limits_0^\infty \operatorname{Im} l(w)(\Gamma(u) - \Gamma^0_\varepsilon(u))^2 \, du +$$

$$+ k(\infty)(\Gamma(0) - \Gamma^0_\varepsilon(0))^2 + O(\varepsilon). \qquad (14)$$

We consider the smoothed estimator $\widetilde{\mu} = \widetilde{\Gamma}^0_\varepsilon(C)\bar{\mathbf{x}}$ defined by the scalar function

$$\widetilde{\Gamma}^0_\varepsilon(u) = \int\limits_{-\infty}^\infty \Gamma^0_\varepsilon(u') \frac{\varepsilon}{\pi} \frac{1}{(u - u')^2 + \varepsilon^2} \, du'.$$

LEMMA 4.9. *If conditions A–E hold and $y \neq 1$, then*

$$(\widetilde{\mu} - \widetilde{\Gamma}^0_\varepsilon(C)\bar{\mathbf{x}})^2 < R^{\text{opt}}_\varepsilon + O(\varepsilon) + \xi_n(\varepsilon), \qquad (15)$$

where $\mathbf{E}\xi_n^2(\varepsilon) \to 0$ *as* $n \to \infty$ *for fixed* $\varepsilon > 0$, *and* $O(\varepsilon)$ *does not depend on* n.

Proof. We pass to the coordinate system where the matrix C is diagonal; let μ_i and $\bar{x}_i$ be components of $\vec{\mu}$ and $\bar{\mathbf{x}}$ therein. We find that

$$(\vec{\mu} - \tilde{\Gamma}_\varepsilon^0(C)\bar{\mathbf{x}})^2 = \vec{\mu}^2 - 2\sum_i \mu_i^2\, \tilde{\Gamma}_\varepsilon^0(\lambda_i) + \sum_i \bar{x}_i^2\, \tilde{\Gamma}_\varepsilon^{02}(\lambda_i) - 2\zeta_n, \quad (16)$$

where $\lambda_j = \lambda_j(C)$, $\zeta_n = \vec{\mu}^T\tilde{\Gamma}_\varepsilon^0(C)(\bar{\mathbf{x}} - \vec{\mu})$, and $\mathbf{E}\zeta_n^2 = O(N^{-1})$. Note that

$$\sum_j \mu_j^2\, \tilde{\Gamma}_\varepsilon^0(\lambda_j) = \frac{1}{\pi} \int_{-\infty}^{\infty} \mathrm{Im}\, a_n(u - i\varepsilon)\, \Gamma_\varepsilon^0(u)du, \quad (17)$$

where $\lambda_j = \lambda_j(C)$. For a fixed $\varepsilon > 0$, $a_n(w) \to a(w)$ as $n \to \infty$ uniformly on $[0, T]$, where $T = 1/\varepsilon$. The contribution of $u \in [0, T]$ to (16) is not larger than

$$\sum_j \mu_j^2 \left(1 - \frac{1}{\pi} \arctan \frac{T - \lambda_j}{\varepsilon} - \frac{1}{\pi} \arctan(\frac{\lambda_j}{\varepsilon})\right) \le$$

$$\le \sum_{\lambda_j > T/2} \mu_j^2 + \frac{\varepsilon^2\mu^2}{2\pi} \le \frac{2}{T}\sum_j \mu_j^2\lambda_j + \frac{\varepsilon^2\vec{\mu}^2}{2\pi} = \frac{2}{T}\vec{\mu}^T C\mu + \frac{\varepsilon^2\vec{\mu}^2}{2\pi}. \quad (18)$$

From Lemma 4.5, it follows that $\mathbf{E}(\vec{\mu}^T C\vec{\mu})^2$ is bounded and the right-hand side of (18) can be expressed in the form $O(\varepsilon) + \xi_n(\varepsilon)$, where $O(\varepsilon)$ is independent on n and $\mathbf{E}\xi_n^2(\varepsilon) \to 0$ as $n \to \infty$ for fixed $\varepsilon > 0$. Thus, the second term of the right-hand side of (16) equals

$$-\frac{2}{\pi} \int_0^{\infty} \mathrm{Im}\, a(u - i\varepsilon)\, \Gamma_\varepsilon^0(u)du + O(\varepsilon) + \xi_n(\varepsilon).$$

We notice that the third term of (16) is

$$\sum_j \bar{\mathbf{x}}_j^2 \tilde{\Gamma}_\varepsilon^{02}(\lambda_j) \leq \sum_j \bar{\mathbf{x}}_j^2 \int_{-\infty}^{\infty} \Gamma_\varepsilon^{02}(u) \frac{\varepsilon/\pi}{(u-\lambda_j)^2 + \varepsilon^2} \, du =$$

$$= \frac{1}{\pi} \int_{-\infty}^{+\infty} \Gamma_\varepsilon^{02}(u) \operatorname{Im} l_n(w) \, du, \tag{19}$$

where the second superscript 2 denotes the square, $w = u - i\varepsilon$, $\varepsilon > 0$, and $0 \leq \Gamma_\varepsilon^0(u) \leq 1$. Note that $l_n(w) \xrightarrow{2} l(w)$ uniformly for $u \in [0, T]$, and the contribution of $u \bar{\in} [0, T]$ is not greater than $2T^{-1}\bar{\mathbf{x}}^T C \bar{\mathbf{x}} + \varepsilon^2 \bar{\mathbf{x}}^2/(2\pi)$. But we have $\mathbf{E}\bar{\mathbf{x}}^2 = O(1)$ and $\mathbf{E}(\bar{\mathbf{x}}^T C \bar{\mathbf{x}})^2 = O(1)$. It follows that the third term of the right-hand side of (16) is not greater than

$$\frac{1}{\pi} \int_0^T \Gamma_\varepsilon^{02}(u) \operatorname{Im} l(u - i\varepsilon) \, du + O(\varepsilon^2) + \xi_n(\varepsilon),$$

where $O(\varepsilon)$ is finite as $\varepsilon \to +0$ and $\xi_n(\varepsilon) \xrightarrow{2} 0$ as $n \to \infty$ for any $\varepsilon > 0$. We substitute $\bar{\mu}^2 = B + o(1)$, $\Gamma_\varepsilon^0(u) = \operatorname{Im} a(w)/\operatorname{Im} l(w)$, where $w = u - i\varepsilon$. Gathering summands we obtain

$$(\bar{\mu} - \tilde{\Gamma}_\varepsilon^0(C)\bar{\mathbf{x}})^2 < B - \frac{1}{\pi} \int_0^T \frac{(\operatorname{Im} a(w))^2}{\operatorname{Im} l(w)} \, du + O(\varepsilon) + \xi_n(\varepsilon), \tag{20}$$

where $w = u - i\varepsilon$. We note that $\operatorname{Im} a(w) = O(\varepsilon)|w|^{-2}$ as $|w| \to \infty$, and, consequently, the integral in (20) from 0 to T can be replaced by the integral from 0 to infinity with the accuracy to $O(\varepsilon)$. The statement of Lemma 4.9 follows. $\square$

Now, we consider the statistics

$$\Gamma_{n\varepsilon}^0(u) = \max\left(0, 1 - nN^{-1}\frac{\operatorname{Im}\ (wg_n(w))}{|s_n(w^{-1})|^2}\right), \qquad \text{where } w = u - i\varepsilon,$$

$$\text{and } \widetilde{\Gamma}^0_{n\varepsilon}(u) = \int_{-\infty}^{\infty} \Gamma^0_{n\varepsilon}(u') \frac{\varepsilon/\pi}{[(u-u')^2 + \varepsilon^2]} \, du'.$$

THEOREM 4.6. *Suppose conditions A–E hold and* $y \neq 1$. *Then,*

1. *for a fixed* $\varepsilon > 0$ *as* $n \to \infty$, *we have* $\widetilde{\Gamma}^0_{n\varepsilon}(u) \xrightarrow{2} \widetilde{\Gamma}^0_{\varepsilon}(u)$ *uniformly on any segment;*

2. *we have*

$$(\vec{\mu} - \widetilde{\Gamma}^0_{n\varepsilon}(C)\bar{\mathbf{x}})^2 < \inf_{\Gamma}(\vec{\mu} - \Gamma(C)\bar{\mathbf{x}})^2 + O(\varepsilon) + \xi_n(\varepsilon), \qquad (21)$$

where $\Gamma = \Gamma(\cdot)$ *is defined by the estimator class* $\mathfrak{K}$, *the quantity* $O(\varepsilon)$ *does not depend on* n, *and* $\mathbf{E}\xi_n^2(\varepsilon) \to 0$ *as* $n \to \infty$ *for any* $\varepsilon > 0$.

Proof. For any fixed $\varepsilon > 0$, we have the uniform convergence in the square mean $\widetilde{\Gamma}^0_{n\varepsilon}(u) \xrightarrow{2} \widetilde{\Gamma}^0_{\varepsilon}(u)$ on any segment by definition of these functions and Lemma 4.5. Denote

$$\rho_n = (\vec{\mu} - \widetilde{\Gamma}^0_{n\varepsilon}(C)\bar{\mathbf{x}})^2 - (\vec{\mu} - \widetilde{\Gamma}^0_{\varepsilon}(C)\bar{\mathbf{x}})^2, \quad \Delta(C) = \widetilde{\Gamma}^0_{n\varepsilon}(C) - \widetilde{\Gamma}^0_{\varepsilon}(C).$$

Let us prove that $\lim_{\varepsilon \to +0} \lim_{n \to \infty} \mathbf{E}\rho_n^2 = 0$. It is suffices to show that $\mathbf{E}(\vec{\mu}^T \Delta \bar{\mathbf{x}})^2 \xrightarrow{2} 0$ and $\mathbf{E}(\bar{\mathbf{x}}^T \Delta \bar{\mathbf{x}})^2 \xrightarrow{2} 0$. We single out a contribution of eigenvalues λ_i of C not exceeding T for some $T > 0$: let $\Delta(u) = \Delta_1(u) + \Delta_2(u)$, where $\Delta_2(u) = \Delta(u)$ for $|u| > T$ and $\Delta_2(u) = 0$ for $|u| \leq T$. Here the scalar argument u stands for eigenvalues of C. By virtue of the first theorem statement, $\Delta_1(u) \xrightarrow{2} 0$ as $n \to \infty$ uniformly on the segment $[0, T]$. The contribution of $|u| > T$ to $\mathbf{E}(\vec{\mu}^T \Delta(C)\bar{\mathbf{x}})^2$ is not greater than

$$\sum_i \mu_j^2 \mathbf{E}\mathbf{x}_j^2 \ \mathrm{ind}(\lambda_j > T) \leq \ T^{-1}\mathbf{E}(\bar{\mathbf{x}}^T C\bar{\mathbf{x}}) = O(T^{-1}).$$

Let $T = 1/\varepsilon$. Then, $\mathbf{E}\rho_n^2 = O(\varepsilon)$ as $n \to \infty$. In view of Lemma 4.9,

$$(\vec{\mu} - \widetilde{\Gamma}^0_{n\varepsilon}(C)\bar{\mathbf{x}})^2 < R^0_\varepsilon + O(\varepsilon) + \xi_n(\varepsilon),$$

where the estimate $O(\varepsilon)$ is uniform in n and $\mathbf{E}\xi_n^2 \to 0$ as $n \to \infty$ for fixed $\varepsilon > 0$. It follows that $R_\varepsilon^0 \le R(\Gamma) + O(\varepsilon)$. This completes the proof of Theorem 4.6. $\square$

Denote $\widehat{\mu}_\varepsilon^0 = \widetilde{\Gamma}_{n\varepsilon}^0(C)\bar{\mathbf{x}}$.

We conclude that, in the sequence of problems $\{\mathfrak{P}_n\}$ of estimation of n-dimensional parameters $\vec{\mu} = \mathbf{E}\mathbf{x}$ for populations restricted by conditions A–E, the family of estimators $\{\widehat{\mu}_\varepsilon^0\}$ is asymptotically ε-dominating over the class of estimators $\widehat{\mu} \in \mathfrak{K}$ of $\vec{\mu}$ as follows: for any $\varepsilon > 0$ and $\delta > 0$, there exists an n_0 such that for any $n > n_0$ for any $\vec{\mu}$ for any estimator $\Gamma(C)\bar{\mathbf{x}}$, the inequality

$$(\vec{\mu} - \widehat{\mu}_\varepsilon^0)^2 < (\vec{\mu} - \widehat{\mu})^2 + \varepsilon \qquad (22)$$

holds with the probability $1 - \delta$.

Under conditions A–E, the estimator $\widehat{\mu}_\varepsilon^0$ provides quadratic losses asymptotically not exceeding $R^{\mathrm{opt}} \le y$ and proves to be ε-unimprovable with $\varepsilon \to 0$.

4.3. MULTIPARAMETRIC SAMPLE LINEAR REGRESSION

Standard procedure of regression analysis starts with a rather artificial assumption that arguments of the regression are fixed, but the response is random due to inaccuracies of observation. In a more successive statistical setting, all empiric data should be considered as random and present a many-dimensional sample from population. Thus, we come to the "statistical regression," or, more specifically, to "sample regression" which actually is a form of regularity hidden amidst random distortion. The quality of the regression may be measured by the quadratic risk of prognoses calculated over the population. The usual minimum square solution minimizing empirical risk measured by residual sum of squares (RSS) has the quadratic risk function that is not minimal (the difference is substantial for large n). In Section 3.3, it was proved that by using Normal Evaluation Principle, standard quality functions of regularized multivariate procedures may be approximately evaluated as in terms of parameters, as in terms of statistics. This fact was used in [68], [69], and [71] for search of regressions asymptotically unimprovable in a wide class. In this section, we develop this approach for finite dimension and finite sample size and present solutions unimprovable with a small inaccuracy that is estimated from the above in an explicit form.

We consider $(n + 1)$-dimensional population $\mathfrak{S}$ in which the observations are pairs $(\mathbf{x}, y)$, where $\mathbf{x} = (x_1, \ldots, x_n)$ is a vector of predictors and y is a scalar response.

Define the centered values $\overset{\circ}{\mathbf{x}} = \mathbf{x} - \mathbf{E}\mathbf{x}$ and $\overset{\circ}{y} = y - \mathbf{E}y$. We restrict the population by a single requirement that the four moments of all variables exist and there exists the moment $M_8 = \mathbf{E}(\overset{\circ}{\mathbf{x}}^2/n)^2\, \overset{\circ}{y}^4$ (here and in the following, squares of vectors denote the squares of lengths). Assume, additionally, that $\mathbf{E}\overset{\circ}{\mathbf{x}}^2 > 0$ (non degenerate case). Denote

$$M_4 = \sup_{|\mathbf{e}|=1} \mathbf{E}(\mathbf{e}^T\overset{\circ}{\mathbf{x}})^4 > 0, \quad M = \max(M_4,\ \sqrt{M_8},\ \mathbf{E}\overset{\circ}{y}^4),$$

$$\text{and}\ \ \gamma = \sup_{\|\Omega\|=1} \mathrm{var}(\overset{\circ}{\mathbf{x}}^T\Omega\, \overset{\circ}{\mathbf{x}}/n)/M, \tag{1}$$

where (and in the following) $\mathbf{e}$ is a nonrandom vector of unit length, and Ω are symmetric, positive, semidefinite matrices with unit spectral norm. We consider the linear regression $y = \mathbf{k}^T\mathbf{x} + l + \Delta$, where $\mathbf{k} \in \mathbb{R}^n$ and $l \in \mathbb{R}^1$. The problem is to minimize the quadratic risk $\mathbf{E}\Delta^2$ by the best choice of $\mathbf{k}$ and l that should be calculated over a sample $\mathfrak{X} = \{(\mathbf{x}_m, y_m)\}$, $m = 1, \ldots, N$, from $\mathfrak{S}$.

We denote
$\lambda = n/N$, $\mathbf{a} = \mathbf{Ex}$, $a_0 = \mathbf{E}y$, $\Sigma = \text{cov}(\mathbf{x}, \mathbf{x})$, $\sigma^2 = \text{var}\, y$, and $\mathbf{g} = \text{cov}(\mathbf{x}, y)$.

If $\sigma > 0$ and the matrix Σ is nondegenerate, then the a priori coefficients $\mathbf{k} = \Sigma^{-1}\mathbf{g}$ and $l = a_0 - \mathbf{k}^T\mathbf{g}$ provide minimum of $\mathbf{E}\Delta^2$, which equals $\sigma^2 - \mathbf{g}^T\Sigma^{-1}\mathbf{g} = \sigma^2(1 - r^2)$, where r is the multiple correlation coefficient.

We start from the statistics

$$\bar{\mathbf{x}} = N^{-1}\sum_{m=1}^{N}\mathbf{x}_m, \quad \bar{y} = N^{-1}\sum_{m=1}^{N}y_m, \quad \widehat{\sigma}^2 = N^{-1}\sum_{m=1}^{N}(y_m - \bar{y})^2,$$

$$S = N^{-1}\sum_{m=1}^{N}\mathbf{x}_m\mathbf{x}_m^T, \quad \widehat{\mathbf{g}}_0 = N^{-1}\sum_{m=1}^{N}\mathbf{x}_m y_m,$$

$$C = N^{-1}\sum_{m=1}^{N}(\mathbf{x}_m - \bar{\mathbf{x}})(\mathbf{x}_m - \bar{\mathbf{x}})^T, \quad \widehat{\mathbf{g}} = N^{-1}\sum_{m=1}^{N}(\mathbf{x}_m - \bar{\mathbf{x}})(y_m - \bar{y}). \tag{2}$$

The standard minimum square "plug-in" procedure with $\widehat{\mathbf{k}} = C^{-1}\widehat{\mathbf{g}}$ and $\widehat{l} = \bar{y} - \mathbf{k}^T\bar{\mathbf{x}}$ has known demerits: this procedure does not guarantee the minimum risk, is degenerate for multicollinear data (for degenerate matrix C), and is not uniform with respect to the dimension.

The quadratic risk of the regression $y = \widehat{\mathbf{k}}^T\mathbf{x} + \widehat{l} + \Delta$, where $\widehat{\mathbf{k}}$ and $\widehat{l}$ are calculated over a sample with the "plug-in" constant term $\widehat{l} = \bar{y} - \widehat{\mathbf{k}}^T\bar{\mathbf{x}}$, is given by

$$\mathbf{E}\Delta^2 = R + \mathbf{E}(\bar{y} - \widehat{\mathbf{k}}^T\bar{\mathbf{x}})^2 = (1 + 1/N)R,$$

where

$$R \overset{def}{=} \mathbf{E}(\sigma^2 - 2\widehat{\mathbf{k}}^T\mathbf{g} + \widehat{\mathbf{k}}^T\Sigma\widehat{\mathbf{k}}). \tag{3}$$

Let us calculate and minimize R. We consider the following class of generalized regularized regressions. Let $H_0 = (I + tS)^{-1}$ and $H = (I + tC)^{-1}$ be the resolvents of the matrices S and C, respectively. We choose the coefficient $\widehat{\mathbf{k}}$ (everywhere below) in the class $\mathfrak{K}$ of statistics of the form $\widehat{\mathbf{k}} = \Gamma\widehat{\mathbf{g}}$, where

$$\Gamma = \Gamma(C) = \int tH(t)\,d\eta(t)$$

and $\eta(t)$ are functions whose variation on $[0, \infty)$ is at most one and that has sufficiently many moments

$$\eta_k \overset{def}{=} \int t^k |d\eta(t)|, \quad k = 1, 2, \ldots$$

The function $\eta(t)$ presenting a unit jump corresponds to the "ridge regression" (see Introduction). The regression with the coefficients $\widehat{\mathbf{k}} \in \mathfrak{K}$ may be called a generalized ridge regression. The quantity (2) depends on $\eta(t)$, $R^1 = R^1(\eta)$, and

$$R(\eta) = \sigma^2 - 2\mathbf{E} \int t\mathbf{g}^T H(t)\widehat{\mathbf{g}}\,d\eta(t) + \iint D(t, u)\,d\eta(t)\,d\eta(u), \quad (4)$$

where

$$D(t, u) \overset{def}{=} tu\,\widehat{\mathbf{g}}^T H(t)\Sigma H(u)\widehat{\mathbf{g}}.$$

Since all arguments of $R(\eta)$ are invariant with respect to the translation of the origin, we assume (everywhere in the following) that $\mathbf{a} = \mathbf{Ex} = 0$ and $a_0 = \mathbf{E}y = 0$.

Our purpose is to single out leading parts of these functionals and to obtain upper bounds for the remainder terms up to absolute constants. To simplify notations of the remainder terms, we define

$$\tau = \sqrt{M}t, \quad \varepsilon = \sqrt{\gamma + 1/N}, \quad c_{lm} = a\,\max(1, \tau^l)\,\max(1, \lambda^m),$$

where a, l, and m are non-negative numbers (for brevity we omit the argument t in $c_{ml} = c_{ml}(t)$). In view of (1), we can readily see that

$$\mathbf{E}(\mathbf{x}^2)^2 \le M, \quad \mathbf{E}(\bar{\mathbf{x}}^2)^2 \le M\lambda^2, \quad \|\Sigma\|^2 \le M,$$
$$\mathbf{g}^2 \le M, \quad \mathbf{E}\widehat{g}^2 \le \mathbf{E}\widehat{g}_0^2 \le M(1+\lambda).$$

As in previous sections, we begin by studying functions of more simple covariance matrices S and then pass to functions of C.

Some Spectral Functions of Sample Covariance Matrices

Our investigation will be based on results of the spectral theory of large sample covariance matrices developed in Chapter 3.

Define

$$H_0 = H_0(t) = (I + tS)^{-1}, \quad \widehat{h}_0(t) = n^{-1}\text{tr } H_0(t),$$
$$h_0(t) = \mathbf{E}\widehat{h}_0(t), \quad s_0 = s_0(t) = 1 - \lambda + \lambda h_0(t),$$
$$V = V(t) = \mathbf{e}^T H_0(t)\bar{\mathbf{x}}, \quad \Phi = \Phi(t) = \bar{\mathbf{x}}^T H_0(t)\bar{\mathbf{x}},$$

and

$$H = H(t) = (I + tC)^{-1}, \quad \widehat{h}(t) = n^{-1}\text{tr } H(t),$$
$$h(t) = \mathbf{E}\widehat{h}(t), \quad s = s(t) = 1 - \lambda + \lambda h(t).$$

We write out some assertions from Section 3.1 in the form of the following lemma.

LEMMA 4.10. *If* $t \ge 0$, *then*

1. $s_0 = \mathbf{E}(1 - t\psi_1) \ge (1 + \tau\lambda)^{-1}$,

$\text{var}(t\psi_1) \le \delta \overset{def}{=} 2\tau^2\lambda^2(\gamma + \tau^2/N) \le c_{42}\varepsilon$;

2. $\mathbf{E}H_0 = (I + ts_0\Sigma)^{-1} + \Omega_0$, *where* $\|\Omega_0\| \le c_{31}\varepsilon$,

$\text{var}(\mathbf{e}^T H_0\mathbf{e}) \le \tau^2/N$;

3. $t(\mathbf{E}V)^2 \le c_{52}\varepsilon^2$, $\text{var}(tV) \le c_{20}/N$;

4. $t\Phi \le 1$, $t\mathbf{E}\Phi = 1 - s_0 + o$, $o^2 \le c_{52}\varepsilon^2$, $\text{var}(t\Phi) \le c_{20}/N$;

5. $\|\mathbf{E}H - \mathbf{E}H_0\| \le c_{74}\varepsilon^2$, $|s - s_0| \le c_{11}/N$;

6. $\mathbf{E}H = (I + ts\Sigma)^{-1} + \Omega$, *where* $\|\Omega\|^2 \le c_{63}\varepsilon^2$.

Functionals of Random Gram Matrices

We consider random Gram matrices of the form S defined in (2) that can be used as sample covariance matrices when the expectation vectors are known a priori and may be set equal to 0.

Let us use the method of the alternative elimination of independent sample vectors. Eliminating one of the sample vectors, say, the vector $\mathbf{x}_1$, we denote

$$\bar{\mathbf{x}}^1 = \bar{\mathbf{x}}_1 - \mathbf{x}_1/N, \quad \widehat{\mathbf{g}}_0^1 = \widehat{\mathbf{g}}_0 - \mathbf{x}_1 y_1/N,$$

$$S^1 = S - \mathbf{x}_1\mathbf{x}_1^T/N \cdot H_0^1 = H_0^1(t) = (I + tS^1)^{-1}.$$

These values do not depend on $\mathbf{x}_1$ and y_1. The identity holds

$$H_0 = H_0^1 - tH_0^1\mathbf{x}_1\mathbf{x}_1^T H_0/N. \tag{5}$$

Also denote

$$v_1 = v_1(t) = \mathbf{e}^T H_0^1(t)\mathbf{x}_1, \quad u_1 = u_1(t) = \mathbf{e}^T H_0(t)\mathbf{x}_1,$$
$$\varphi_1 = \varphi_1(t) = \mathbf{x}_1^T H_0^1(t)\,\mathbf{x}_1/N, \quad \psi_1 = \psi_1(t) = \mathbf{x}_1^T H_0(t)\,\mathbf{x}_1/N.$$

We have the identities

$$u_1 = (1-t\psi_1)v_1, \quad (1+t\varphi_1)(1-t\psi_1) = 1, \quad H_0\mathbf{x}_1 = (1-t\psi_1)H_0^1\mathbf{x}_1. \tag{6}$$

Obviously, $0 \le t\psi_1 \le 1$. It can be readily seen that

$$1 - s_0 = t\mathbf{E}\psi_1 = t\mathbf{E}N^{-1}\mathrm{tr}(H_0 S). \tag{7}$$

From (1) it follows that

$$\|H_0\| \le \|H_0^1\| \le 1, \quad \mathbf{E}u_1^4 \le \mathbf{E}v_1^4 \le M. \tag{8}$$

Remark 1.

$$\mathbf{E}\widehat{\mathbf{g}}_0^2 \le M(1 + \lambda), \quad \mathbf{E}|\widehat{\mathbf{g}}_0|^4 \le 2M^2(1 + \lambda)^2,$$
$$\mathbf{E}|\widehat{\mathbf{g}}_0^1|^2 \le M(1 + \lambda), \quad \mathbf{E}|\widehat{\mathbf{g}}_0^1|^4 \le 2M^2(1 + \lambda)^2. \tag{9}$$

Indeed, the value $\widehat{\mathbf{g}}_0^1$ does not depend on $\mathbf{x}_1$ and y_1, and

$$\mathbf{E}\widehat{\mathbf{g}}_0^2 = \mathbf{E}y_1\mathbf{x}_1^T\widehat{\mathbf{g}}_0 = \mathbf{E}y_1\mathbf{x}_1^T\widehat{\mathbf{g}}_0^1 + \mathbf{E}y_1^2\mathbf{x}_1^2/N.$$

Here, the first summand equals $\mathbf{g}^2(1-N^{-1}) \le M$. The second one is not greater than $(\mathbf{E}y_1^4(\mathbf{x}_1^2/N)^2)^{1/2} \le M\lambda$. Further,

$$\mathbf{E}(\widehat{\mathbf{g}}_0^2)^2 = \mathbf{E}y_1\widehat{\mathbf{g}}^2\mathbf{x}_1^T\widehat{\mathbf{g}}_0 = \mathbf{E}y_1\widehat{\mathbf{g}}_0^2\ \mathbf{x}_1^T\widehat{\mathbf{g}}_0^1 + \mathbf{E}y_1^2\widehat{\mathbf{g}}_0^2\mathbf{x}_1^2/N.$$

Using the Schwarz inequality, we find

$$\mathbf{E}(\widehat{\mathbf{g}}_0^2)^2 \le 2\mathbf{E}(y_1\ \mathbf{x}_1^T\widehat{\mathbf{g}}_0^1)^2 + 2\mathbf{E}y_1^4(\mathbf{x}_1^2)^2/N^2.$$

In the first summand here, for fixed $\widehat{\mathbf{g}}_0^1$, we have $\mathbf{E}y_1^2(\mathbf{x}_1^T\mathbf{e})^2 \le M$, where $\mathbf{e} = \widehat{\mathbf{g}}_0^1/|\widehat{\mathbf{g}}_0^1|$. Therefore, the first summand is not greater than $2M\mathbf{E}\widehat{\mathbf{g}}_0^2 \le 2M^2(1 + \lambda)$. By (1) the second summand is not greater than $2M^2\lambda^2$. The second inequality in (9) follows. The same arguments may be used to establish the second pair of inequalities (9).

LEMMA 4.11. *If $t \ge 0$, then*

$$|t\mathbf{E}\bar{\mathbf{x}}^T H_0(t)\ \widehat{\mathbf{g}}_0| \le M^{1/4}c_{32}\ \varepsilon, \qquad \mathrm{var}(t\ \bar{\mathbf{x}}^T H_0(t)\ \widehat{\mathbf{g}}_0) \le \sqrt{M}c_{42}/N.$$

Proof. Eliminating $(\mathbf{x}_1, y_1)$, we substitute $\widehat{\mathbf{g}}_0 = \overset{\sim 1}{\mathbf{g}}_0 + \mathbf{x}_1 y_1/N$. We have $\mathbf{E}\mathbf{x}_1^T H_0^1\widehat{\mathbf{g}}_1^1 = 0$, $\mathbf{E}y_1 = 0$. It follows:

$$\begin{aligned}
t\mathbf{E}\bar{\mathbf{x}}^T H_0\ \widehat{\mathbf{g}}_0 &= t\mathbf{E}\mathbf{x}_1^T H_0\ \widehat{\mathbf{g}}_0 = t\mathbf{E}\mathbf{x}_1^T H_0\widehat{\mathbf{g}}_0^1 + \mathbf{E}t\psi_1 y_1 = \\
&= t\mathbf{E}(1 - t\psi_1)\mathbf{x}_1^T H_0^1\ \widehat{\mathbf{g}}_0^1 + \mathbf{E}(1 - s_0 - \Delta_1)y_1 = \\
&= t\mathbf{E}\Delta_1\ \mathbf{x}_1^T H_0^1\ \widehat{\mathbf{g}}_0^1 - \mathbf{E}\Delta_1 y_1,
\end{aligned}$$

where $\Delta_1 = 1 - t\psi_1 - s_0$. By Statement 1 of Lemma 4.10, $\mathbf{E}\Delta_1^2 \le \delta \le c_{42}\varepsilon^2$. Applying the Schwarz inequality, (1), and (9), we obtain that

$$\begin{aligned}
(t\mathbf{E}\bar{\mathbf{x}}^T H_0\ \widehat{\mathbf{g}}_0)^2 &\le [t^2\mathbf{E}(\mathbf{x}_1^T H_0^1\ \widehat{\mathbf{g}}_0^1)^2 + \mathbf{E}y_1^2]\ c_{42}\varepsilon^2 \le \\
&\le \sqrt{M}\ [t^2\mathbf{E}(\widehat{\mathbf{g}}_0^1)^2 + 1]\ c_{42}\varepsilon^2 \le \sqrt{M}c_{63}\ \varepsilon^2.
\end{aligned}$$

The first statement follows.

To estimate the variance, we use Lemma 3.2. Let us eliminate the variables $\mathbf{x}_1$ and y_1. Denote $\widetilde{\mathbf{x}} = \bar{\mathbf{x}} - \mathbf{x}_1/N$. Then, $\bar{\mathbf{x}}^T H_0 \, \widehat{\mathbf{g}}_0$ equals

$$\widetilde{\mathbf{x}}^T H_0^1 \, \widehat{\mathbf{g}}_0^1 - t \, \widetilde{\mathbf{x}}^T H_0^1 \mathbf{x}_1 \, \mathbf{x}_1^T H_0 \, \widehat{\mathbf{g}}_0^1/N + \mathbf{x}_1^T H_0 \, \widehat{\mathbf{g}}_0^1/N + y_1 \, \bar{\mathbf{x}}^T H_0 \mathbf{x}_1/N.$$

The first term in the right-hand side does not depend on $\mathbf{x}_1$ and y_1. In view of the identical dependence on sample vectors, we conclude that $t^2 \mathrm{var}(\bar{\mathbf{x}}^T H_0 \, \widehat{\mathbf{g}}_0)$ is not greater than

$$3[\mathbf{E}(t^2 \, \widetilde{\mathbf{x}}_1^T H_0^1 \mathbf{x}_1 \mathbf{x}_1^T H_0 \widehat{\mathbf{g}}_0^1)^2 + \mathbf{E}(t \mathbf{x}_1^T H_0 \, \widehat{\mathbf{g}}_0^1)^2 + \mathbf{E}(t y_1 \, \bar{\mathbf{x}}^T H_0 \mathbf{x}_1)^2]/N.$$

In view of (6), this inequality remains valid if H_0 is replaced by H_0^1. After this replacement, we use (1). The square of the sum of the first two summand in the bracket is not greater than

$$at^4 \, \mathbf{E}(\mathbf{x}_1^T H_0^1 \, \widehat{\mathbf{g}}_0^1)^4 \, \mathbf{E} \, (t^2 (\widetilde{\mathbf{x}}^T H_0^1 \mathbf{x}_1)^2 + 1)^2 \leq$$
$$\leq aMt^4 \mathbf{E}|\widehat{\mathbf{g}}_0^1|^4 \mathbf{E}(\sqrt{M}t^2 \, \widetilde{\mathbf{x}}_1^T (H_0^1)^2 \widetilde{\mathbf{x}}_1 + 1),$$

where a is a numerical coefficient. From the definition, one can see that $t\Phi \leq 1$ and $t\widetilde{\mathbf{x}}_1^T (H_0^1)^2 \widetilde{\mathbf{x}}_1 \leq 1$. The square of the sum of the first two summands in the bracket is not greater than $aM^3 t^4 (\sqrt{M} t + 1) \, (1 + \lambda)^2 \leq M c_{52}$. In the third summand, $\bar{\mathbf{x}}^T H_0 \mathbf{x}_1 = \widetilde{\mathbf{x}}^T H_0 \mathbf{x}_1 + \psi_1$. By (1) and (6), the square of this summand is not greater than

$$2(t^4 \mathbf{E} y_1^4 \mathbf{E}(\widetilde{\mathbf{x}}^T H_0^1 \mathbf{x}_1)^4 + M) \leq 2M(Mt^4 \mathbf{E}|\widetilde{\mathbf{x}}|^4 + 1) \leq M c_{42}.$$

We conclude that the variance in Statement 2 is not greater than $\sqrt{M} c_{42}/N$. The proof of Lemma 4.11 is complete. $\square$

LEMMA 4.12.

$$t \mathbf{E} \mathbf{e}^T H_0(t) \, \widehat{\mathbf{g}}_0 = t s_0(t) \mathbf{E} \mathbf{e}^T H_0(t) \mathbf{g} + o,$$

$$\textit{where} \quad |o| \leq c_{31}\varepsilon; \quad \mathrm{var}(t \mathbf{e}^T H_0(t) \, \widehat{\mathbf{g}}_0) \leq c_{41}/N. \tag{10}$$

Proof. Denote $\Delta_1 = t\psi_1 - t\mathbf{E}\psi_1$. Using (6), we find

$$
\begin{aligned}
t\mathbf{E}\mathbf{e}^T H_0 \widehat{\mathbf{g}}_0 &= t\mathbf{E} y_1 \mathbf{e}^T H_0 \mathbf{x}_1 = t\mathbf{E} y_1 (1 - t\psi_1) \mathbf{e}^T H_0^1 \mathbf{x}_1 = \\
&= t s_0 \mathbf{E} y_1 \mathbf{e}^T H_0^1 \mathbf{x}_1 - t\mathbf{E} y_1 \Delta_1 \ \mathbf{e}^T H_0^1 \mathbf{x}_1 = \\
&= t s_0 \mathbf{E}\mathbf{e}^T H_0^1 \mathbf{g} - t\mathbf{E} v_1 y_1 \Delta_1 = \\
&= t s_0 \mathbf{E}\mathbf{e}^T H_0 \mathbf{g} + t^2 s_0 \mathbf{E} u_1 \mathbf{x}_1^T H_0^1 \mathbf{g}/N - t\mathbf{E} v_1 y_1 \Delta_1.
\end{aligned}
$$

The last two terms present the remainder term o in the lemma formulation. We estimate it by the Schwarz inequality:

$$
\begin{aligned}
|o| &\le t^2 (\mathbf{E} u_1^2 \ \mathbf{E}(\mathbf{x}_1^T H_0 \mathbf{g})^2)^{1/2}/N + t\sqrt{\delta}(\mathbf{E} v_1^2 y_1^2)^{1/2} \le \\
&\le \tau^2/N + t\sqrt{\delta M} \le c_{31}\varepsilon.
\end{aligned}
$$

The first statement in (10) is proved.

Now we estimate the variance eliminating independent variables. Denote $f = t\mathbf{e}^T H_0 \ \widehat{\mathbf{g}}_0$. Let $f = f^1 + \Delta_1$, where f^1 does not depend on $\mathbf{x}_1$ and y_1. We have

$$
f = \mathbf{e}^T H_0^1 \ \widehat{\mathbf{g}}_0^1 + t u_1 y_1/N + t^2 u_1 (\mathbf{x}_1^T H_0^1 \ \widehat{\mathbf{g}}_0^1)/N.
$$

By Lemma 4.2, we obtain var $f \le N\mathbf{E} \ \Delta_1^2$.

Therefore,

$$
\begin{aligned}
\text{var } f &\le 2N^{-1}[t^4 \mathbf{E} u_1^2 (\mathbf{x}_1^T H_0^1 \ \widehat{\mathbf{g}}_0^1)^2 + t^2 \mathbf{E} u_1^2 y_1^2] \le \\
&\le 2N^{-1}(\mathbf{E} u_1^4)^{1/2} \ M^{1/2} t^2 \ [t^2 (\mathbf{E}|\widehat{\mathbf{g}}_0^1|^4)^{1/2} + 1] \le \\
&\le 2N^{-1}\tau^2 (2\tau^2(1 + \lambda) + 1) \le c_{41}/N.
\end{aligned}
$$

Lemma 4.12 is proved. $\square$

LEMMA 4.13. *If $t \ge 0$, then*

$$
\begin{aligned}
t\mathbf{E}\widehat{\mathbf{g}}_0^T H_0 \widehat{\mathbf{g}}_0 &= \sigma^2 (1 - s_0) + t s_0 \mathbf{E}\mathbf{g}^T H_0 \widehat{\mathbf{g}}_0 + o_1 = \\
&= \sigma^2 (1 - s_0) + t s_0^2 \mathbf{E}\mathbf{g}^T H_0 \mathbf{g} + o_2,
\end{aligned}
$$

where

$$
|o_1| \le \sqrt{M} c_{32}\varepsilon, \quad |o_2| \le \sqrt{M} c_{32}\varepsilon; \quad t^2 \text{var}(\widehat{\mathbf{g}}_0^T H_0 \widehat{\mathbf{g}}_0) \le M c_{42}/N.
$$

Proof. Using (6), we find that $t\mathbf{E}\widehat{\mathbf{g}}_0 H_0 \widehat{\mathbf{g}}_0 = t\mathbf{E}y_1 \mathbf{x}_1^T H_0 \widehat{\mathbf{g}}_0$ equals

$$t\mathbf{E}y_1\mathbf{x}_1^T H_0\mathbf{x}_1/N + ty_1\mathbf{E}\mathbf{x}_1^T H_0 \widehat{\mathbf{g}}_0^1 =$$
$$= t\mathbf{E}y_1^2\psi_1 + ty_1\mathbf{E}(1 - t\psi_1)\mathbf{x}_1^T H_0^1\widehat{\mathbf{g}}_0^1.$$

Substituting $t\psi_1 = 1 - s_0 - \Delta_1$, we find that $\mathbf{E}t\widehat{\mathbf{g}}_0^T H_0\widehat{\mathbf{g}}_0$ equals

$$\sigma^2(1 - s_0) - \mathbf{E}\Delta_1 y_1^2 + ts_0\mathbf{E}y_1\mathbf{x}_1^T H_0^1\widehat{\mathbf{g}}_0^1 + t\mathbf{E}\Delta_1 y_1 \; \mathbf{x}_1^T H_0^1 \; \widehat{\mathbf{g}}_0^1. \quad (11)$$

The square of the second term in the right-hand side is not greater than $M\delta \leq Mc_{42}\varepsilon^2$, where δ is defined in Lemma 4.10. Using the Schwarz inequality, we obtain that the square of the fourth term in (11) is not greater than

$$\delta t^2\mathbf{E}y_1^2(\mathbf{x}_1^T H_0^1 \; \widehat{\mathbf{g}}_0^1)^2 \leq$$
$$\leq \delta M t^2(\mathbf{E}|\widehat{\mathbf{g}}_0^1|^4)^{1/2} \leq 2\delta M\tau^2(1 + \lambda) \leq Mc_{63}\varepsilon^2.$$

In the third term of the right-hand side of (11), we have $\mathbf{E}y_1\mathbf{x}_1^T = \mathbf{g}^T$, and this term equals

$$ts_0\mathbf{E}\mathbf{g}^T H_0^1\widehat{\mathbf{g}}_0^1 = ts_0\mathbf{E}\mathbf{g}^T H_0\widehat{\mathbf{g}}_0^1 + t^2 s_0\mathbf{E}\mathbf{g}^T H_0\mathbf{x}_1 \; \mathbf{x}_1^T H_0^1 \; \widehat{\mathbf{g}}_0^1/N =$$
$$= ts_0\mathbf{E}\mathbf{g}^T H_0\widehat{\mathbf{g}}_0 - ts_0\mathbf{E}y_1\mathbf{g}^T H_0\mathbf{x}_1/N + t^2 s_0 \; \mathbf{E}\mathbf{g}^T H_0\mathbf{x}_1 \; \mathbf{x}_1^T H_0^1 \; \widehat{\mathbf{g}}_0^1/N.$$

Here in the right-hand side, the first summand is that included in the lemma formulation. The square of the sum of the second and third summands on the right-hand side is not greater than

$$2t^2\mathbf{E}(\mathbf{g}^T \; H_0\mathbf{x}_1)^2 \; [\mathbf{E}y_1^2 + t^2\mathbf{E}(\mathbf{x}_1^T H_0^1 \; \widehat{\mathbf{g}}_0^1)^2]/N^2 \leq$$
$$\leq 2t^2\mathbf{g}^2\mathbf{E}u_1^2\sqrt{M} \; (1 + t^2\mathbf{E}(\widehat{\mathbf{g}}_0^1)^2)/N^2 \leq$$
$$\leq 2M\tau^2 \; [1 + 2\tau^2(1 + \lambda)]/N^2 \leq Mc_{41}/N^2.$$

We conclude that $o_1^2 \leq Mc_{63}\varepsilon^2$ and $|o_1| \leq \sqrt{M}c_{32}\varepsilon$.

In view of the first statement of Lemma 4.12, we have

$$t\mathbf{E}\mathbf{g}^T H_0\widehat{\mathbf{g}} = ts_0\mathbf{E}\mathbf{g}^T H_0\mathbf{g} + o,$$

where $|o| \leq |\mathbf{g}|c_{31}\varepsilon$. Consequently, $|o_2| \leq \sqrt{M}c_{32}\varepsilon$.

Now we estimate the variance of $f \overset{def}{=} t\, \widehat{\mathbf{g}}_0^T H_0\, \widehat{\mathbf{g}}_0$. Using Lemma 4.12 and taking into account the identical dependence on sample vectors, we have $\operatorname{var} f \leq N\Delta_1^2$, where $\Delta_1 = f - f^1$, and f^1 does not depend on $\mathbf{x}_1$ and y_1. We rewrite f in the form

$$f = t\widehat{\mathbf{g}}_0^{1T} H_0^1\, \widehat{\mathbf{g}}_0^1 + 2t\widehat{\mathbf{g}}_0^{1T} H_0\mathbf{x}_1\, y_1/N +$$
$$+ t\mathbf{x}_1^T H_0\mathbf{x}_1\, y_1^2/N^2 - t^2\widehat{\mathbf{g}}_0^T H_0^1\mathbf{x}_1\, \mathbf{x}_1^T H_0\widehat{\mathbf{g}}_0/N.$$

Here the first summand does not depend on $(\mathbf{x}_1, y_1)$. We find that $\operatorname{var} f$ is not greater than

$$a\left[t^2\mathbf{E}(y_1\widehat{\mathbf{g}}_0^{1T} H_0\mathbf{x}_1)^2 + t^2\mathbf{E}\psi_1^2 y_1^4 + t^4\mathbf{E}(\widehat{\mathbf{g}}_0^{1T} H_0^1\mathbf{x}_1)^2(\mathbf{x}_1^T H_0\, \widehat{\mathbf{g}}_0^1)^2\right]/N,$$

where a is a numerical constant. We apply the Schwarz inequality. The square of the first summand in the square bracket is not greater than

$$Mt^4\mathbf{E}|\widehat{\mathbf{g}}_0^1|^4\, \mathbf{E}y_1^4 \leq 2M^4t^4(1+\lambda)^2 \leq M^2c_{42}.$$

It follows that the first summand is not greater than Mc_{21} in absolute value. The second summand is not greater than Mt^2 by (1). Using (6) and (1), we obtain that the third summand is not greater than

$$t^4\mathbf{E}(\widehat{\mathbf{g}}_0^T H_0^1\mathbf{x}_1)^4/N^2 \leq Mt^4\mathbf{E}|\widehat{\mathbf{g}}_0|^4 \leq 2M^3t^4(1+\lambda)^2 \leq Mc_{42}.$$

Consequently, $\operatorname{var} f \leq Mc_{42}/N$. This completes the proof of Lemma 4.13. □

LEMMA 4.14. *If $t \geq u \geq 0$, then*

$$tu\, \mathbf{E}\widehat{\mathbf{g}}_0^T H_0(t)\Sigma H_0(u)\, \widehat{\mathbf{g}}_0 = tu\, \mathbf{E}\widehat{\mathbf{g}}_0^{1T} H_0^1(t)\Sigma H_0^1(u)\, \widehat{\mathbf{g}}_0^1 + o,$$

where $|o| \leq \sqrt{M}c_{31}/N$; the inequality holds

$$t^2u^2\, \operatorname{var}(\widehat{\mathbf{g}}_0^T H_0(t)\Sigma H_0(u)\, \widehat{\mathbf{g}}_0) \leq \delta \overset{def}{=} M^2t^2c_{42}/N.$$

Proof. Substituting $\widehat{\mathbf{g}}_0 = \widehat{\mathbf{g}}_0^1 + \mathbf{x}_1 y_1/N$, we obtain

$$f \overset{\text{def}}{=} \widehat{\mathbf{g}}_0^T H_0(t)\Sigma H_0(u)\,\widehat{\mathbf{g}}_0 = tu\,\widehat{\mathbf{g}}_0^{1T} H_0(t)\Sigma H_0(u)\,\widehat{\mathbf{g}}_0^1 +$$
$$+ tu\,\widehat{\mathbf{g}}_0^{1T} H_0(t)\Sigma H_0(u)\mathbf{x}_1\,y_1/N + tuy_1\,\mathbf{x}_1^T H_0(t)\Sigma H_0(u)\widehat{\mathbf{g}}_0^1/N +$$
$$+ tuy_1^2\,\mathbf{x}_1^T H_0(t)\Sigma H_0(u)\mathbf{x}_1/N^2. \tag{12}$$

To prove the lemma statement, it suffices to show that the three last summands in (12) are small and the difference

$$d = tu\,\widehat{\mathbf{g}}_0^{1T} H_0^1(t)\Sigma H_0^1(u)\,\widehat{\mathbf{g}}_0^1 - tu\,\widehat{\mathbf{g}}_0^{1T} H_0(t)\Sigma H_0(u)\,\widehat{\mathbf{g}}_0^1$$

is small.

In the second summand in (12), we use the Schwarz inequality. First, we single out the dependence on y_1. We use (6) to replace H_0 by H_0^1. It follows that the expected square of the second summand in the right-hand side of (12) that is not greater than

$$t^2 u^2\,\sqrt{M}\mathbf{E}|\widehat{\mathbf{g}}_0^{1T} H_0^1(t)\Sigma H_0(u)\mathbf{x}_1|^2/N^2 \leq M^3 t^4 (1+\lambda)/N^2\,Mc_{41}/N^2.$$

The third summand in (12) is estimated likewise. To estimate the quantity d, we present in the form

$$d = t^2 u\widehat{\mathbf{g}}_0^{1T} H_0^1(t)\mathbf{x}_1\,\mathbf{x}_1^T H_0(t)\Sigma H_0^1(u)\,\widehat{\mathbf{g}}_0^1/N +$$
$$+ tu^2\widehat{\mathbf{g}}_0^{1T} H_0^1(u)\mathbf{x}_1\,\mathbf{x}_1^T H_0(u)\Sigma H_0^1(t)\,\widehat{\mathbf{g}}_0^1/N +$$
$$+ t^2 u^2\widehat{\mathbf{g}}_0^{1T} H_0^1(t)\mathbf{x}_1\,\mathbf{x}_1^T H_0(t)\Sigma H_0(u)\mathbf{x}_1\,\mathbf{x}_1^T H_0^1\widehat{\mathbf{g}}_0^1/N^2.$$

Let us estimate $\mathbf{E}d^2$. Note that

$$\sqrt{tu}\,\mathbf{x}_1^T H_0(t)\Sigma H_0(u)\mathbf{x}_1 \leq \sqrt{tu\Phi(t)\Phi(u)} \leq 1.$$

It follows that

$$\mathbf{E}d^2/3 \leq t^4 u^2\,\mathbf{E}|\widehat{\mathbf{g}}_0^{1T} H_0^1(t)\mathbf{x}_1|^2\,\mathbf{E}|\mathbf{x}_1^T H_0(t)\Sigma H_0^1(u)\widehat{\mathbf{g}}_0^1|^2/N^2 +$$
$$+ t^2 u^4\mathbf{E}|\widehat{\mathbf{g}}_0^{1T} H_0^1(u)\mathbf{x}_1|^2\,\mathbf{E}|\mathbf{x}_1^T H_0(u)\Sigma H_0^1(t)\,\widehat{\mathbf{g}}_0^1|^2/N^2 +$$
$$+ Mt^3 u^3\mathbf{E}|\widehat{\mathbf{g}}_0^{1T} H_0^1(t)\mathbf{x}_1|^2\,|\widehat{\mathbf{g}}_0^{1T} H_0^1(u)\mathbf{x}_1|^2/N^2. \tag{13}$$

Substituting H_0^1 for H_0, using the relation $H_0\mathbf{x}_1 = (1 - t\psi_1)H_0^1\mathbf{x}_1$ and (1), we obtain

$$\mathbf{E}d^2/3 \leq 2M^2t^6(\mathbf{E}|\widehat{\mathbf{g}}_0^1|^2)^2/N^2 + M^2t^6\mathbf{E}|\widehat{\mathbf{g}}_0^1|^4/N^2 \leq$$
$$\leq 2M^4t^6(1 + \lambda)^2/N^2 \leq Mc_{62}/N^2.$$

It follows that $\mathbf{E}|d| \leq \sqrt{M}c_{31}/N$.

The expected square of the second summand is not greater than

$$+ 2t^4u^2\sqrt{M}\mathbf{E}(\widehat{\mathbf{g}}_0^{1T}H_0^1(t)\mathbf{x}_1)^2(\mathbf{x}_1^T H_0(t)\Sigma H_0(u)\mathbf{x}_1)^2/N^4.$$

We use (6) to replace H_0 by H_0^1. It follows that the expected square of the second summand that is not greater than

$$4t^4M^2\mathbf{E}|\widehat{\mathbf{g}}_0^1|^2/N^2 \leq 4M^2t^2\tau^2(1 + \lambda)/N^2 \leq M^2t^2c_{21}/N^2.$$

The contribution of the fourth summand in (12) to f is not greater than $\sqrt{M}t\mathbf{E}y_1^2/N \leq \sqrt{M}c_{10}/N$.

Thus, the right-hand side of (12) presents the leading term in the lemma formulation with the accuracy up to c_{31}/N. The first statement of the lemma is proved.

Further, we estimate $\text{var}f$ similarly using Lemma 3.2. We find that $\text{var}f \leq N\mathbf{E}\Delta_1^2$, where $f - \Delta_1$, does not depend on $\mathbf{x}_1$ and y_1. The value Δ_1 equals the sum of last three terms in (12) minus d. The expectation of squares of the second and third terms in (12) is not greater than $M^2t^2c_{21}/N^2$. By (1), the square of the fourth term in (12) contributes no more than $Mt^2\mathbf{E}y_1^4/N^2 \leq M^2t^2/N^2$. We have the inequality $\mathbf{E}d^2 \leq 2M^2t^2c_{42}/N^2$. It follows that $\mathbf{E}\Delta_1^2 \leq M^2t^2c_{42}/N^2$. This is the second lemma statement. The proof of Lemma 4.14 is complete. $\square$

LEMMA 4.15. *If* $t \geq u \geq 0$, *then*

$$tu\mathbf{E}\widehat{\mathbf{g}}_0^T H_0(t)SH_0(u)\widehat{\mathbf{g}}_0 = tus_0(t)s_0(u)\mathbf{E}\widehat{\mathbf{g}}_0^{1T}H_0^1(t)\Sigma H_0^1(u)\ \widehat{\mathbf{g}}_0^1 +$$
$$+ (1 - s_0(u))ts_0(t)\ \mathbf{E}\widehat{\mathbf{g}}_0^{1T}H_0^1(t)\mathbf{g} + (1 - s_0(t))us_0(u)\mathbf{E}\widehat{\mathbf{g}}_0^{1T}H_0^1(u)\mathbf{g} +$$
$$+ \sigma^2(1 - s_0(t))(1 - s_0(u)) + o, \tag{14}$$

and $\mathbf{E}|o| \le \sqrt{M}c_{42}\varepsilon; \quad t^2u^2\mathrm{var}(\widehat{\mathbf{g}}_0^T H_0(t)SH_0(u)\widehat{\mathbf{g}}_0) \le Mc_{62}/N.$

Proof. We notice that

$$f \overset{def}{=} tu\mathbf{E}\widehat{\mathbf{g}}_0^T H_0(t)SH_0(u)\,\widehat{\mathbf{g}}_0 = tu\mathbf{E}\widehat{\mathbf{g}}_0^T H_0(t)\mathbf{x}_1\,\mathbf{x}_1 H_0(u)\,\widehat{\mathbf{g}}_0.$$

Substituting $\widehat{\mathbf{g}}_0 = \widehat{\mathbf{g}}_0^1 + \mathbf{x}_1 y_1/N$, we find

$$f = tu\mathbf{E}\widehat{\mathbf{g}}_0^{1T} H_0(t)\mathbf{x}_1\,\mathbf{x}_1^T H_0(u)\widehat{\mathbf{g}}_0^1 + tu\mathbf{E}y_1\widehat{\mathbf{g}}_0^{1T} H_0(t)\psi_1(u)\mathbf{x}_1 +$$
$$+ tu\mathbf{E}y_1\widehat{\mathbf{g}}_0^{1T} H_0(u)\psi_1(t)\mathbf{x}_1 + tu\mathbf{E}y_1^2\psi_1(t)\psi_1(u)/N^2.$$

In the first summand of the right-hand side, we substitute $H_0\mathbf{x}_1$ from (6) and $t\psi_1(t) = 1 - s_0(t) + \Delta_1$, $\mathbf{E}\Delta_1^2 \le \delta$. We find that the first summand is $tu\mathbf{E}s_0(t)s_0(u)\,\widehat{\mathbf{g}}_0^{1T} H_0^1(t)\Sigma H_0^1(u)\,\widehat{\mathbf{g}}_0^1 + o$, where the leading term is involved in the formulation of the lemma, and the remainder term is such that

$$\mathbf{E}o^2 \le at^2u^2\mathbf{E}[\widehat{\mathbf{g}}_0^{1T} H_0^1(t)\mathbf{x}_1\,\mathbf{x}_1^T H_0^1(u)\,\widehat{\mathbf{g}}_0^1]^2\delta,$$

where a is numerical coefficient and δ is defined in Lemma 4.14. In view of (1), we have

$$\mathbf{E}o^2 \le Mt^4\mathbf{E}|\widehat{\mathbf{g}}_0^1|^4c_{42}\varepsilon^2 \le Mc_{84}\varepsilon^2.$$

We transform the last three summands of f substituting $t\psi_1(t) = = 1 - s_0(t) + \Delta_1$, $\mathbf{E}\Delta_1^2 \le \delta$. The sum of these terms is equal to

$$(1- s_0(u))ts_0(t)\mathbf{E}\widehat{\mathbf{g}}_0^{1T} H_0^1(t)\mathbf{x}_1 y_1 +$$
$$+ (1 - s_0(t))us_0(u)\mathbf{E}\widehat{\mathbf{g}}_0^{1T} H_0^1(u)\mathbf{x}_1 y_1 +$$
$$+ (1 - s_0(t))(1 - s_0(u))\mathbf{E}y_1^2 + o, \tag{15}$$

where the remainder term o is such that $\mathbf{E}o^2$ is not greater than

$$a\big[u^2\mathbf{E}(\widehat{\mathbf{g}}_0^{1T} H_0^1(t)\mathbf{x}_1)^2 y_1^2\delta + t^2\mathbf{E}(\widehat{\mathbf{g}}_0^{1T} H_0^1(u)\mathbf{x}_1)^2 y_1^2\delta + \mathbf{E}y_1^4\delta\big] \le$$
$$\le aM(2\tau^2(1 + \lambda) + 1)\,\delta \le Mc_{63}\varepsilon^2,$$

where a is a numerical coefficient. The leading part of (15), as is readily seen, coincides with three terms in (14). The weakest upper estimate for the squares of the remainder terms is $M c_{84} \varepsilon^2$. Consequently, the first lemma statement holds with the remainder term $\sqrt{M} c_{42} \varepsilon$.

In the second statement, we first substitute $t S H_0(t) = I - H_0(t)$. It follows

$$\text{var } f = \text{var}(u\ \widehat{\mathbf{g}}_0^T H_0(t)\ \widehat{\mathbf{g}}_0 - u\ \widehat{\mathbf{g}}_0^T H_0(t) H_0(u)\ \widehat{\mathbf{g}}_0).$$

Here the variance of the minuend is not greater than $M c_{42}/N$ by Lemma 4.13. The variance of the subtrahend is not greater than $M c_{62}/N$ by Lemma 4.14. The last statement of Lemma 4.15 follows. $\square$

THEOREM 4.7. *If* $t \geq u \geq 0$, *then*

$$tu\mathbf{E}\widehat{\mathbf{g}}_0^T H_0(t) S H_0(u)\widehat{\mathbf{g}}_0 = tu s_0(t) s_0(u)\mathbf{E}\widehat{\mathbf{g}}_0^T H_0(t) \Sigma H_0(u)\widehat{\mathbf{g}}_0 +$$

$$+ (1 - s_0(u)) t s_0(t)\mathbf{E}\widehat{\mathbf{g}}_0^T H_0(t)\mathbf{g} + (1 - s_0(t)) u s_0(u)\mathbf{E}\widehat{\mathbf{g}}_0^T H_0(u)\mathbf{g} +$$

$$+ \sigma^2 (1 - s_0(t))(1 - s_0(u)) + o, \qquad (16)$$

where $|o| \leq \sqrt{M} c_{42} \varepsilon$.

Proof. First, we apply Lemma 4.15. The left-hand side can be transformed by (14) with the remainder term $\sqrt{M} c_{42} \varepsilon$. We obtain the first summand in (16). Now we compare the right-hand sides of (14) and (16). By (6), the difference between the second summands does not exceed

$$t|\mathbf{E}\widehat{\mathbf{g}}_0^{1T} H_0^1(t)\mathbf{g} - \mathbf{E}\widehat{\mathbf{g}}_0^T H_0(t)\mathbf{g}| \leq$$

$$\leq t\ |\mathbf{E}t\ \widehat{\mathbf{g}}_0^{1T} H_0^1(t)\mathbf{x}_1\ \mathbf{x}_1^T H_0(t)\mathbf{g}/N| + t|\mathbf{E}y_1\ \mathbf{x}_1^T H_0(t)\mathbf{g}/N| \leq$$

$$\leq [|\mathbf{E}t\widehat{\mathbf{g}}_0^{1T} H_0^1(t)\mathbf{x}_1|^2\ \mathbf{E}|t\mathbf{x}_1^T H_0(t)\mathbf{g}|^2 + \mathbf{E}y_1^2\ \mathbf{E}|t\mathbf{x}_1^T H_0(t)\mathbf{g}|^2]^{1/2}/N \leq$$

$$\leq \left(\mathbf{E}|\widehat{\mathbf{g}}_0^1|^2 M^2 t^4 + M^2 t^2\right)^{1/2}/N \leq \sqrt{M} c_{21}/N.$$

The difference between the third summands also does not exceed this quantity. The fourth summands coincide. We conclude that the equality in the formulation of the theorem holds with the inaccuracy at most $\sqrt{M} c_{42} \varepsilon$. Theorem 4.7 is proved. $\square$

Functionals in the Regression Problem

To pass to matrices C, $H = H(t)$, and $\widehat{\mathbf{g}}$, we use the identities $C = S - \bar{\mathbf{x}}\bar{\mathbf{x}}^T$, $\widehat{\mathbf{g}} = \widehat{\mathbf{g}}_0 - \bar{\mathbf{x}}\bar{y}$, and the identity $H = H_0 - tH_0\bar{\mathbf{x}}\bar{\mathbf{x}}^T H$, where $\bar{\mathbf{x}}$ and $\bar{y}$ are centered sample averages of $\mathbf{x}$ and y.

Remark 2. $\mathbf{E}|\widehat{\mathbf{g}}|^4 \leq aM^2(1 + \lambda)^2$, where a is a numerical coefficient.

LEMMA 4.16.

$$t\mathbf{E}\mathbf{g}^T H(t)\widehat{\mathbf{g}} = ts(t)\mathbf{E}\mathbf{g}^T H(t)\mathbf{g} + o_1,$$
$$t\mathbf{E}\widehat{\mathbf{g}}^T H(t)\widehat{\mathbf{g}} = \sigma^2(1 - s(t)) + ts(t)\mathbf{E}\mathbf{g}^T H(t)\widehat{\mathbf{g}} + o_2, \qquad (17)$$

where $|o_1|, |o_2| \leq \sqrt{M}c_{43}\varepsilon$.

Proof. We have

$$t\mathbf{E}\mathbf{g}^T H\widehat{\mathbf{g}} = t\mathbf{E}\mathbf{g}^T H_0\widehat{\mathbf{g}}_0 + t^2\mathbf{E}\mathbf{g}^T H\bar{\mathbf{x}}\ \bar{\mathbf{x}}^T H\widehat{\mathbf{g}}_0 - t\mathbf{E}\mathbf{g}^T H_0\bar{\mathbf{x}}\ \bar{y}.$$

By Lemma 4.12, the first summand equals $ts_0\mathbf{E}\mathbf{g}^T H_0\mathbf{g} + o$, where $|o| \leq \sqrt{M}c_{31}\varepsilon$. We estimate the remaining terms using the Schwarz inequality. By Statement 3 of Lemma 4.10 with $\mathbf{e} = \mathbf{g}/|\mathbf{g}|$, we find that the second term is not greater in absolute value than

$$|\mathbf{g}|(\mathbf{E}tV^2t^3\ \mathbf{E}\bar{\mathbf{x}}^2\widehat{\mathbf{g}}^2)^{1/2} \leq \sqrt{M}c_{42}\varepsilon.$$

The third summand does not exceed $|\mathbf{g}|t(\mathbf{E}\bar{\mathbf{x}}^2\mathbf{E}\bar{y}^2)^{1/2} \leq \sqrt{M}c_{11}$. We conclude that

$$|t\mathbf{E}\mathbf{g}^T H\widehat{\mathbf{g}} - t\mathbf{E}\mathbf{g}^T H_0\widehat{\mathbf{g}}_0| \leq \sqrt{M}c_{42}\varepsilon.$$

In view of Lemma 4.10, we can replace s_0 by s with an accuracy up to $t\mathbf{g}^2c_{11}/N \leq \sqrt{M}c_{21}/N$. It follows that $t\mathbf{E}\mathbf{g}^T H\widehat{\mathbf{g}} = ts\mathbf{g}^T H_0\mathbf{g} + o$, where $|o| \leq \sqrt{M}c_{42}\varepsilon$. In view of Lemma 4.10, replacing H_0 by H in the right-hand side we produce an inaccuracy of the same order. The first statement of our lemma is proved.

Further, from (16) it follows

$$t\mathbf{E}\widehat{\mathbf{g}}^T H\widehat{\mathbf{g}} = t\mathbf{E}\widehat{\mathbf{g}}_0^T H_0\widehat{\mathbf{g}}_0 + t^2\mathbf{E}\widehat{\mathbf{g}}_0^T H_0\bar{\mathbf{x}}\ \bar{\mathbf{x}}^T H\widehat{\mathbf{g}}_0 - 2t\mathbf{E}\bar{y}\bar{\mathbf{x}}^T H\widehat{\mathbf{g}},$$

where by Lemma 4.11, the second summand is not greater in absolute value than

$$t^2\big(\mathbf{E}|\widehat{\mathbf{g}}_0^T H_0\bar{\mathbf{x}}|^2\ \mathbf{E}\bar{\mathbf{x}}^2\widehat{\mathbf{g}}_0^2\big)^{1/2} \le \sqrt{M}c_{43}\varepsilon.$$

The third summand in absolute value is not greater than

$$t(\mathbf{E}\bar{y}^2\ \mathbf{E}\bar{\mathbf{x}}^2\widehat{\mathbf{g}}^2)^{1/2} \le \sqrt{M}c_{11}/\sqrt{N}.$$

Applying Lemma 4.13, we recall that

$$t\mathbf{E}\widehat{\mathbf{g}}_0^T H\widehat{\mathbf{g}}_0 = \sigma^2(1 - s_0) + ts_0^2\mathbf{E}\mathbf{g}^T H_0\widehat{\mathbf{g}}_0 + o,$$

where $\mathbf{E}|o| \le \sqrt{M}c_{32}\varepsilon$. The difference between s and s_0 contributes no more than $\sqrt{M}c_{21}/N$. Now we have

$$t\mathbf{E}\mathbf{g}^T H_0\widehat{\mathbf{g}}_0 = t\mathbf{E}\mathbf{g}^T H\widehat{\mathbf{g}} - t^2\mathbf{E}\mathbf{g}^T H_0\bar{\mathbf{x}}\ \bar{\mathbf{x}}^T H\widehat{\mathbf{g}}_0 + t\mathbf{E}\bar{y}\mathbf{g}^T H\bar{\mathbf{x}},$$

where the first summand is written out in the lemma formulation. By Lemma 4.11 with $\mathbf{e} = \mathbf{g}/|\mathbf{g}|$, the second term is not greater in absolute value than

$$|\mathbf{g}|(\mathbf{E}tV^2\mathbf{E}t^3\bar{\mathbf{x}}^2\widehat{\mathbf{g}}_0^2)^{1/2} \le \sqrt{M}c_{42}\varepsilon.$$

The third term in the absolute value is not greater than

$$t(\mathbf{E}\bar{\mathbf{x}}^2\mathbf{E}\bar{y}^2)^{1/2} \le \sqrt{M}c_{11}/N.$$

We conclude that the first part of Statement 2 is valid. The second equation in Statement 2 follows from Statement 1, Lemma 4.10, and Lemma 4.13. This proves Lemma 4.16. $\square$

LEMMA 4.17. *If $t \ge u \ge 0$, then*

$$tu|\mathbf{E}\widehat{\mathbf{g}}^T H(t)\Sigma H(u)\widehat{\mathbf{g}} - \mathbf{E}\widehat{\mathbf{g}}_0^T H_0(t)\Sigma H_0(u)\widehat{\mathbf{g}}_0| \le \sqrt{M}c_{63}\varepsilon.$$

Proof. Replacing H by $H_0 - tH_0\bar{\mathbf{x}}\,\bar{\mathbf{x}}^T H$, we obtain

$$
\begin{aligned}
tu\widehat{\mathbf{g}}^T H(t)\Sigma H(u)\widehat{\mathbf{g}} &= \\
&= tu\widehat{\mathbf{g}}^T H_0(t)\Sigma H_0(u)\widehat{\mathbf{g}} + t^2 u\widehat{\mathbf{g}}^T H_0(t)\bar{\mathbf{x}}\,\bar{\mathbf{x}}^T H(t)\Sigma H_0(u)\,\widehat{\mathbf{g}} + \\
&\quad + tu^2\,\widehat{\mathbf{g}}^T H_0(u)\bar{\mathbf{x}}\,\bar{\mathbf{x}}^T H(u)\Sigma H_0(t)\,\widehat{\mathbf{g}} + \\
&\quad + t^2 u^2\,\widehat{\mathbf{g}}^T H_0(t)\bar{\mathbf{x}}\,\bar{\mathbf{x}}^T H(t)\Sigma H(u)\bar{\mathbf{x}}\,\bar{\mathbf{x}}^T H_0(u)\,\widehat{\mathbf{g}}. \qquad (18)
\end{aligned}
$$

Here the first summand provides the required expression in the formulation of the lemma with an accuracy to the replacement of vectors $\widehat{\mathbf{g}}$ by $\widehat{\mathbf{g}}_0$, i.e., to an accuracy up to

$$
\begin{aligned}
tu\mathbf{E}\widehat{\mathbf{g}}_0^T H_0(t)\Sigma H_0(u)\bar{\mathbf{x}}\bar{y} + tu\mathbf{E}\widehat{\mathbf{g}}_0^T H_0(u)\Sigma H_0(t)\bar{\mathbf{x}}\bar{y} + \\
+ tu\mathbf{E}\bar{\mathbf{x}}^T H_0(t)\Sigma H_0(u)\bar{\mathbf{x}}\,\bar{y}^2.
\end{aligned}
$$

Let us obtain upper estimates for these three terms. We single out first the dependence on $\bar{y}$. The square of the first term does not exceed

$$
Mt^4 \mathbf{E}\bar{y}^2 \mathbf{E}(\widehat{\mathbf{g}}^2\bar{\mathbf{x}}^2) \le 2M^3 t^4 \lambda(1+\lambda)/N \le M c_{42}/N.
$$

The square of the second term can be estimated likewise. The square of the third term is not larger than

$$
Mt^2 u^2 \mathbf{E}\Phi(t)\,\Phi(u)\bar{y}^2 \le Mt^2 \mathbf{E}\bar{y}^2 \le \sqrt{M}\tau^2/N.
$$

It remains to estimate the sum of last three terms in (18). We have $\|\Sigma\| \le M$. By Lemma 4.11 and Remark 1, for $u \le t$, the expectation of the second summand in (18) is not greater than

$$
\sqrt{M}\left(\mathbf{E}|t\,\widehat{\mathbf{g}}_0^T H_0\bar{\mathbf{x}}|^2\,t^4\mathbf{E}\bar{\mathbf{x}}^2\widehat{\mathbf{g}}^2\right)^{1/2} \le c_{53}\varepsilon.
$$

The third summand can be estimated likewise. To estimate the fourth summand in (18), we note that

$$
|\sqrt{u}H_0(u)\bar{\mathbf{x}}|^2 \le |u\,\bar{\mathbf{x}}^T H_0^2(u)\bar{\mathbf{x}}| \le |u\Phi(u)| \le 1.
$$

In view of Lemma 4.11, the expectation of the fourth summand in (18) is not larger than

$$
\sqrt{M}\left(\mathbf{E}|t\,\widehat{\mathbf{g}}^T H_0(t)\bar{\mathbf{x}}|^2 \mathbf{E}t^5|\bar{\mathbf{x}}|^4\widehat{\mathbf{g}}^2\right)^{1/2} \le \sqrt{M}c_{63}\varepsilon.
$$

We conclude that lemma statement holds.

LEMMA 4.18. *If* $t \geq u \geq 0$, *then*

$$tu|\mathbf{E}\widehat{\mathbf{g}}^T H(t)CH(u)\widehat{\mathbf{g}} - \mathbf{E}\widehat{\mathbf{g}}_0^T H_0(t)SH_0(u)\,\widehat{\mathbf{g}}_0| \leq \sqrt{M}c_{43}\varepsilon.$$

Proof. We substitute $\widehat{\mathbf{g}} = \widehat{\mathbf{g}}_0 - \bar{\mathbf{x}}\,\bar{y}$. It follows that

$$\begin{aligned}
f &\overset{def}{=} tu\mathbf{E}\widehat{\mathbf{g}}^T H(t)CH(u)\widehat{\mathbf{g}} = \\
&= tu\mathbf{E}\widehat{\mathbf{g}}_0^T H(t)CH(u)\,\widehat{\mathbf{g}}_0 - tu\mathbf{E}\widehat{\mathbf{g}}_0^T H(t)CH(u)\bar{\mathbf{x}}\,\bar{y} - \\
&\quad - tu\mathbf{E}\widehat{\mathbf{g}}_0^T H(u)CH(t)\bar{\mathbf{x}}\,\bar{y} + tu\mathbf{E}\bar{y}^2\bar{\mathbf{x}}^T H_0(t)CH_0(u)\bar{\mathbf{x}}.
\end{aligned}$$
$$(19)$$

Substituting $uCH(u) = I - H(u)$ in the last three summands, we find that the square of the second term is not greater than

$$t^2\mathbf{E}\bar{y}^2\mathbf{E}\widehat{\mathbf{g}}_0^2\bar{\mathbf{x}}^2 \leq 2M^2t^2\lambda(1+\lambda)/N \leq Mc_{22}/N.$$

The square of the third summand can be estimated likewise. The square of the fourth summand does not exceed

$$t^2\mathbf{E}\bar{y}^4\mathbf{E}|\bar{\mathbf{x}}|^4 \leq M^2t^2\lambda^2/N^2 \leq Mc_{22}/N^2.$$

Thus, the quantity f is equal to the first summand of the right-hand side of (19) to an accuracy up to $\sqrt{M}c_{11}/\sqrt{N}$.

It remains to estimate the contribution of the difference

$$\begin{aligned}
uH_0(t)CH(u) &- uH_0(t)SH_0(u) = \\
&= H(t) - H(t)H(u) - H_0(t) + H_0(t)H_0(u).
\end{aligned}$$

Using (16), we find that within an accuracy up to $\sqrt{M}c_{11}/\sqrt{N}$,

$$\begin{aligned}
|f - tu\mathbf{E}\widehat{\mathbf{g}}_0^T H_0(t)SH_0(u)\widehat{\mathbf{g}}_0| &\leq \\
\leq t\mathbf{E}|\widehat{\mathbf{g}}_0^T (H(t) - H_0(t))\,(H(u) + H_0(u))\widehat{\mathbf{g}}_0| &= \\
= t^2\mathbf{E}|\widehat{\mathbf{g}}_0^T H_0(t)\bar{\mathbf{x}}\,\bar{\mathbf{x}}^T G\widehat{\mathbf{g}}_0| \leq 2\big(\mathbf{E}|t\,\widehat{\mathbf{g}}_0^T H_0(t)\bar{\mathbf{x}}|^2\mathbf{E}t^2\bar{\mathbf{x}}^2\widehat{\mathbf{g}}_0^2\big)^{1/2},&
\end{aligned}$$

where $\|G\| \leq 2$. By Lemma 4.11, the right-hand side of the last inequality is not larger than $2Mt\sqrt{\lambda(1+\lambda)}c_{32}\varepsilon \leq \sqrt{M}c_{43}\varepsilon$. We conclude that the statement of our lemma is to an accuracy up to $\sqrt{M}c_{43}\varepsilon$. The lemma is proved. $\square$

THEOREM 4.8. *If* $t \geq u \geq 0$, *then*

$$tu\, \mathbf{E}\widehat{\mathbf{g}}^T H(t)CH(u)\widehat{\mathbf{g}} = tus(t)s(u)\mathbf{E}\widehat{\mathbf{g}}^T H(t)\Sigma H(u)\widehat{\mathbf{g}} +$$
$$+ (1 - s(u))ts^2(t)\mathbf{E}\mathbf{g}^T H(t)\mathbf{g} +$$
$$+ (1 - s(t))us^2(u)\mathbf{E}\mathbf{g}^T H(u)\mathbf{g} + \sigma^2(1 - s(t))(1 - s(u)) + o,$$

(20)

where $|o| \leq \sqrt{M}c_{63}\varepsilon$.

Proof. We transform the left-hand side. First, we apply Lemma 4.18 and obtain the leading term $tu\mathbf{E}\widehat{\mathbf{g}}_0^T H_0(t)SH_0(u)\widehat{\mathbf{g}}_0$ with a correction not greater than $\sqrt{M}c_{43}\varepsilon$. Then, we apply Theorem 4.7. Up to the same accuracy, this terms equals

$$tus_0(t)s_0(u)\mathbf{E}\widehat{\mathbf{g}}_0^T H_0(t)\Sigma H_0(u)\widehat{\mathbf{g}}_0 + (1 - s_0(u))ts_0(t)\mathbf{E}\widehat{\mathbf{g}}_0^T H_0(t)\mathbf{g} +$$
$$+ (1 - s_0(t))us_0(u)\mathbf{E}\widehat{\mathbf{g}}_0^T H_0(u)\mathbf{g} + \sigma^2(1 - s_0(t))(1 - s_0(u)).$$

(21)

We transform the first summand in (21) using Lemma 4.18 and Theorem 4.17. This lemma gives a correction not greater than $\sqrt{M}c_{31}/N$. The first summand in the right-hand side of (20) is obtained with a correction not greater than $\sqrt{M}c_{63}\varepsilon$. Next, we transform the second summand in (21). By Lemma 4.12, the equality holds $t\mathbf{E}\widehat{\mathbf{g}}^T H_0(t)\mathbf{g} = ts_0(t)\mathbf{E}\mathbf{g}^T H_0(t)\mathbf{g} + o$, where o is not greater in absolute value than $Mtc_{31}\varepsilon \leq \sqrt{M}c_{41}\varepsilon$. The difference between s_0 and s yields a lesser correction. We obtain the second summand of the right-hand side of (20). Similarly, we transform the expression with argument u. We obtain the third summand in (20). The substitution of s for s_0 gives a correction in the last summand that is not larger than $\sqrt{M}c_{11}/N$. We conclude that the right-hand sides of (21) and (20) coincide with an accuracy up to $\sqrt{M}c_{63}\varepsilon$. This proves Theorem 4.8. $\square$

Minimization of Quadratic Risk

We first express the leading part of the quadratic risk R in terms of sample characteristics, that is, as a function of $\mathbf{C}$ and $\widehat{\mathbf{g}}$. Our problem is to construct reliable estimators for the functions $t\mathbf{E}\widehat{\mathbf{k}}^T\mathbf{g}$ and $D(t, u) = tu\mathbf{E}\widehat{\mathbf{k}}^T\Sigma\widehat{\mathbf{k}}$ that are involved in the expression (3) for the quadratic risk.

We consider the statistics

$$\widehat{s}(t) = 1 - nN^{-1} + N^{-1}\text{tr}(I + tC)^{-1}, \quad \widehat{\kappa}(t) = t\,\widehat{\mathbf{g}}^T H(t)\widehat{\mathbf{g}},$$

$$\widehat{K}(t, u) \overset{def}{=} tu\widehat{\mathbf{g}}^T H(t)CH(u)\widehat{\mathbf{g}} = \frac{t\widehat{\kappa}(u) - u\widehat{\kappa}(t)}{t - u},$$

$$\widehat{\Delta}(t, u) = \widehat{K}(t, u) - (1 - \widehat{s}(t))\widehat{\kappa}(u) - \\ - (1 - \widehat{s}(u))\widehat{\kappa}(t) + \widehat{\sigma}^2(1 - \widehat{s}(t))(1 - \widehat{s}(u)),$$

where $\widehat{K}(t, u)$ is extended by continuity to $t = u$.

Remark 3. If $t \geq u \geq 0$, then

$$s(t)s(u)\mathbf{E}D(t, u) = \mathbf{E}\widehat{\Delta}(t, u) + o, \quad \text{where} \quad \mathbf{E}|o| \leq \sqrt{M}\,c_{63}\varepsilon.$$

It is convenient to replace the dependence of the functionals on $\eta(t)$ by that on a function $\rho(t)$ of the form

$$\rho(t) \overset{def}{=} \int_{0 \leq x \leq t} \frac{1}{s(x)}\,d\eta(x).$$

We note that the variation of the function $t^k\rho(t)$ on $[0, \infty)$ does not exceed $\sqrt{M}\,\eta_{k+1}\lambda$. Let us consider the quadratic risk (3) defined as a function of $\rho(t)$, $R = R(\eta) = R(\rho)$.

THEOREM 4.9. *The statistic*

$$\widehat{R}(\rho) \overset{def}{=} \widehat{\sigma}^2 - 2\int \left[\widehat{\kappa}(t) - \widehat{\sigma}^2(1 - \widehat{s}(t))\right] d\rho(t) + \iint \widehat{\Delta}(t, u)\, d\rho(t)\, d\rho(u)$$

is an estimator of $R = R(\rho)$ for which $\mathbf{E}\widehat{R}(\rho) = R(\rho) + o$, where $|o| \leq \sqrt{M}\,\eta_8\,c_{05}\varepsilon$.

Proof. The expression (3) equals $R(\rho)$, with an accuracy up to $\sqrt{M}/N$. Let us compare (3) with the right-hand side of the expression for $R(\widehat{\rho})$ in the formulation of our theorem. We have $|\sigma^2 - \mathbf{E}\widehat{\sigma}^2| \leq \leq 2\sqrt{M/N}$, $s(t) \geq (1 + \tau\lambda)^{-1}$. By Lemma 4.12,

$$|\mathbf{E}t\mathbf{g}^T H(t)\widehat{\mathbf{g}} - \mathbf{E}\left(\widehat{\kappa}(t) - \sigma^2(1 - s(t))\right)/s(t)| \leq \sqrt{M}c_{54}\varepsilon.$$

The differential $d\rho(t) = s^{-1}(t)d\eta(t)$, and the variation of $\eta(t)$ is not larger than 1. One can see that the second summand in (3) equals the second term of the expression for $\widehat{R}(\rho)$, with an accuracy up to $\sqrt{M}c_{54}\varepsilon$. The third summand in (3) by Theorem 4.8 is equal to the third term of $\widehat{R}(\rho)$, with an accuracy up to $\sqrt{M}c_{85}\varepsilon$. The coefficient c_{85} increases not faster than t^8 with t. We arrive to the theorem statement. $\square$

Now we pass to the calculation of the nonrandom leading part of the quadratic risk.

Define

$$\phi(t) = t\widehat{\mathbf{g}}^T(I + t\Sigma)^{-1}\mathbf{g}, \quad \kappa(t) = \sigma^2(1 - s(t))^2 + s^2(t)\phi(ts(t)),$$
$$K(t, u) = \frac{t\kappa(u) - u\kappa(t)}{t - u},$$
$$\Delta(t, u) = K(t, u) - (1 - s(t))\kappa(u) - (1 - s(u))\kappa(t) + \\ + \sigma^2(1 - s(t))(1 - s(u)),$$

where the function $K(t, u)$ is extended by continuity to $t = u$.

Remark 4. $\mathbf{E}\widehat{\kappa}(t) = \kappa(t) + o$, where $|o| \leq \sqrt{M}c_{43}\varepsilon$.

LEMMA 4.19. *If $t \geq u \geq 0$, then*

$$|\mathbf{E}\widehat{K}(t, u) - K(t, u)| \leq c_{43}\sqrt{M}\varepsilon,$$
$$|\mathbf{E}\widehat{\Delta}(t, u) - \Delta(t, u)| \leq c_{34}\sqrt{M}\varepsilon.$$

Proof. First, for some $d > 0$, let $|t - u| \geq d$. For these arguments, by Remark 4, we have $|\mathbf{E}\widehat{K}(t, u) - K(t, u)| \leq \tau c_{43}/d$. Let $|t - u| \leq d$,

$d > 0$. We expand the functions $\kappa(u)$ and $\widehat{\kappa}(u)$ to the Taylor series up to the second derivatives

$$
\begin{aligned}
\mathbf{E}\widehat{K}\,(t,u) - K(t,u) = \\
= d^{-1}\mathbf{E}[u\widehat{\kappa}(u) + ud\widehat{\kappa}'(u) + ud^2\widehat{\kappa}''(\xi)/2 + d\widehat{\kappa}(u)] - \\
- d^{-1}[u\kappa(u) + ud\kappa'(u) + ud^2\kappa''(\zeta)/2 + d\kappa(u)], \quad (22)
\end{aligned}
$$

where ξ and ζ are intermediate values of the arguments, $u \leq \xi$, $\zeta \leq 1$. Here

$$
\begin{aligned}
|\mathbf{E}\widehat{\kappa}''(\xi)| \leq 2\mathbf{E}\widehat{\mathbf{g}}^T HCH\widehat{\mathbf{g}} + t|\mathbf{E}\widehat{\mathbf{g}}^T HCHCH\widehat{\mathbf{g}}| \leq \\
\leq 3\mathbf{E}\widehat{\mathbf{g}}^T C\widehat{\mathbf{g}} \leq aM^{3/2}\lambda(1+\lambda),
\end{aligned}
$$

where a is a numerical coefficient. We also find that

$$
\begin{aligned}
|s'(u)| \leq N^{-1}\mathbf{E}\,\mathrm{tr}(HCH) \leq \sqrt{M}\lambda, \\
|s''(u)| \leq N^{-1}\mathbf{E}\,\mathrm{tr}(HCHCH) \leq N^{-1}\mathbf{E}\mathrm{tr}\,C^2 \leq M\lambda, \\
|\phi'(us(u))| \leq \mathbf{g}^T R\mathbf{g} + t(1+\tau\lambda)\,\mathbf{g}^T R\Sigma R\mathbf{g} \leq Mc_{21}, \\
|\phi''(us(u))| \leq a(1+\tau\lambda)^2\,\mathbf{g}^T R\Sigma R\mathbf{g} \leq M^{3/2}c_{22},
\end{aligned}
$$

where $R = (I + us(u)\Sigma)^{-1}$. We conclude that $t|\mathbf{E}\kappa''(\xi)| \leq Mc_{32}$. Thus, the terms with the second derivatives contribute to (22) no more than $Mc_{32}d$. Further, we estimate $|u\mathbf{E}\widehat{\kappa}'(u) - u\kappa'(u)|$. We have

$$
\mathbf{E}\widehat{\kappa}(u+d) - \mathbf{E}\widehat{\kappa}(u) = d\widehat{\kappa}'(u) + d^2\mathbf{E}\widehat{\kappa}''(\xi).
$$

Analogously we substitute $\kappa(u+d) - \kappa(u)$. Subtracting these expressions we find that,

$$
d\mathbf{E}\widehat{\kappa}'(u) + d^2\mathbf{E}\widehat{\kappa}''(\xi) - d\mathbf{E}\kappa'(u) - d^2\mathbf{E}\kappa''(\zeta)
$$

is not greater than $\sqrt{M}c_{43}\varepsilon$ in absolute value. Consequently,

$$
|\mathbf{E}u\widehat{\kappa}'(u) - u\kappa'(u)| \leq \sqrt{M}c_{43}\varepsilon + Mc_{32}d.
$$

It remains to estimate the summand $(\mathbf{E}u\widehat{\kappa}(u) - u\kappa(u))/d$ in the right-hand side of (22). This difference is not greater than $c_{53}\varepsilon/d$ in absolute value, and consequently all the right-hand sides of (22) do not exceed $\sqrt{M}c_{43}\varepsilon + c_{32}(Md + c_{22}\varepsilon/d)$. Let us choose $d = \sqrt{c_{22}\varepsilon/M}$. Then (since $\varepsilon \le 1$), the right-hand side of (22) is not greater in absolute value than $\sqrt{M\varepsilon}\, c_{43}$. The first statement is proved.

Further, we have

$$|\mathbf{E}\widehat{\Delta}(t, u) - \Delta(t, u)| \le |\mathbf{E}\widehat{K}(t, u) - K(t, u)| +$$
$$+ |r(t)\kappa(u) - \widehat{r}(t)\,\widehat{\kappa}(u) + r(u)\kappa(t) - \widehat{r}(u)\widehat{\kappa}(t)| +$$
$$+ |\sigma^2 r(t)r(u) - \widehat{\sigma}^2\widehat{r}(t)\,\widehat{r}(u)|,$$

where $r(t) = 1 - s(t)$, $\widehat{r}(t) = 1 - \widehat{s}(t)$. The first summand is estimated in Lemma 4.19. Note that the variance of $n^{-1}\mathrm{tr}\, H$ is not greater than c_{20}/N and therefore $|\mathbf{E}\widehat{s}(t) - s(t)| \le c_{11}\varepsilon$. By Remark 4, the upper estimate $c_{43}\sqrt{M\varepsilon}$ also holds for $|\mathbf{E}\widehat{\Delta}(t, u) - \Delta(t, u)|$. Lemma 4.19 is proved. $\square$

THEOREM 4.10. *The quadratic risk* (3) *is* $R = R(\rho) = R_0(\rho) + o$, *where*

$$R_0 = R_0(\rho) \stackrel{def}{=} \sigma^2 - 2\int s(t)\phi(ts(t))\, d\rho(t) + \iint \Delta(t, u)\, d\rho(t)\, d\rho(u),$$
$$(23)$$

and $|o| \le \sqrt{M}\eta_6\, c_{05}\sqrt{\varepsilon}$. *If some function of bounded variation* $\rho^{\mathrm{opt}}(t)$ *exists satisfying the equation*

$$\int \Delta(t, u)\, d\rho^{\mathrm{opt}}(u) = \kappa(t) - \sigma^2(1 - s(t)), \quad t \ge 0,$$

then $R_0(\rho)$ *reaches the minimum for* $\rho(t) = \rho^{\mathrm{opt}}(t)$ *and*

$$\min_\rho R_0(\rho) = \sigma^2 - \int s(t)\phi(ts(t))\, d\rho^{\mathrm{opt}}(t).$$

Proof. We start from Theorem 4.9. The difference between σ^2 and $\mathbf{E}\widehat{\sigma}^2$ is not greater than $2\sqrt{M/N}$. The difference between $s(t)$

and $\mathbf{E}\widehat{s}(t)$ is not larger than $c_{11}/\sqrt{N}$. By Lemma 4.16,

$$\left|\phi(ts(t)) - \mathbf{E}\left[\widehat{\kappa}(t) - \widehat{\sigma}^2(1 - \widehat{s}(t))\right]\right| \le \sqrt{M}c_{43}\varepsilon.$$

We obtain the two first summands in (23). Further, by Lamma 4.19, $|\Delta(t,u) - \mathbf{E}\widehat{\Delta}(t,u)| \le c_{43}\sqrt{M}\varepsilon$. The statement of Theorem 4.10 follows. $\square$

Usually, to estimate the efficiency of the linear regression, the RSS is used, which presents an empirical quadratic risk estimated over the same sample $\mathfrak{X}$

$$R^{\mathrm{emp}} = \widehat{\sigma}^2 - 2\widehat{\mathbf{k}}^T\widehat{\mathbf{g}} + \widehat{\mathbf{k}}^T C\widehat{\mathbf{k}}.$$

THEOREM 4.11. *For the linear regression with* $\mathbf{k} = \Gamma(C)\widehat{\mathbf{g}}$ *and* $l = \bar{y} - \widehat{\mathbf{k}}^T\bar{\mathbf{x}}$, *the empiric quadratic risk* $R^{\mathrm{emp}} = R^{\mathrm{emp}}(\eta)$ *may be written in the form*

$$R^{\mathrm{emp}}(\eta) = \sigma^2 - 2\int \kappa(t)\,d\eta(t) + \iint K(t,u)\,d\eta(t)d\eta(u) + o,$$

where $|o| \le c_{43}\sqrt{M}\varepsilon$.

Proof. By Lemmas 4.12 and 4.19, we have

$$|\sigma^2 - \mathbf{E}\widehat{\sigma}^2| \le 2\sqrt{M}\varepsilon \quad |\kappa(t) - \mathbf{E}\widehat{\kappa}(t)| \le \sqrt{M}c_{43}\varepsilon,$$

$$|K(t,u) - \mathbf{E}\widehat{K}(t,u)| \le c_{43}\sqrt{M}\varepsilon.$$

The variation of $\eta(t)$ on $[0,\infty)$ is not larger than 1. We can easily see that the statement of Theorem 4.11 holds. $\square$

Special Cases

We consider "shrinkage-ridge estimators" defined by the function $\eta(x) = \alpha\,\mathrm{ind}(x \ge t)$, $t \ge 0$. The coefficient $\alpha > 0$ is an analog of the shrinkage coefficient in estimators of the Stein estimator type, and $1/t$ presents a regularization parameter. In this case, by Theorem 4.10, the leading part of the quadratic risk (3) is

$$R_0(\rho) = R_0(\alpha, t) = \sigma^2 - 2\alpha\phi(ts(t)) + \alpha^2\Delta(t, t)/s^2(t).$$

If $\alpha = 1$, we have

$$R_0(\rho) = R_0(1, t) = \frac{1}{s^2(t)}\frac{d}{dt}[t\,(\sigma^2 - \kappa(t))].$$

In this case, the empirical risk is $R^{\mathrm{emp}}(t) = s^2(t)R_0(t)$. For the optimum value $\alpha = \alpha^{\mathrm{opt}} = s^2(t)\phi(ts(t))/\Delta(t, t)$, we have

$$R_0(\rho) = R_0(d^{\mathrm{opt}}, t) = \sigma^2\left(1 - \frac{s^2(t)\phi^2(ts(t))}{\Delta(t, t)}\right).$$

Example 1. Let $\lambda \to 0$ (the transition to the case of fixed dimension under the increasing sample size $N \to \infty$). To simplify formulas, we write out only leading terms of the expressions. If $\lambda = 0$, then $s(t) = 1$, $\quad h(t) = n^{-1}\mathrm{tr}(I + t\Sigma)^{-1}$, $\quad \kappa(t) = \phi(t)$, $\Delta(t, t) = \phi(t) - t\phi'(t)$. Set $\Sigma = I$. We have

$$\phi(t) \approx \frac{\sigma^2 r^2 t}{1 + t}, \quad h(t) \approx \frac{1}{1 + t}, \quad \Delta(t, t) \approx \frac{\sigma^2 r^2 t^2}{(1 + t)^2},$$

where $r^2 = \mathbf{g}^2/\sigma^2$ is the square of the multiple correlation coefficient. The leading part of the quadratic risk (3) is

$$R_0 = \sigma^2[1 - 2\alpha^2 t/(1 + t) + \alpha^2 r^2 t^2/(1 + t)^2].$$

For the optimal choice of d, as well as for the optimal choice of t, we have $\alpha = (1 + t)/t$ and $R^{\mathrm{opt}} = \sigma^2(1 - r^2)$, i.e., the quadratic risk (3) asymptotically attains its a priori minimum.

Example 2. Let $N \to \infty$ and $n \to \infty$ so that the convergence holds $\lambda = n/N \to \lambda_0$. Assume that the matrices Σ are nondegenerate for each n, $\sigma^2 \to \sigma_0^2$, $r^2 = \mathbf{g}^T\Sigma^{-1}\mathbf{g}/\sigma^2 \to r_0^2$, and the parameters $\gamma \to 0$. Under the limit transition, for each fixed $t \geq 0$, the remainder terms in Theorems 4.8–4.11 vanish. Let $d = 1$ and $t \to \infty$ (the transition to the standard nonregularized regression under the increasing dimension asymptotics). Under these conditions,

$$s(t) \to 1 - \lambda_0, \quad s'(t) \to 0, \quad \phi(ts(t)) \to \sigma_0^2 r_*^2,$$

$$\kappa(t) \to \kappa(\infty) \stackrel{def}{=} \sigma_0^2 r_0^2 (1 - \lambda_0) + \sigma_0^2 \lambda_0, \quad t\kappa'(t) \to 0.$$

The quadratic risk (3) tends to R_0 so that

$$\lim_{t \to \infty} \lim_{\gamma \to 0} \lim_{N \to \infty} |\mathbf{E}R(t) - R_0| = 0,$$

where $R_0 \stackrel{def}{=} \sigma_0^2 (1 - r_0^2)/(1 - \lambda_0)$. This limit expression was obtained by I. S. Yenyukov (see in [2]). It presents an explicit dependence of the quality of the standard regression procedure on the dimension of observations and the sample size. Note that under the same conditions, the empirical risk $R^{\mathrm{emp}} \to \sigma_0^2 (1 - r_0^2)(1 - \lambda_0)$ that is less than $\sigma_0^2 (1 - r_0^2)$.

Example 3. Under the same conditions as in Example 2, let the coefficients d be chosen optimally and then $t \to \infty$. We have $\alpha = \alpha^{\mathrm{opt}}(t) = s^2(t)\phi(ts(t))/\Delta(t,t)$ and $t \to \infty$. Then,

$$s(t) \to 1 - \lambda_0, \quad \phi(ts(t)) \to \sigma_0^2 r_0^2,$$
$$\Delta(t,t) \to \sigma_0^2 (1 - \lambda_0)[\lambda_0(1 - r_0^2) + (1 - \lambda_0)r_0^2],$$
$$\alpha^{\mathrm{opt}} \to r_0^2 (1 - \lambda_0)[\lambda_0(1 - r_0^2) + (1 - \lambda_0)r_0^2].$$

By (23), the quadratic risk (3) $R_0(t, \alpha^{\mathrm{opt}}) \to R_0$ as $t \to \infty$, where

$$R_0 = \frac{\sigma_0^2 (1 - r_0^2)[\lambda_0 + (1 - \lambda_0)r_0^2]}{\lambda_0(1 - r_0^2) + (1 - \lambda_0)r_0^2} \leq \frac{\sigma_0^2 (1 - r_0^2)}{1 - \lambda_0}.$$

If $\lambda_0 = 1$, the optimal shrinkage coefficient $\alpha^{\mathrm{opt}} \to 0$ and the quadratic risk remains finite (tends to σ_0^2) in spite of the absence of a regularization, whereas the quadratic risk for the standard linear regression tends to infinity.

CHAPTER 5

MULTIPARAMETRIC DISCRIMINANT ANALYSIS

Many-dimensional recognition problems arise when a small number of leading features do not provide satisfactory discrimination and the statistician should additionally invoke a large number of less informative discriminating variables. Wald [86] characterized the set of variables in the discriminant analysis as a finite number of well-discriminating "structure variables" and as an infinite number of poorly discriminating "incident variables." The small number of well-discriminating variables may be successfully treated by standard methods. But the technique of extracting discriminative information from a large number of incident variables was not developed until recently.

In 1976, Meshalkin [46] noticed (see Introduction) that linear discriminant function may be approximately normal if a large number of addends in the discriminant functions are uniformly small and the concept of distributions uniformly approaching each other in parametric space may be used. He applied the Kolmogorov asymptotics and derived concise limit formulas for the probabilities of errors.

In 1979 the author of this book used the same asymptotics for the development of an extended theory [60], [61] of improving discriminant by introducing weights of independent contributions of variables. This theory is presented in Section 5.1.

In Section 5.2, we consider the general case of dependent variables. The standard Wald discriminant function is modified by introducing weights of variables in the coordinate system of where the (pooled) sample covariance is diagonal. As in [63], we start from the assumption of normality of variables, prove limit theorems, and find the "best-in-the-limit" linear discriminant function. Then, we extend these results to a wide class of distributions

193

using Normal Evaluation Principle proposed in Section 3.3. In the general case, linear discriminant function is no more normal. We return to the Fisher criterion of discriminant analysis quality and find solutions unimprovable in the meaning of this criterion.

5.1. DISCRIMINANT ANALYSIS OF INDEPENDENT VARIABLES

In this section, we solve the problem of discriminating large-dimensional vectors $\mathbf{x} = (x_1, x_2, \ldots, x_n)$ from two populations under assumption that the variables $x_1, x_2, \ldots, x_n$ are independent and normally distributed. We consider a generalized family of linear discriminant functions different by introduction of weights for independent addends and apply the multiparametric technique to find the best weights for asymptotically improved linear discriminant function and asymptotically best rules for the selection of variables.

Asymptotical Problem Setting.

Let $\mathfrak{P} = \{\mathfrak{P}_n\}$ be a sequence of the discrimination problems

$$(\mathfrak{S}_\nu, \, \mathfrak{X}_\nu, \, N, \, \bar{\mathbf{x}}_\nu, \, g(\mathbf{x}), \, \widehat{\alpha}_\nu, \, \nu = 1, 2)_n \quad n = 1, 2, \ldots, \quad (1)$$

where $\mathfrak{S}_1$ and $\mathfrak{S}_2$ are two populations; $\mathfrak{X}_\nu$ are samples from $\mathfrak{S}_\nu$, of the same size $N > 2$, $\bar{\mathbf{x}}_\nu$ are sample mean vectors for $\mathfrak{X}_\nu$, $\nu = 1, 2$; $g(\mathbf{x}) = g(\mathbf{x}, \mathfrak{X}_1, \mathfrak{X}_2)$ is the discriminant function used in the classification rule $g(\mathbf{x}) > \theta$ against $g(\mathbf{x}) \geq \theta$,

$$\widehat{\alpha}_1 = \mathbf{P}(g(\mathbf{x}) < \theta | \mathfrak{S}_1), \quad \widehat{\alpha}_2 = \mathbf{P}(g(\mathbf{x}) \geq \theta | \mathfrak{S}_2)$$

are sample-dependent probabilities of errors of two kinds and θ is a threshold (we do not write out the subscripts n for arguments of (1)).

Let us restrict (1) with the following assumptions.

A. For each n, the vectors $\mathbf{x}$ in populations $\mathfrak{S}_\nu$ from $\mathfrak{P}$ are distributed normally with the density

$$f(\mathbf{x}, \vec{\mu}_\nu) = (2\pi)^{-n/2} \, \exp(\mathbf{x} - \vec{\mu}_\nu)^2/2, \quad \nu = 1, 2,$$

where $\vec{\mu}_1 = (\mu_{11}, \mu_{12}, \ldots, \mu_{1n})$ and $\vec{\mu}_2 = (\mu_{21}, \mu_{22}, \ldots, \mu_{2n})$.
Denote $\vec{\mu} = \vec{\mu}_1 - \vec{\mu}_2 = (\mu_1, \mu_2, \ldots, \mu_n)$.

B. The populations $\mathfrak{S}_\nu$ in (1) are contigual so that

$$\max_{i=1,..,n} N\,\mu_i^2/2 \le c,$$

where c does not depend on n (here and in the following, squares of vectors denote squares of their length), and the ratio $n/N \to \lambda > 0$

We introduce a description of the set $\{\mu_i^2\}$ in terms of an empirical distribution function

$$R_n(v) = n^{-1}\sum_{i=1}^{n} \mathrm{ind}(N\mu_i^2/2 \le v), \quad R_n(c) = 1.$$

C. For any $v > 0$ in $\mathfrak{P}$, the limits exist

$$\lim_{n\to\infty} R_n(v) = R(v), \quad J = \lim_{n\to\infty} \bar\mu^2 = 2\lambda \int v dR(v). \qquad (2)$$

This condition does not restrict applications of the limit theory to finite-dimensional problems and is introduced in order to provide the limit form of results. Under Assumption B, $R(c) = 1$.

Define $\hat{\mathbf{a}} = (\hat a_1, \hat a_2, \ldots, \hat a_n) = (\bar{\mathbf{x}}_1 + \bar{\mathbf{x}}_2)/2$,
$\bar{\mathbf{x}} = (\bar x_1, \bar x_2, \ldots, \bar x_n) = \bar{\mathbf{x}}_1 - \bar{\mathbf{x}}_2 \sim \mathbf{N}(\bar\mu, 2/N)$.

D. The discriminant function is of the form

$$g(\mathbf{x}) = \sum_{i=1}^{n} \eta_i \bar x_i (x_i - \hat a_i), \qquad (3)$$

where the weights η_i have one of two forms: $\eta_i = \eta(v_i),\ v_i = N\mu_i^2/2$ (weighting by a priori data) or $\eta_i = \eta(u_i),\ u_i = N\bar x_i^2/2$ (weighting by sample data), $i = 1, 2, \ldots, n$. Assume that the function $\eta(\cdot)$ does not depend on n and is of bounded variation on $[0, \infty)$ and continuous everywhere, perhaps, except a finite number of discontinuity points not coinciding with discontinuity points of $R(v)$.

A Priori Weighting of Variables

Denote

$$\widehat{G}_{n\nu}(\eta) = \int g(\mathbf{x}) f(\mathbf{x}, \vec{\mu}_\nu) \, d\mathbf{x} = 1/2 \sum_i \eta(v_i) \bar{x}_i (\mu_{\nu i} - \widehat{a}_i),$$

$$\widehat{D}_n(\eta) = \int \eta^2(v_i) [g(\mathbf{x}) - \widehat{G}_{n\nu}(\eta)]^2 f(\mathbf{x}, \vec{\mu}_\nu) \, d\mathbf{x} = \sum_i \eta^2(v_i) \, \bar{x}_i^2,$$

where $v_i = N\mu_i^2/2$ (here and in the following in sums over i, the subscript $i = 1, 2, \ldots, n$).

Probabilities of errors of two kinds depend on samples and are

$$\widehat{\alpha}_1 = \Phi\left(-\frac{\widehat{G}_{n1}(\eta) - \theta}{\sqrt{\widehat{D}_n(\eta)}} \right), \quad \widehat{\alpha}_2 = \Phi\left(-\frac{\widehat{G}_{n2}(\eta) + \theta}{\sqrt{\widehat{D}_n(\eta)}} \right). \quad (4)$$

LEMMA 5.1. *Under Assumptions A–D for $\eta_i = \eta(n\mu_i^2/2)$ in (3), $i = 1, 2, \ldots, n$,*

1. *the limits in the square mean exist*

$$\text{l.i.m.}_{n\to\infty} \widehat{G}_{n\nu}(\eta) = (-)^{\nu+1} G(\eta), \; \nu = 1, 2, \; \text{l.i.m.}_{n\to\infty} \widehat{D}_n(\eta) = D(\eta),$$

where

$$G(\eta) = \lambda \int \eta(v) v dR(v), \quad D(\eta) = 2\lambda \int \eta^2(v)(v+1) dR(v); \quad (5)$$

2. *if $D(\eta) > 0$, then*

$$\text{plim}_{n\to\infty} \widehat{\alpha}_1(\eta) = \Phi\left(-\frac{G(\eta) - \theta}{\sqrt{D(\eta)}} \right), \quad \text{plim}_{n\to\infty} \widehat{\alpha}_2(\eta) = \Phi\left(\frac{G(\eta) - \theta)}{D(\eta)} \right).$$

Proof. Let $v_i = N\mu_i^2/2$, $i = 1, 2, \ldots, n$. The expectation

$$\mathbf{E}\widehat{G}_{n1}(\eta) = 1/2 \sum_i \eta(v_i) \, \mu_i^2 = n/N \int \eta(v) v dR_n(v).$$

The quantity $\mathbf{E}\widehat{G}_{n2}(\eta)$ is different by sign. The domain of integration is bounded, and the function in the integrand is piecewise continuous almost everywhere. We conclude that the integral above converges, and $\widehat{G}_{n1}(\eta) \to G(\eta)$, whereas $\widehat{G}_{n2}(\eta) \to -G(\eta)$.

The expectation

$$\mathbf{E}\widehat{D}_n(\eta) = \sum_i \eta(v_i)\mathbf{E}\bar{x}_i^2 = \sum_i \eta(v_i)\left(\mu_i^2 + \frac{2}{N}\right) =$$

$$= \frac{2n}{N}\int \eta(v)(v+1)\,dR_n(v).$$

By the same reasoning, this integral converges to the integral over $dR(v)$ in the lemma formulation.

Let us show that variances of $\widehat{G}_{n\nu}(\eta)$ and $\widehat{D}_n(\eta)$ vanish. These two quantities present sums of independent addends and their variance is not greater than sums of expectations of squares. We note that the function $\eta(\cdot)$ is bounded, $\mu_i^4 \le 4c^2/n^2$. All fourth central moments of $\bar{x}_i$ are equal to $3/N^2$. It follows that var $\widehat{G}_{n\nu}(\eta)$ and var $\widehat{D}_n(\eta)$ are not greater than sums of $O(n^{-2})+O(N^{-2})$. In view of condition B, these variances are $O(n^{-1})$. The first statement of Lemma 5.1 is proved. The second statement follows immediately. The proof is complete. $\square$

Example 1. If $\eta(v) = 1$ for all $v > 0$, then $\widehat{G}_{n1}(1) \to J/2$ and $\widehat{D}_n(1) \to J+2\lambda$ as $n \to \infty$ in the square mean, where J is defined by (2).

Minimum of Errors by A Priori Weighting

It is easy to check that the minimum of the limit value of $(\widehat{\alpha}_1 + \widehat{\alpha}_2)/2$ defined by (4) is achieved for the threshold $\theta = 0$ (nontrivial optimum thresholds may be obtained for essentially different sample sizes, see Introduction). Lemma 5.1 states that if $D(\eta) > 0$, then

$$\min_\theta \ \operatorname*{plim}_{n\to\infty}(\widehat{\alpha}_1 + \widehat{\alpha}_2)/2 = \Phi(-\rho(\eta)/2),$$

where the "effective limit Mahalanobis distance" $\rho(\eta) = 2G(\eta)/\sqrt{D(\eta)}$.

THEOREM 5.1. *Under Assumptions A–D, the variation of $\eta(\cdot)$ in $g(\mathbf{x})$ of the form (3) with $\eta_i = \eta(N\mu_i^2/2)$, $i = 1, 2, \ldots, n$, leads to the minimum half-sum of limit error probabilities*

$$\inf_\eta \min_\theta \ \text{plim} \ \frac{\widehat{\alpha}_1 + \widehat{\alpha}_2}{2} = \Phi(-\rho(\eta_{\text{opt}})/2)$$

with $\eta(v) = \eta_{\text{opt}}(v) = v/(v+1)$ and

$$\rho^2(\eta_{\text{opt}}) = 2\lambda \int \frac{v^2}{v+1} \, dR(v).$$

Proof. Let us vary $\eta(v)$ fixing $\theta = 0$. We obtain the necessary condition of the extremum

$$D(\eta) \int v \delta\eta(v) \, dR(v) = 2G(\eta) \int (v+1)\eta(v)\delta\eta(v) \, dR(v).$$

It follows that $\eta(v) = \text{const } v/(v+1)$. The proportionality coefficient does not affect the value of $\rho(\eta)$. Set $\eta(v) = \eta_{\text{opt}}(v) = v/(v+1)$. Let us show that the value $\rho(\eta_{\text{opt}})$ is not less than $\rho(\eta)$ for any other $\eta(t)$. Using the Cauchy–Bunyakovskii inequality, we obtain

$$2G(\eta) = 2 \int v\eta(v) \, dR(v) \leq$$

$$\leq 2 \left[\int \frac{v^2}{v+1} \, dR(v) \int (v+1)\eta^2(v) \, dR(v) \right]^{1/2} =$$

$$= \rho(\eta_{\text{opt}}) \sqrt{D(\eta)}.$$

This completes the proof of Theorem 5.1. □

The optimal weighting of the form $\eta_{\text{opt}}(v) = v/(v+1)$ was first found by Deev in [18].

Example 2. (the case of a portion r of noninformative variables). Let $R(v) = r \geq 0$ for $0 \leq v \leq b$, and $R(b) = 1$ for $v > b$. If $\eta(t) = 1$ for all t (no weighting), then $G(1) = J/2 = (1 - r)b$, $D(1) = J + 2\lambda = 2[(1 - r) \, b + \lambda]$, and $\rho^2(1) = 2(1 - r)^2 b^2/$

$[(1-r)b+\lambda)]$. If $\eta(v) = \eta_{\mathrm{opt}}(v) = v/(v+1)$ (optimum weighting), then $\rho^2(\eta_{\mathrm{opt}}) = 2(1-r)b^2/(b+\lambda) \geq \rho^2(1)$.

Note that under Assumptions A–D, the contributions of variables to the discrimination are $\mu_i^2 = O(n^{-1})$, $i = 1, 2, \ldots, n$, while the bias of their estimators $\bar{x}_i^2$ is of the order $1/N$, that is, of the same order magnitude. This means that true contributions are substantially different from their estimators. Thus, we have the problem of improving discrimination by weighting variables using random $\bar{x}_i^2$.

Empirical Weighting of Variables

We consider now a more realistic problem when the weights in (3) are calculated by the estimators $\bar{x}_i$ of the quantities μ_i, $i = 1, 2, \ldots, n$.

Let $F_m^\beta(u)$ be the standard biased χ^2 distribution function with m degrees of freedom and the bias parameter $\beta \geq 0$. Denote

$$f_m^\beta(u) = \frac{\partial F_m^\beta(u)}{\partial u}, \quad f_{m+2}^\beta(u) = -\frac{\partial F_m^\beta(u)}{\partial \beta^2}.$$

THEOREM 5.2. *Let Assumptions A–D be valid. For the discriminant function (3) with weights $\eta_i = \eta(N\bar{x}_i^2/2)$, $i = 1, 2, \ldots, n$, the limits in the square mean exist*

$$\mathrm{l.i.m.}_{n\to\infty} \widehat{G}_{n1}(\eta) = G(\eta) \stackrel{def}{=} \lambda \int [\int \eta(u) f_3^\beta(u)\, du]\, \beta^2\, dR(\beta^2),$$

$$\mathrm{l.i.m.}_{n\to\infty} \widehat{D}_n(\eta) = D(\eta) \stackrel{def}{=} 2\lambda \int [\int \eta^2(u) u f_1^\beta(u)\, du]\, dR(\beta^2). \quad (6)$$

Proof. We begin with the second statement. Denote $v_i = N\mu_i^2/2$, $u_i = N\bar{x}_i^2/2$. The latter variable can be expressed as $u_i = (v_i + \xi)^2$, where $\xi \sim \mathbf{N}(0,1)$. Therefore, u_i is distributed as biased χ^2 with the density $f_1^\beta(\cdot)$ with one degree of freedom and bias $\beta^2 = Nv_i^2/2$. Thus, the expectation

$$\mathbf{E}\widehat{D}_n(\eta) = 2/N\mathbf{E}\sum_i u_i\eta^2(u_i) = 2n/N\int [\int u\eta(u) f_1^\beta(u)\, du]\, dR_n(\beta^2).$$

By conditions A and C, the left-hand side of this expression converges to the limit in the theorem formulation.

The expectation

$$\mathbf{E}\widehat{G}_{n1}(\eta) = \mathbf{E}\sum_i \eta(u_i)\,\bar{x}_i(\mu_{1i} - \widehat{a}_i) = 1/2\mathbf{E}\sum_i \eta(u_i)\,\mu_i\bar{x}_i =$$

$$= \frac{1}{2}\int \sum_i \eta(u_i)\mu_i\left(\mu_i + \frac{2}{N}\frac{\partial}{\partial\mu_i}\right)f_1(\bar{x}_i)\,d\bar{x}_i =$$

$$= \frac{1}{N}\sum_i \int \eta(u)(\beta^2 + \beta\frac{\partial}{\partial\beta})\,f_i^\beta(u)\,du,$$

where $f_i(\bar{x}_i) = \sqrt{N/4\pi}\,\exp(-N(\bar{x}_i - \mu_i)^2/4)$ and $\beta^2 = \beta^2(i) = N\mu_i^2/2$ for each i. Using the identity

$$f_3^\beta(u) - f_1^\beta(u) = 2\frac{\partial f_1^\beta(u)}{\partial\beta^2},$$

we obtain the expression that tends to the limit in the theorem formulation.

It is left to check that variances of both $\widehat{G}_{n1}$ and $\widehat{D}_n$ vanish. These variances present sums of addends that can be estimated from the above by second moments. Each of these moments does not exceed $O(n^{-2}) + O(N^{-2})$. In view of condition A, it follows that the variances of both $\widehat{G}_{n1}(\eta)$ and $\widehat{D}_n(\eta)$ are $O(n^{-1})$. The proof is complete. $\square$

Denote

$$\sigma(u) = \int \beta^2 f_3^\beta(u)\,dR(\beta^2), \quad \pi(u) = \int u f_1^\beta(u)\,dR(\beta^2). \quad (7)$$

The limits in (6) may be rewritten in the form

$$G(\eta) = \lambda\int_0^\infty \eta(u)\sigma(u)\,du, \quad D(\eta) = 2\lambda\int_0^\infty \eta^2(u)\pi(u)\,du. \quad (8)$$

THEOREM 5.3. *Let Assumptions A–D hold and $D(\eta) > 0$. Then, using $g(\mathbf{x})$ in (3) with the weight coefficients $\eta_i = \eta(n\bar{x}_i^2/2)$, we have*

$$\operatorname*{plim}_{n\to\infty} \widehat{\alpha}_1 = \Phi\left(-\frac{G(\eta) - \theta}{\sqrt{D(\eta)}}\right), \quad \operatorname*{plim}_{n\to\infty} \widehat{\alpha}_2 = \Phi\left(-\frac{G(\eta) + \theta}{\sqrt{D(\eta)}}\right),$$

where $G(\eta)$ and $D(\eta)$ are defined by (8).

Proof. Let $\widehat{F}(\cdot)$ denote the conditional distribution function of the random value $\widehat{g} = g(\mathbf{x})$ with fixed $\bar{\mathbf{x}}_1$ and $\bar{\mathbf{x}}_2$. Obviously, $\widehat{\alpha}_1 = \widehat{F}(\theta)$. The function $g(\mathbf{x})$ is a weighted sum of independent random values. Denote by $\widehat{G}_{n1}$, $\widehat{D}_n$, and $\widehat{T}_{n1}$ sums of the first moments, variances, and third absolute moments (conditional under fixed samples) of the addends in (3). Applying the Esseen inequality, we find that the distribution function of $\widehat{g} = g(\mathbf{x})$ differs from the distribution function of $\mathbf{N}(\widehat{G}_{n1}, \widehat{D}_n)$ by $\omega = O(\widehat{T}_{n1}/\widehat{D}_n^{3/2})$. Here $\widehat{D}_n \to D(\eta) > 0$ in the square mean. The expectation

$$|\mathbf{E}\,\widehat{T}_{n1}| \le \sum_i \mathbf{E}|\eta_i \bar{x}_i \,(x_i - \widehat{a}_i)|^3,$$

where coefficients η_i and moments $\mathbf{E}|x_i - \widehat{a}_i|^6$ are bounded. Hence, $\mathbf{E}\,\widehat{T}_{n1}$ is of order of magnitude of the sum of $\sqrt{|\bar{x}_i|^6}$. In view of conditions A and B, in this sum, each addend is $O(n^{-3/2}) + O(N^{-3/2})$ as $n \to \infty$. By Theorem 5.2, we have $\widehat{G}_{n1} \xrightarrow{\mathbf{P}} G(\eta)$, and $\widehat{D}_n(\eta) \xrightarrow{\mathbf{P}} D(\eta) > 0$. We conclude that the random value $\omega \xrightarrow{\mathbf{P}} 0$ and $\widehat{F}(\theta)$ tends in probability to the distribution function of $\mathbf{N}(G(\eta), D(\eta))$. The symmetric conclusion for $\nu = 2$ follows from assumptions. Theorem 5.3 is proved. $\square$

The limit half-sum of error probabilities is reached for $\theta = 0$ and is

$$\alpha(\eta) \stackrel{def}{=} \operatorname*{plim}_{n\to\infty} (\widehat{\alpha}_1 + \widehat{\alpha}_2)/2 = \Phi(-\rho(\eta)/2), \tag{9}$$

where $\rho^2(\eta) = 4G^2(\eta)/D(\eta)$.

Example 3. Suppose that the contributions μ_i^2 of all variables to the distance between the populations are identical. Then, for some β^2, $R(v) = 0$ for $v < \beta^2$ and $R(v) = 1$ for $v \ge \beta^2$. We have

$$\sigma(u) = \beta^2 f_3^\beta(u), \quad \pi(u) = u f_1^\beta(u),$$

$$G(\eta) = \lambda\beta^2 \int \eta(u) f_3^\beta(u)\, du, \quad D(\eta) = 2\lambda \int u\eta^2(u) f_1^\beta(u)\, du,$$

where

$$f_1^\beta(t^2) = \frac{1}{\sqrt{2\pi}\, t} \exp(-(t^2 + \beta^2)/2)\mathrm{ch}\, \beta t,$$

$$f_3^\beta(t^2) = \frac{1}{\sqrt{2\pi}\, \beta} \exp(-(t^2 + \beta^2)/2)\, \mathrm{sh}\, \beta t, \qquad \beta, t > 0.$$

If $\eta(u) = 1$ for all $u > 0$, then $G(1) = \lambda\beta^2 = J/2$, and using (6), we find that $D(1) = 2(\beta^2 + \lambda) = J + 2\lambda$, and $\rho(\eta_{\mathrm{opt}}) = \rho(1)$.

Minimum Error Probability for Empirical Weighting

We find the maximum of the function $\rho(\eta)$ defined in (9).

THEOREM 5.4. *Let Assumptions A–D hold. Varying the threshold θ and the function $\eta(\cdot)$ under fixed λ and $R(v)$ in $g(\mathbf{x})$ of the form (3) with $\eta_i = \eta(N\bar{x}_i^2/2)$, $i = 1, 2, \ldots, n$, we obtain that*

$$\inf_\eta \min_\theta \plim_{n\to\infty} (\alpha_1 + \alpha_2)/2 = \Phi(-\rho(\eta_{\mathrm{opt}})/2),$$

where

$$\eta_{\mathrm{opt}}(u) = \frac{\sigma(u)}{\pi(u)} \quad and \quad \rho^2(\eta_{\mathrm{opt}}) = 2\lambda \int_{+0}^{\infty} \frac{\sigma^2(u)}{\pi(u)}\, du. \qquad (10)$$

Proof. We seek the extremum of (9) by variation of $\eta(u)$. The necessary condition of the extremum is

$$D(\eta) \int_0^\infty \sigma(u)\delta\eta(u)\, du = G(u) \int_0^\infty \pi(u)\eta(u)\delta\eta(u)\, du.$$

Hence, $\eta(u) = \mathrm{const}\, \sigma(u)/\pi(u)$. The constant coefficient does not affect the value of $\rho(\eta)$. Set $\eta_{\mathrm{opt}}(u) = \sigma(u)/\pi(u)$. Let us prove that $\eta_{\text{opt}}(u)$ is bounded. By Assumption B, the supports of the distributions $R_n(\beta^2)$ and $R(\beta^2)$ are bounded. For χ^2 densities, we have the inequality $u f_1^\beta(u) \geq f_3^\beta(u)$. It follows that

$$\sigma(u) = \int \beta^2 f_3^\beta(u)\, dR(\beta^2) \leq u \int \beta^2 f_1^\beta(u)\, dR(\beta^2) \leq c\pi(u)/2,$$

$u \geq 0$. We see that the function $\eta_{\text{opt}}(u)$ is bounded and continuous for $u \geq 0$. Substituting this function in (10), we obtain the second relation in (7). For any other $\eta(u)$ using the integral Cauchy–Bunyakovskii inequality, we find

$$2G(\eta) = 2\lambda \int\limits_0^\infty \sigma(u)\eta(u)\, du \leq 2\lambda \left[\int\limits_{+0}^\infty \frac{\sigma^2(u)}{\pi(u)}\, du \int\limits_0^\infty \pi(u)\eta^2(u)\, du \right]^{1/2} =$$

$$= \sqrt{D(\eta)}\rho(\eta_0).$$

This completes the proof of Theorem 5.4. $\square$

Example 4. Let $\gamma > 0$. Consider a special distribution of $\beta > 0$ with

$$dR(\beta^2) = \sqrt{\frac{\gamma}{2\pi}} \int \exp\left(-\frac{\gamma \vec{\beta}^2}{2} \right) \delta(\beta^2 - \vec{\beta}^2) d\vec{\beta} =$$

$$= \left(\frac{\gamma}{2} \right)^{1/2} \Gamma^{-1}(1/2) \exp\left(-\frac{\gamma\beta^2}{2} \right) \beta^{-1}\, d\beta^2.$$

Using the integral representation of $f_1^\beta(u)$, we find that

$$\int f_1^\beta(u)\, dR(\beta^2) = \left(\frac{g}{2} \right)^{1/2} \Gamma^{-1}(1/2)\ \exp\left(-\frac{gu}{2} \right) u^{-1/2},$$

where $g = \gamma/(1+\gamma)$ and

$$\int \beta^2 f_3^\beta(u)\, dR(\beta^2) = \frac{u}{1+\gamma} \int f_1^\beta(u)\, dR(\beta^2).$$

Hence, $\sigma(u) = \pi(u)/(1 + \gamma)$. If $\eta(u) = 1$ for all $u > 0$, we have

$$D(1) = 2(1 + \gamma)\, G(1), \quad \rho^2(1) = \frac{2\gamma}{1 + \gamma} \int v dR(v).$$

For the best weight function $\eta(u) = \eta_{\mathrm{opt}}(u)$, we obtain

$$\rho^2(\eta_{\mathrm{opt}}) = \frac{2\lambda}{1 + \gamma} \int \sigma(u)\, du = \frac{2\lambda}{1 + \gamma} \int v dR(v).$$

Thus, $\rho(\eta_{\mathrm{opt}}) = \rho(1)$. We conclude that, for this special case, the optimal weighting does not diminish the limit half-sum of the error probabilities.

Example 5. Let all variables contribute identically to the distance between populations, $\mu_i^2 = \vec{\mu}^2/n$, $i = 1, \ldots, n$. Then, $dR(\beta^2) \neq 0$ only at the point $\beta^2 = N\vec{\mu}^2/2n$ and $\rho^2(1) = J\beta^2/(\beta^2 + 1)$. The best weighting function is

$$\eta_{\mathrm{opt}}(u) = \frac{\beta^2 f_3^\beta(u)}{u f_1^\beta(u)} = \beta\, \frac{\mathrm{th}\,(\beta\sqrt{u})}{\sqrt{u}}.$$

The following inequality is valid

$$\int \frac{[f_3^\beta(u)]^2}{u f_1^\beta(u)}\, du \geq \frac{1}{1 + \beta^2}.$$

It is easy to examine that $\rho(\eta_{\mathrm{opt}}) \geq \rho(1)$. It is remarkable that in spite of identical contributions of variables to J, the optimal weighting provides the increase of $\rho(\eta)$ and the decrease of $\mathop{\mathrm{plim}}\limits_{n \to \infty} (\hat{\alpha}_1 + \hat{\alpha}_2)/2$ owing to the effect of the suppression of large deviations of estimators.

THEOREM 5.5. *If Assumptions A–D hold and $\rho_{\mathrm{opt}}(\eta) = \rho(1)$, then there exists $\gamma > 0$ such that*

$$\frac{dR(v)}{dv} = \left(\frac{\gamma}{2}\right)^{1/2} \Gamma^{-1}\left(\frac{1}{2}\right) \exp\left(-\frac{\gamma v}{2}\right) v^{-1/2}. \tag{11}$$

Proof. We compare (9) and (10). The inequality $\rho(\eta_{\text{opt}}) \geq \rho(1)$ is the Cauchy–Bunyakovskii inequality for the functions $\sqrt{\pi(u)}$ and $\sigma(u)/\sqrt{\pi(u)}$. The case of the equality implies $\sqrt{\pi(u)} = \sigma(u)/\sqrt{\pi(u)}$ almost for all $u > 0$. In view of the continuity, we obtain $\sigma(u) = C_1\pi(u)$, where $C_1 > 0$. Substituting this relation to (9), we find that

$$C_1\pi(u) = \int (1+u)f_1^\beta(u)\,dR(\beta^2) + 2u \int \frac{\partial f_1^\beta(u)}{\partial u}\,dR(\beta^2)$$
$$= \pi(u) + \pi(u)/u + 2\pi'(u), \quad u > 0.$$

We integrate this differential equation and obtain that

$$\pi(u) = C_2 u^{1/2} \exp(-C_3 u), \quad C_2, C_3 > 0. \tag{12}$$

Let us substitute $\pi(u)$ from definition, divide by u, and perform the Fourier transformation of the both parts of this equality. It follows that

$$\int \frac{\pi(u)}{u} \exp(iut)\,dt = \int \chi_1^\beta(t)\,dR(\beta^2) = \frac{\text{const}}{(C_3 - it)^{1/2}},$$

$t \geq 0$. Denote $s = t/(1 - 2it)$. Substituting (12), we obtain

$$\int \exp(is\beta^2)\,dR(\beta^2) = \frac{\text{const}}{C_3 + is\,(2C_3 - 1)^{1/2}}.$$

This relation holds, in particular, at the interval $\{s : \text{Im } s = 1/4$ and $|\text{Re } s| < 1/4\}$. The analytical continuation to $\text{Im } s \geq 0$ makes it possible to perform the inverse Fourier transformation

$$\frac{dR(v)}{dv} = C_4 \int_{-\infty}^{+\infty} \frac{\exp(-ivs)}{[C_3 - is(2C_3 - 1)]^{1/2}}\,ds =$$
$$= C_5\, v^{-1/2} \exp\left(-\frac{\gamma v}{2}\right),$$

where $C_4, C_5 > 0$ and $\gamma = 2C_3/(1 - 2C_3) > 0$. Normalizing, we obtain (11). The proof is complete. $\square$

Thus, the distribution (11) is the only limit distribution for which the effect of weighting by Theorem 5.4 produces no gain.

Statistics to Estimate Probabilities of Errors

Usually, the observer does not know, even approximately, the true values of either the parameters $\vec{\mu}_1\ \vec{\mu}_2$, or the law of distribution $R(\cdot)$. The functions $\sigma(u)$ and $\pi(u)$ involved in Theorem 5.4 also are not known. Let us construct their estimators. We describe the set of the observed values $\bar{x}_i^2$ in terms of the sample-dependent function

$$\widehat{Q}_n(u) = n^{-1} \sum_i \text{ind}(N\bar{x}_i^2/2 \leq u).$$

Remark 1. For each $u \geq 0$, the limit exists

$$Q(u) = \underset{n\to\infty}{\text{l.i.m.}}\ \widehat{Q}_n(u) = \int F_1^{\beta}(u)\, dR(\beta^2), \quad u \geq 0 \qquad (13)$$

(here and in the following, l.i.m. stands for the limit in the square mean). Indeed, since the random values u_i are distributed as $F_1^{\beta}(u)$ with $\beta^2 = N\mu_i^2/2$, we find that the expectation of $\widehat{Q}_n(u)$ tends to the right-hand side of (13). It is easy to see that var $\widehat{Q}_n(u) \leq 1$. Statement (13) is grounded.

From (10), it follows that the function $Q(u)$ is monotone, continuous, increases as $u^{1/2}$ for small $u > 0$, and $Q(u) \to 1$ exponentially as $u \to \infty$. The following integral relation holds:

$$\int\limits_0^\infty (1 - Q(u))\, du = \int (1 + \beta^2)\, dR(\beta^2).$$

LEMMA 5.2. *The functions $\sigma(u)$ and $\pi(u)$ can be expressed in terms of the derivatives of $Q(u)$ as follows:*

$$\sigma(u) = 2u\, Q''(u) + (1 + u)\, Q'(u), \quad \pi(u) = uQ'(u). \qquad (14)$$

These relations follow from (13), (7), and (6).

Remark 2. If the function $\eta(u)$ is twice differentiable, $u \geq 0$, then

$$G(\eta) = \lambda \int [(u-1)\eta(u) - 2u\eta'(u)] \, dQ(u) =$$

$$= \lambda \int_0^\infty [\eta(u) + (u-3)\eta'(u) - 2u\eta''(u)] \, (1 - Q(u)) \, du - \lambda\eta(0);$$

$$D(\eta) = 2\lambda \int u\eta^2(u) \, dQ(u) =$$

$$= 2\lambda \int_0^\infty [\eta(u) + 2u\eta'(u)] \, \eta(u)(1 - Q(u)) \, du.$$

For a special case when $\eta(u) = 1$ for all $u \geq 0$, we obtain

$$G(1) = J/2 = \lambda \int (u-1) \, dQ(u),$$

$$D(1) = J = 2\lambda \int_0^\infty (1 - Q(u)) \, du.$$

Remark 3. The weighting function $\eta_{\mathrm{opt}}(u)$ can be written in the form

$$\eta_{\mathrm{opt}}(u) = \frac{2uQ''(u) + (1+u)Q'(u)}{uQ'(u)}, \tag{15}$$

and the corresponding limit "effective Mahalanobis distance" is

$$\rho(\eta_{\mathrm{opt}}) = 2\kappa \int_{+0}^\infty \frac{[2uQ''(u) + (1+u)Q'(u)]^2}{uQ'(u)} \, du. \tag{16}$$

As a corollary of Theorem 5.4, we can formulate the following assertion.

THEOREM 5.6. *Under Assumptions A–D for $g(\mathbf{x})$ of the form* (3) *with weights $\eta_i = \eta(N\bar{x}_i^2/2)$, $i = 1, 2, \ldots, n$, we have*

$$\operatorname*{plim}_{n \to \infty} \frac{\widehat{\alpha}_1 + \widehat{\alpha}_2}{2} \geq \Phi\left(-\frac{\rho(\eta_{\mathrm{opt}})}{2}\right),$$

where $\rho(\eta_{\mathrm{opt}})$ is defined by (16), and the equality holds for $\eta(u)$ defined by (15).

This theorem presents the limit form of asymptotically unimprovable discriminant function that can be constructed using the empiric contributions $\bar{x}_i^2$ of variables to the square distance between samples.

Contribution of a Small Number of Variables

If the observer knows parameters of the populations exactly, then to minimize the discrimination errors, one needs all variables. For a sample discrimination rule with no weighting of variables, the limit probability of the discrimination error is not minimum, and the problem arises of choosing the best subset of variables minimizing the discrimination error.

Under the problem setting accepted above, sums of the increasing number of variables produce nonrandom limit contributions to the discrimination. However, the contributions of separate variables remain essentially random. To obtain stable recommendations concerning the selection, we gather variables with neighboring values of $\bar{\mu}_i^2$ and $\bar{x}_i^2$ into groups that are sufficiently large to have stable characteristics. Let us investigate the effect of an exclusion of an increasing number k of variables from the consideration. More precisely, consider two sequences (1) of the discrimination problems such that the second sequence is different by $\tilde{k} = n - k$ variables with the same other arguments. Assume that $k/n \to \gamma \geq 0$ as $n \to \infty$. In this section, we suppose that γ is small. Let us mark the characteristics of the second sequence by the sign tilde.

Contribution of Variables in Terms of Parameters

Suppose that all the excluded variables have close values of $\mu_i^2 \approx 2/N\, v_0$, so that the new limit function is

$$\tilde{R}(v) = \frac{R(v) - \gamma \,\mathrm{ind}(v > v_0)}{1 - \gamma}. \tag{17}$$

Let $\delta z = \tilde{z} - z$ denote the change of a value z when we pass from the first sequence to the second one. Assume that γ is small, and let us study the effect of the exclusion of variables, keeping only first-order terms in $\gamma \to 0$. Then,

$$\gamma = -\delta\gamma/\gamma, \quad \delta R(v) = [\text{ind}(v - v_0) - R(v)]\delta\gamma/\gamma,$$
$$\delta G(\eta) = v_0\eta(v_0)\delta\gamma, \quad \delta D(\eta) = 2(v_0 + 1)\,\eta^2(v_0)d\gamma,$$

where $G(\eta)$ and $D(\eta)$ are defined by (5). We set $\eta(v) = 1$ for all $v \geq 0$. For a problem with the known set $\{\mu_i^2\}$ of contributions, we obtain

$$\delta\rho^2(\eta) = \frac{2J}{(J + 2\gamma)^2}\,[(J + 4\gamma)\,v_0 - J]\,\,\delta\gamma,$$

where J are defined by (2).

Proposition 1. Let Assumptions A–D hold, $\eta(v) = 1$ for all $v \geq 0$, and a small portion of variables be excluded by a priori values of μ_i^2 so that (17) holds. Then, the limit error probability $\alpha(\eta) = \Phi(-\rho(\eta)/2)$ is decreased by the exclusion of variables (17) if and only if the inequality holds $v_0 < J/(J + 4\lambda)$. In order to decrease the limit probability of the discrimination error, it should be recommended to exclude variables with

$$\mu_i^2 < \frac{2}{N}\frac{J}{J + 4\lambda}.$$

Contribution of Variables in Terms of Statistics

Now we investigate the contribution of variables when a number q of variables with close sample characteristics are excluded. We assume that $q/n \to \gamma \geq 0$, where γ is small. Let δ be the symbol of variation when we pass to the decreased number of variables, so that $\gamma = -\delta\lambda/\lambda$, and

$$\delta Q(u) = [\text{ind}(u \geq u_0) - Q(u)]\,\delta\lambda/\lambda, \qquad (18)$$

where $u_0 = \plim_{n\to\infty} N\bar{x}_i^2/2$ for all excluded variables. Keeping only first-order terms, we have

$$\delta\,G(\eta) = [(u_0 - 1)\eta(u_0) - 2u_0\,\eta'(u_0)]\delta\gamma,$$
$$\delta D(\eta) = 2\delta\gamma \int_0^{u_0} [\eta(u) + 2u\eta'(u)]\eta(u)\,du = 2u_0\eta^2(u_0)\delta\gamma.$$

Let $\eta(u) = 1$ for all $u > 0$. Then, we have

$$\delta J = 2\delta G(1) = 2(u_0 - 1)\,\delta\gamma, \quad \delta D(1) = 2u_0\delta\gamma.$$

We obtain

$$\delta\rho^2(1) = \frac{2J}{(J+2\gamma)^2}[u_0(J+4\gamma) - 2(J+2\gamma)]\delta\gamma.$$

Proposition 2. Let Assumptions A–D hold, $\eta(u) = 1$ for all $u > 0$, and a small portion of variables be excluded by $\bar{x}_i^2$ so that (18) holds. Then, $\alpha(\eta) = \Phi(-\rho(\eta)/2)$ decreases if and only if

$$u_0 < 2\frac{J+2\gamma}{J+4\gamma},$$

and to decrease the limit probability of the discrimination error, it should be recommended to exclude variables with

$$\bar{x}_i^2 < \frac{2}{N}\frac{K}{K+2\gamma},$$

where $K = J + 2\lambda$.

Selection of Variables by Threshold

We now consider the selection of a substantial portion of variables. Let q be the number of variables left for the discrimination that were selected by the rule $v_i = N\mu_i^2/2 > \tau^2$ and $u_i = N\bar{x}_i^2/2 > \tau^2$, where τ^2 is the selection threshold, $k \leq n$. To investigate the selection effect in our consideration, we apply the weighting function of the form $\eta(u) = \text{ind}(u \geq \tau^2)$.

Selection by A Priori Threshold

Proposition 3. Given τ under Assumptions A–D for the weighting function,
$\eta(v) = \text{ind}(v \geq \tau^2)$ with $v = v_i = N\bar{x}_i^2/2$, the limit exists

$$\delta = \lim_{n\to\infty}\frac{k}{n} = 1 - R(\tau^2).$$

Indeed, by (5), we have

$$\frac{k}{n} = n^{-1} \sum_i \eta(v_i) = \int \eta(v)\, dR_n(v) = 1 - R_n(\tau^2),$$

where $R_n(v) \to R(v)$ for all $v > 0$.

To treat the selection problem, let us redefine

$$G(\delta) = G(\eta), \quad D(\delta) = D(\eta), \quad \rho(\delta) = \rho(\eta),$$
$$J(\delta) = 2\gamma \int v\eta(v)\, dR(v).$$

By (5), we have

$$G(\delta) = \lambda \int_{v > \tau^2} v\, dR(v), \quad D(\delta) = 2\gamma \int_{v > \tau^2} (v + 1)\, dR(v).$$

Let us find the threshold τ^2 and the limit portion δ of variables left for the discrimination that minimize $\alpha(\delta) = \Phi(-\rho(\delta)/2)$.

Denote by b_1 and b_2 the left and right boundaries of the distribution $R(v)$: $b_1 = \inf\{v \geq 0 : R(v) > 0\}$, $b_2 = \inf\{v \geq 0 : R(v) = 1\}$.

THEOREM 5.7. *Suppose Assumptions A–D hold and the discrimination function* (3) *is used with nonrandom weighting coefficients of the form* $\eta_i = \mathrm{ind}(N\mu_i^2/2 \geq \tau^2)$. *Then, under the variation of* δ, *the condition*

$$b_1 < J(1)/(J(1) + 4\gamma) \tag{19}$$

is sufficient for $\alpha(\delta) = \Phi(-\rho(\delta)/2)$ *to reach a minimum for* δ *such that* $0 < \delta \leq 1$. *The derivative* $\alpha'(\delta)$ *exists almost everywhere for* $0 < \delta \leq 1$ *and its sign coincides with the sign of the difference*

$$J(\delta) - \tau^2(J(\delta) + 4\gamma\delta).$$

Proof. For $\delta = 1$, we have $\tau^2 = 0$ and $\eta(v) = 1$ for all $v \geq 0$. Hence,

$$(J(\delta) - J(1))/2\gamma = -\int (1 - \eta(v))\, v\, dR(v) \geq -(1 - \delta)\tau^2.$$

The minimum of $\alpha(\delta)$ is attained if $\rho^2(\delta) = J^2(\delta)/(J(\delta) + 2\gamma\delta)$. We find

$$\rho^2(\delta) - \rho^2(1) \geq c[J^2(1) - \tau^2 J(\delta)\,(J(1) + 2\gamma) - 2\gamma\tau^2 J(\delta)], \quad c > 0.$$

If $\delta \to 1$, then $\tau^2 \to b_1$, $J(\delta) \to J(1)$, and if (19) holds, then there exists $\delta < 1$ such that $\rho^2(\delta) \geq \rho^2(1)$. On the other hand, $\delta \to 0$ implies that $J(\delta) \to 0$ and $\rho^2(\delta) \leq J(\delta) \to 0$. The function $J(\delta)$ is continuous for all δ, $0 < \delta \leq 1$. Therefore, $\rho^2(\delta)$ reaches a minimum for $0 < \delta < 1$. The derivatives $\alpha'(\delta)$ and $\rho'(\delta)$ exist at all points where $\dfrac{d\tau^2}{d\delta}$ exists, i.e., at all points where $dR(\tau^2) > 0$. For these δ, we have

$$\frac{dJ(\delta)}{d\delta} = 2\lambda\tau^2,$$

$$\frac{d\rho^2(\delta)}{d\delta} = \frac{2\lambda J(\delta)}{(J(\delta) + 2\lambda\delta)^2}\,[\tau^2(J(\delta) + 4\lambda\delta) - J(\delta)].$$

The second statement of the theorem follows. The proof of the theorem is complete. $\square$

Example 6. Consider the case of a portion $r \geq 0$ of noninformative variables. The function $R(v) = r \geq 0$ for $0 \leq v < b_2$ and $R(b_2) = 1$. The value $J(1) = 2\lambda(1 - r)\,b_2$. For $\delta < 1 - r$, we have $\tau^2 = b_2$, $J(\delta) = 2\lambda\delta b_2$, $\rho^2(\delta) = \rho^2(1)\delta$, and the increase of δ decreases $\alpha(\delta)$. For $r > 0$ and $\delta \geq 1 - r$, we have $b_1 = 0$, and (19) is valid. We have $J(\delta) = 2\lambda(1 - r)b_2$ independently on δ, and the decrease of δ decreases $\alpha(\delta)$. For $r = 0$, we have $b_1 = b_2$, relation (19) does not hold, $J(\delta) = 2\lambda\,\delta b_2$, and the decrease of δ increases $\alpha(\delta)$. The selection is not purposeful if $r = 0$.

Empirical Selection of Variables

We consider a selection of $k \leq n$ variables with sufficiently large values of the statistics $\widehat{u}_i = \bar{x}_i^2 \geq 2\tau^2/N$, where τ^2 is a fixed threshold. This problem is equivalent to the discrimination with the weight coefficients $\eta(u) = \text{ind}(u \geq \tau^2)$, where $u = u_i$, $i = 1, 2, \ldots, n$. The number of variables left in the discriminant function is $k = \eta(u_1) + \eta(u_2) + \cdots + \eta(u_n)$.

LEMMA 5.3. *If Assumptions A–D hold and the discriminant function* (3) *is used with the weights* $\eta_i = \mathrm{ind}(u_i \geq \tau^2)$, $i = 1, \ldots, n$, *then the limit exists*

$$\delta = \operatorname*{plim}_{n \to \infty} \frac{k}{n} = 1 - Q(\tau^2),$$

where $Q(\cdot)$ *is defined by* (13).

Proof. We find that

$$\mathbf{E} \frac{k}{n} = \int [\int \eta(u) \, dF_1^\beta(u) \, du] \, dR(\beta^2) + o(1) =$$
$$= \int \eta(u) \, dQ(u) + o(1) = 1 - Q(u) + o(1).$$

The ratio k/n is a sum of independent random values, and its variance is not greater than $1/n \to 0$. This proves the lemma. $\square$

One can see that $\tau^2 = \tau(\delta)$ is a function decreasing monotonously with the increase of δ. The function $\eta(\cdot)$ is determined by the value of δ uniquely.

Let us redefine $G(\delta) = G(\eta)$, $D(\delta) = D(\eta), \rho(\delta) = \rho(\eta)$, and $\alpha(\delta) = \alpha(\eta)$.

Remark 4. Under Assumptions A–D with $\eta(u) = \mathrm{ind}(u \geq \tau^2)$, we have

$$G(\delta) = \lambda \int_{\tau^2}^\infty \sigma(u) \, du = \lambda \int \beta^2 (1 - F_3^\beta(\tau^2)) \, dR(\beta^2),$$
$$D(\delta) = 2\lambda \int_{\tau^2}^\infty \pi(u) \, du = 2\lambda \int [\int_{u > \tau^2} u f_1^\beta(u) \, du] \, dR(\beta^2)$$
$$= 2\lambda \int \left[(1 - F_3^\beta(\tau^2)) + \beta^2 (1 - F_5^\beta(\tau^2)) \right] dR(\beta^2).$$

We use (13) to express these functions in terms of the function $Q(u)$.

Remark 5. Under Assumptions A–D, we have

$$G(\delta) = \lambda(\tau^2 - 1)(1 - Q(\tau^2)) + \lambda \int\limits_{\tau^2}^{\infty} (1 - Q(u))\, du;$$

$$D(\delta) = 2\lambda(1 - Q(\tau^2)) + \lambda \int\limits_{\tau^2}^{\infty} (1 - Q(u))\, du.$$

By virtue of Theorem 5.3, random values $\widehat{\alpha}_1$ and $\widehat{\alpha}_2$ converge in probability to the limits depending on δ.

Remark 6. Consider the problem of the influence of an informational noise on the discrimination. We modify the sequence of problems (1) by adding a block of independent variables number $i = 0$ and assume that the random vector $\mathbf{x}^0$ is distributed as $\mathbf{N}(\vec{\mu}_\nu^0, I_r)$, where $\vec{\mu}_\nu^0 \in \mathbb{R}^r$, $\nu = 1, 2$, and I_r is the identity matrix of size $r \times r$. To simplify formulas, suppose that $\vec{\mu}_1^0$ and $\vec{\mu}_2^0$ do not depend on n. Denote $J^0 = (\vec{\mu}_1^0 - \vec{\mu}_2^0)^2$. The discriminant function is modified by an addition of a normal variable distributed as $\mathbf{N}(\pm J^0/2, J^0)$. Suppose that all remaining variables are noninformative: let $R(v) = 1$ for all $v > 0$. In this case, as $n \to \infty$,

$$\alpha(\delta) \to \Phi\left(-\frac{J^0}{2\sqrt{J^0 + D(\delta)}}\right),$$

where $D(\delta)$ is defined by Remark 4. We also have $Q(u) = F_1^0(u)$, $\delta = 1 - F_1^0(\tau^2)$, and $D(\rho) = 2\lambda\delta + 4\lambda f_3^0(\tau^2)$. Here the second summand is added to the variance of $g(\mathbf{x})$; this additional term is produced by the selection of those variables that have the greater deviations of estimators (this effect was analyzed in detail in [62]). For a small portion δ of variables left, the selection substantially increases the effect of informational noise (as $\ln 1/\delta$).

THEOREM 5.8. *Suppose Assumptions A–D hold and the discriminant function* (3) *is used with the weighting coefficients of the form* $\eta_i = \mathrm{ind}(N\bar{x}_i^2/2 \geq \tau^2)$. *Then, under variation of* δ

1. *the condition*

$$2 \int (v+1)\, dR(v) \int v \exp(-v/2)\, dR(v) <$$
$$< \int \exp(-v/2)\, dR(v) \int v\, dR(v) \tag{20}$$

is sufficient for the value $\alpha(\delta) = \Phi\left(-G(\eta)/\sqrt{D(\eta)}\right)$ *to have a minimum for* $0 < \delta < 1$;

2. *the derivative* $\alpha'(\delta)$ *exists for* $0 < \delta \leq 1$ *and the sign of* $\alpha'(\delta)$ *coincides with the sign of the difference*

$$\pi(\tau^2) \int\limits_{\tau^2}^{\infty} \sigma(u)\, du - 2\sigma(\tau^2) \int\limits_{\tau^2}^{\infty} \pi(u)\, du. \tag{21}$$

Proof. If $\delta = 1$, then $\tau^2 = 0$, $G(1) = \lambda \int \beta^2\, dR(\beta^2) = J/2$, and $D(1) = 2\lambda \int (\beta^2 + 1)\, dR(\beta^2) = J + 2\lambda$. Using relations (6), we calculate the derivatives as $\tau^2 \to +0$

$$\tau \frac{df_1^{\beta}(\tau^2)}{d\tau} = a\, \exp\left(-\beta^2/2\right),$$

$$\frac{dG(\delta)}{d\tau^3} = -\lambda \frac{2a}{3} \int \beta^2\, \exp\left(-\beta^2/2\right) dR(\beta^2),$$

$$\frac{dD(\delta)}{d\tau^3} = -2\lambda \frac{2a}{3} \int \exp\left(-\beta^2/2\right) dR(\beta^2),$$

where $a = 2[\sqrt{2}\Gamma(1/2)]^{-1}$.

Calculating the derivative $\dfrac{d\alpha(\delta)}{d\tau^3}$ at the point $\tau^2 = 0$, we find that this derivative is positive if condition (20) holds. It follows that τ^2 is monotone depending on δ and $\alpha(\delta) < \alpha(1)$ for some $\delta < 1$, $\tau^2 > 0$. The first statement is proved.

Further, we notice that

$$\rho^2(\delta) = 2\lambda \left[\int_{\tau^2}^{\infty} \sigma(u) \, du \right]^2 \left[\int_{\tau^2}^{\infty} \pi(u) \, du \right]^{-1}.$$

Differentiating $\rho^2(\delta)$ we obtain the second statement of the theorem. This completes the proof. $\square$

Example 7. The function $R(v) = r$ for $0 \le v < b = \beta^2$ and $R(b) = 1$. The condition (19) takes the form

$$r > (2b + \exp(b/2) - 1)^{-1} (2b + 1).$$

The selection is purposeful for sufficiently large r and large b.

For variables with identical nonzero contributions, we have $r = 0$ and the inequality does not hold. In this case, we find that

$$\frac{\rho^2(\delta)}{\rho^2(1)} = \frac{(1 - F_3^\beta(\tau^2))^2 (1 + \beta)^2}{(1 - F_1^\beta(\tau^2)) + \beta^2 (1 - F_5^\beta(\tau^2))}.$$

But $F_5^\beta(\tau^2) \le F_3^\beta(\tau^2)$. Replacing F_5^β by F_3^β, we obtain the inequality

$$\frac{\rho^2(\delta)}{\rho^2(1)} \le 1 - F_3^\beta(\tau^2) < 1.$$

The minimum of $\alpha(\delta)$ is attained for $\delta = 1$, that is, using all variables.

Example 8. Consider the special limit distribution (11) when the derivative $R'(v)$ exists. In this case, $\sigma(u) = \pi(u)/(1 + \gamma)$. In the inequality (20), the left-hand side equals the right-hand side, and the sufficient condition for the selection to be purposeful is not satisfied. The value

$$\rho^2(\delta) = \frac{2\lambda}{1 + \gamma} \int_{\tau^2}^{\infty} \sigma(u) \, du.$$

One can see that $\alpha(\delta)$ is strictly monotone, decreasing with the decrease of δ and the increase of τ^2. The minimum of $\alpha(\delta)$ is attained when all variables are used.

Remark 7. Let us rewrite the selection conditions (19) and (20) in the form

$$b < J/(J + 4\lambda),$$

$$2\int v \exp(-v/2)\, dR(v) < J/(J + 2\lambda) \int \exp(-v/2)\, dR(v).$$

The left-hand side of the second inequality has the meaning of the mean contribution of weakly discriminating variables. It can be compared with the first inequality. One may see that for the "good" discrimination when $\int v dR(v) \gg 1$, the boundary of the purposefulness of the selection using estimators is twice as less than under the selection by parameters. The sufficient condition (20) of the purposeful selection involves quantities that are usually unknown to the observer. Let us rewrite it in the form of limit functions of estimators. Denote

$$w(u) = 2\ln\left[u^{-1/2}\exp(u/2)Q'(u)\right].$$

THEOREM 5.9. *Suppose conditions A–D are fulfilled and the discrimination function* (3) *is used with the weights* $\eta_i = \text{ind}\,(\bar{x}_i^2/2 \geq \tau^2)$ *of variables. Then, under the variation of* δ, *the condition*

$$(1 - 2w'(0))\int_0^\infty u dQ(u) > 1 \tag{22}$$

is sufficient for the $\alpha(\delta)$ *to attain a minimum for* $0 < \delta < 1$. *The minimum of* $\alpha(\delta)$ *is attained for* $\delta = \delta_{\text{opt}}$ *and* $\tau = \tau_{\text{opt}}$ *and is such that*

$$\delta_{\text{opt}} = 1 - Q(\tau_{\text{opt}}^2) \quad and \quad w'(\tau_{\text{opt}}^2) = G(\delta_{\text{opt}})/D(\delta_{\text{opt}}).$$

Proof. Indeed, from (13) it follows that as $u \to 0$, we have

$$w'(0) = \int \beta^2 \exp(-\beta^2/2) \, dR(\beta^2) \bigg/ \int \exp(-\beta^2/2) \, dR(\beta^2).$$

The relation (22) readily follows. The second assertion of the theorem follows from Theorem 5.8. $\square$

Thus, the investigation of the empirical distribution function $\widehat{Q}_n(u)$ makes it possible to estimate the effect of selection of variables from sample data. If inequality (20), holds, then the selection is purposeful (in the limit). Using (20), we are able to approximate the best limit selection threshold τ_{opt}^2 and the best limit portion δ_{opt} of chosen variables.

5.2.   DISCRIMINANT ANALYSIS OF DEPENDENT VARIABLES

In this section, we present the construction of asymptotically unimprovable, essentially multivariate procedure of linear discriminant analysis of vectors with dependent components. For normal variables, this investigation was first carried out in 1983 and is presented first in [63] and then in [71]. In this section, we extend results of [63] to a wide class of distributions using the Normal Evaluation Principle described in Section 3.3.

We consider the problem of discriminating n-dimensional normal vectors $\mathbf{x} = (\mathbf{x}_1, \ldots, \mathbf{x}_n)$ from one of two populations $\mathfrak{S}_1$ and $\mathfrak{S}_2$ with common unknown covariance matrix $\Sigma = \text{cov}(\mathbf{x}, \mathbf{x})$. We begin with the case of normal distributions. The discrimination rule is $w(\mathbf{x}) \geq \theta$ against $w(\mathbf{x}) < \theta$, where $w(\mathbf{x})$ is a linear discriminant function, and θ is a threshold. The quality of the discriminant analysis is measured by probabilities of errors of two kinds (classification errors)

$$\alpha_1 = \mathbf{P}(w(\mathbf{x}) < \theta | \mathfrak{S}_1) \quad \text{and} \quad \alpha_2 = \mathbf{P}(w(\mathbf{x}) \geq \theta | \mathfrak{S}_2) \qquad (1)$$

for observations $\mathbf{x}$ from $\mathfrak{S}_1$ and $\mathfrak{S}_2$. If the populations are normal $\mathbf{N}(\vec{\mu}_\nu, \Sigma)$, $\nu = 1, 2$, with nondegenerate matrix Σ, then it is well known that, by the Neumann–Pearson lemma, the minimum of $(\alpha_1 + \alpha_2)/2$ is attained with $\theta = 0$ and $w(\mathbf{x})$ of the form

$$w^{\text{NP}}(\mathbf{x}) = \ln \frac{f_1(\mathbf{x})}{f_2(\mathbf{x})} = (\vec{\mu}_1 - \vec{\mu}_2)^T \Sigma^{-1}(\mathbf{x} - (\vec{\mu}_1 + \vec{\mu}_2)/2),$$

where

$$f_\nu(\mathbf{x}) = (2\pi \det \Sigma)^{-1/2} \exp(-(\mathbf{x} - \vec{\mu}_\nu)^T \Sigma^{-1}(\mathbf{x} - \vec{\mu}_\nu)/2),$$

are normal distribution densities, $\nu = 1, 2$. The minimum of the half-sum $(\alpha_1 + \alpha_2)/2$ equals $\Phi(-\sqrt{J}/2)$, where $J = (\vec{\mu}_1 - \vec{\mu}_2)^T \Sigma^{-1}(\vec{\mu}_1 - \vec{\mu}_2)$ is "the square of the Mahalanobis distance." The estimator of $w(\mathbf{x})$ is constructed over samples $\mathfrak{X}_1 = \{\mathbf{x}_m\}$, $m = 1, 2, \ldots, N_1$ and $\mathfrak{X}_2 = \{\mathbf{x}_m\}$, $m = 1, 2, \ldots, N_2$, of size $N_1 > 1$ and

$N_2 > 1$, $N = N_1 + N_2$, from $\mathfrak{S}_1$ and $\mathfrak{S}_2$. To construct an estimator of $w(\mathbf{x})$, we use sample means

$$\bar{\mathbf{x}}_\nu = N_\nu^{-1} \sum_{m=1}^{N_\nu} \mathbf{x}_m, \quad \mathbf{x}_m \in \mathfrak{X}_\nu, \quad \nu = 1, 2$$

and the standard unbiased "pooled" sample covariance matrix C of the form

$$C = \frac{N_1 - 1}{N - 2} C_1 + \frac{N_2 - 1}{N - 2} C_2, \tag{2}$$

where

$$C_\nu = \frac{1}{N_\nu - 1} \sum_{\mathfrak{X}_\nu} (\mathbf{x}_m - \bar{\mathbf{x}}_\nu)(\mathbf{x}_m - \bar{\mathbf{x}}_\nu)^T, \ \nu = 1, 2.$$

In applications, traditionally the standard "plug-in" Wald discriminant function is used

$$w(\mathbf{x}) = (\bar{\mathbf{x}}_1 - \bar{\mathbf{x}}_2)^T C^{-1} (\mathbf{x} - (\bar{\mathbf{x}}_1 + \bar{\mathbf{x}}_2)/2).$$

However, it is well known that this function may be illconditioned, and obviously not the best even for low-dimensional problems (see Introduction). In this section, we choose a class of generalized, always stable discriminant functions replacing Σ^{-1} by a "generalized ridge estimator" of the inverse covariance matrices and develop a limit theory that can serve for the construction of improved and unimprovable discriminant procedures. This development is a continuation of researches initiated by A. N. Kolmogorov in 1968–1970 (see Introduction).

Asymptotical Setting

We apply the multiparametric technique developed in Chapter 3 and use the Kolmogorov asymptotics to isolate principal parts of the expected probabilities of errors with the purpose to find linear discriminant function, minimizing the probability of errors under the assumption that the dimension $n \to \infty$ together with sample sizes $N_1 \to \infty$ and $N_2 \to \infty$ so that $n/N_\nu \to y_\nu$, $\nu = 1, 2$.

Consider a sequence of discrimination problems

$$\mathfrak{P} = \{(\mathfrak{S}_\nu,\ \vec{\mu}_\nu,\ \Sigma,\ N_\nu,\ \mathfrak{X}_\nu,\ \bar{\mathbf{x}}_\nu,\ C,\ w(\mathbf{x}),\ \alpha_\nu)_n\},\quad \nu = 1, 2,$$

$n = 1, 2, \ldots$ (we will not write out the subscripts n for the arguments of $\mathfrak{P}$). The n-dimensional observation vectors $\mathbf{x} = (\mathbf{x}_1, \ldots, \mathbf{x}_n)$ are taken from one of two populations $\mathfrak{S}_1$ and $\mathfrak{S}_2$; the population centers are $\vec{\mu}_1 = \mathbf{E}_1\mathbf{x}$ and $\vec{\mu}_2 = \mathbf{E}_2\mathbf{x}$, where (and in the following) $\mathbf{E}_1$ and $\mathbf{E}_2$ are expectation operators in $\mathfrak{S}_1$ and $\mathfrak{S}_2$, respectively.

Suppose $\mathfrak{P}$ is restricted by the following assumptions (A–E).

A. The populations $\mathfrak{S}_\nu$ are normal $\mathbf{N}(\vec{\mu}_\nu, \Sigma)$, $\nu = 1, 2$.

B. For each n in $\mathfrak{P}$, all eigenvalues of Σ are located on the segment $[c_1, c_2]$, where $c_1 > 0$ and c_2 do not depend on n.

C. For each n in $\mathfrak{P}$, the numbers $N > n+2$ and as $n \to \infty$, the limits exist $y_\nu = \lim\limits_{n\to\infty} n/N_\nu > 0$, $\nu = 1, 2$, and $y = \lim\limits_{n\to\infty} n/(N_1 + N_2) = y_1 y_2/(y_1 + y_2)$, where $y < 1$.

Denote

$$\vec{\mu} = \vec{\mu}_1 - \vec{\mu}_2,\quad \bar{\mathbf{x}} = \bar{\mathbf{x}}_1 - \bar{\mathbf{x}}_2,$$
$$\widehat{G} = 1/2\ \vec{\mu}^T \Gamma(C)\bar{\mathbf{x}},\quad \widehat{D} = \bar{\mathbf{x}}^T \Gamma(C)\Sigma\Gamma(C)\bar{\mathbf{x}}.$$

Consider the empiric distribution functions

$$F_{n\Sigma}(u) = n^{-1} \sum_{i=1}^{n} \mathrm{ind}\,(\lambda_i \leq u)$$

of eigenvalues λ_i of Σ, $i = 1, \ldots, n$, and the function

$$B_n(u) = \sum_{i=1}^{n} \mu_i^2/\lambda_i\ \mathrm{ind}(\lambda_i \leq u),$$

where μ_i are components of $\vec{\mu}$ in the system of coordinates, in which Σ is diagonal, $i = 1, 2, \ldots, n$.

D. For $u > 0$ as $n \to \infty$, $F_{n\Sigma}(u) \to F_{\Sigma}(u)$ and $B_n(u) \to B(u)$ almost everywhere.

Under conditions A–D, the limit exists

$$J = B(c_2) = \lim_{n \to \infty} \vec{\mu}^T \Sigma^{-1} \vec{\mu}, \quad \vec{\mu}^2 \leq c_2 B(c_2).$$

E. We consider the generalized discriminant function of the form

$$w(\mathbf{x}) = (\bar{\mathbf{x}}_1 - \bar{\mathbf{x}}_2)^T \Gamma(C) \big(\mathbf{x} - (\bar{\mathbf{x}}_1 + \bar{\mathbf{x}}_2)/2\big), \tag{3}$$

where the matrix $\Gamma(C)$ depends on C so that $\Gamma(C)$ is diagonalized together with C and has eigenvalues $\Gamma(\lambda)$ corresponding to the eigenvalues λ of C, where the scalar function $\Gamma : \mathbb{R}^1 \to \mathbb{R}^1$ is

$$\Gamma = \Gamma(u) = \int_{t \geq 0} (1 + ut)^{-1} \, d\eta(t),$$

and $\eta(t)$ is a function of finite variation on $[0, \infty)$ not depending on n. In addition, we assume that the function $\eta(t)$ is differentiable at some neighborhood of the point $t = 0$ with $t|\eta'(t)| \leq b$, and a sufficient number of moments exist

$$\beta_j = \int t^j |d\eta(t)|, \quad j = 1, 2, \ldots$$

Under Assumptions A–E, the probabilities of discrimination errors are

$$\alpha_1 = \Phi\left(-\frac{\mathbf{E}_1 w(\mathbf{x}) - \theta}{\sqrt{\mathrm{var}\ w(\mathbf{x})}}\right), \quad \alpha_2 = \Phi\left(-\frac{\theta - \mathbf{E}_2 w(\mathbf{x})}{\sqrt{\mathrm{var}\ w(\mathbf{x})}}\right),$$

where the conditional expectation operators $\mathbf{E}_1$ and $\mathbf{E}_2$ and conditional variance $\mathrm{var}\ w(\mathbf{x})$ (identical in both populations) are calculated for fixed samples $\mathfrak{X}_\nu$, $\nu = 1, 2$.

The half-sum $\alpha = (\alpha_1 + \alpha_2)/2$ reaches the minimum for

$$\theta = \theta_n^{\text{opt}} = 1/2\,(\mathbf{E}_1\,w(\mathbf{x}) + \mathbf{E}_2\,w(\mathbf{x}))$$
$$= 1/2\,(\overset{\circ}{\mathbf{x}}_2^T\Gamma(C)\overset{\circ}{\mathbf{x}}_2 - \overset{\circ}{\mathbf{x}}_1^T\Gamma(C)\overset{\circ}{\mathbf{x}}_1),$$

where $\overset{\circ}{\mathbf{x}}_\nu = \bar{\mathbf{x}}_\nu - \vec{\mu}_\nu$, $\nu = 1, 2$, and for $\theta = \theta_n^{\text{opt}}$

$$\alpha = \alpha(\theta) = \frac{\alpha_1(\theta) + \alpha_2(\theta)}{2} = \Phi(-\widehat{G}/\sqrt{\widehat{D}}). \qquad (4)$$

We study sample-dependent probability of error (4). Let us show that random values $\widehat{G}$ and $\widehat{D}$ have decreasing variance, find their limit expressions, and minimize the principal part of the ratio $\widehat{G}^2/\widehat{D}$.

Moments of Generalized Discriminant Function

Let us use resolvents $H(C) = (I + tC)^{-1}$ of the matrices C and matrices $\Gamma = \Gamma(C) = \int (I + tC)^{-1}\,d\eta(t)$, presenting linear transformation of $H(C)$. We consider spectral functions

$$h_n(t) = n^{-1}\text{tr}\,H(t), \quad b_n(t) = \vec{\mu}^T H(t)\,\vec{\mu},$$
$$k_n(t) = \bar{\mathbf{x}}^T H(t)\,\bar{\mathbf{x}}, \quad \varphi_n(t) = n^{-1}\text{tr}\,\Sigma H(t), \quad t \geq 0,$$
$$\Phi_n(t, t') = \vec{\mu}^T H(t)\Sigma H(t')\,\vec{\mu}, \quad \Psi_n(t, t') = \bar{\mathbf{x}}^T H(t)\Sigma H(t')\,\bar{\mathbf{x}},$$
$$t, t' \geq 0.$$

Remark 1. By the well-known Helmert transformation, the matrices (2) can be transformed to the form

$$S = (N - 2)^{-1}\sum_{m=1}^{N-2}\mathbf{x}_m\mathbf{x}_m^T, \qquad (5)$$

where independent $\mathbf{x}_m \sim \mathbf{N}(0, \Sigma)$, $m = 1, 2, \ldots, N - 2$.

Denote $y = n/(N - 2)$, $N_0 = (N_1 - 1)(N_2 - 1)/(N_1 + N_2 - 2)$, $y_0 = \lim_{n\to\infty} n/N_0$.

LEMMA 5.4. *Under Assumptions A–D as* $n \to \infty$, *the variances of functions* $h_n(t)$, $b_n(t)$, $k_n(t)$, $\Phi_n(t, t')$, *and* $\Psi_n(t, t')$ *decrease as* $O(N^{-1})$.

Proof. The assertions of this lemma follows immediately from Theorem 3.13 of Chapter 3 since all these functions present functionals from the class $\mathcal{L}_3$. $\square$

Denote $s_n = s_n(t) = 1 - y\,(1 - h_n(t)) > 0$.

Remark 2. Under Assumptions A–D for each $t > 0$ as $n \to \infty$, we have

$$\mathbf{E}H(t) = (I + ts_n(t)\Sigma)^{-1} + \Omega_n,$$

$$h_n(t) = \int (1 + ts_n(t)u)^{-1}\,dF_{n\Sigma}(u) + w_n(t), \tag{6}$$

where $w_n(t)$ and $\|\Omega_n\|$ are polynomials of fixed degree with vanishing coefficients. This is a corollary of Theorem 3.2 from Chapter 3.

Remark 3. Under Assumptions A–D for each $t \geq 0$

$$h(t) = \underset{n\to\infty}{\text{l.i.m.}}\ h_n(t), \quad s(t) = \underset{n\to\infty}{\text{l.i.m.}}\ s_n(t) = 1 - y + yh(t)$$

(where and in the following, l.i.m. denotes the limit in the square mean) and the equation holds

$$h(t) = \int (1 + ts(t)u)^{-1}\,dF_\Sigma(u) \tag{7}$$

Remark 4. If $y < 1$, then the following is true.

1. Equation (7) has a unique solution $h(t)$ that presents an analytical function $h(z)$ regular everywhere except for $z < 0$.
2. The function $h(z)$ satisfies the Hölder condition on the plane of complex z and decreases as $O(1/|z|)$ when $|z| \to \infty$.
3. For real $z < 0$, the quantity $\text{Im}\ h(z) \neq 0$ everywhere, except the segment $[v_1, v_2]$, where $v_1 = 1/(c_2(1 + \sqrt{y}))$, $v_2 = 1/(c_1(1 - \sqrt{y}))$.

These statements follow from Theorem 3.2 in Chapter 3.

LEMMA 5.5. *Under Assumptions A–D for each $t \geq 0$, the limits exist*

$$b(t) = \underset{n \to \infty}{\text{l.i.m.}} \, b_n(t) = \int (1 + ts(t)u)^{-1} \, dB(u),$$

$$k(t) = \underset{n \to \infty}{\text{l.i.m.}} \, k_n(t) = b(t) + y_0 \, \frac{1 - h(t)}{ts(t)},$$

$$\varphi(t) = \underset{n \to \infty}{\text{l.i.m.}} \, \varphi_n(t) = \underset{n \to \infty}{\text{l.i.m.}} \, n^{-1}\text{tr}\,(\Sigma H(t)) = \frac{y\,(1 - h(t))}{ts(t)}.$$

Proof. The first assertion follows readily from (6) and Assumption D. Further, in view of (6) we have

$$\mathbf{E}N^{-1}\text{tr}\Sigma H(t) = \mathbf{E}N^{-1}\text{tr}\Sigma(I + ts(t)\Sigma)^{-1} + \omega_n =$$

$$= \frac{n - \text{tr}(I + ts(t)\Sigma)^{-1}}{Nts(t)} + \omega_n = y\frac{1 - h(t)}{ts(t)} + \omega_n$$

with different $\omega_n \to 0$. This is the second statement.

For normal distributions, the matrix $H(t) = (I + tC)^{-1}$ is independent of $\bar{\mathbf{x}}$. Using the previous lemma statement, we obtain

$$\mathbf{E}\,\bar{\mathbf{x}}^T H(t)\bar{\mathbf{x}} = \mathbf{E}\,\frac{1}{N_0}\text{tr}(\Sigma H(t)) = y_0\frac{1 - h(t)}{ts(t)}, \quad t > 0.$$

Lemma 5.5 is proved. □

LEMMA 5.6. *Under Assumptions A–D for any fixed $t, t' \geq 0$, we have*

$$s(t)s(t')\mathbf{E}\,\bar{\mu}^T H(t)\Sigma H(t')\bar{\mu} = \frac{b_n(t) - b_n(t')}{t' - t} + \omega_n(t, t'), \quad (8)$$

$$s(t)s(t')\mathbf{E}\,\bar{\mathbf{x}}^T H(t)\Sigma H(t')\bar{\mathbf{x}} = \frac{k_n(t) - k_n(t')}{t' - t} + \omega_n(t, t'), \quad (9)$$

where the expressions in the right-hand sides are extended by continuity to $t = t'$, and $\omega_n(t, t')$ are some polynomials of fixed degree with coefficients vanishing as $n \to \infty$.

Proof. In view of Remark 1, we calculate the expressions in the left-hand sides of (8) and (9) for $H(t) = (I + tS)^{-1}$ and $H(t') = (I + t'S)^{-1}$ with S defined by (5). In the asymptotic approach, it is convenient to prove the lemma for samples $\mathfrak{X}_\nu$ with excluded vectors $\mathbf{x}_\nu \in \mathfrak{X}_\nu$, $\nu = 1, 2$. Set $t > 0$, $t \neq t'$. Denote

$$S^\nu = S - (N-2)^{-1}\bar{\mathbf{x}}_\nu \bar{\mathbf{x}}_\nu^T, \quad H^\nu(t) = (I + tS^\nu)^{-1},$$

$$\psi_\nu(t) = \bar{\mathbf{x}}_\nu^T H(t)\,\bar{\mathbf{x}}_\nu, \quad s_n(t) = \mathbf{E}\,(1 - t\psi_\nu(t)), \quad \nu = 1, 2.$$

Then, $H(t) = (1 - t\psi_\nu)H^\nu(t)$, $\nu = 1, 2$. By Lemma 3.1, var $\psi_\nu(t) \to 0$ as $n \to \infty$, $\nu = 1, 2$. Using the independence of $H(t)$ from sample means $\bar{\mathbf{x}}_\nu$, $\nu = 1, 2$, we obtain

$$t\,(H(t) - H(t'))/(t' - t) = t\,\mathbf{E}\,H(t)SH(t') = t\,\mathbf{E}\,H(t)\bar{\mathbf{x}}_\nu\bar{\mathbf{x}}_\nu^T H(t') =$$

$$= t\,\mathbf{E}\,(1 - t\psi_\nu(t))\,(1 - t'\psi(t'))H^\nu(t)\bar{\mathbf{x}}_\nu\,\bar{\mathbf{x}}_\nu^T H^\nu(t') =$$

$$= ts_n(t)s_n(t')\,\mathbf{E}\,H^\nu(t)\Sigma H^\nu(t') + \Omega_n, \tag{10}$$

where (different) matrices Ω_n are such that $\|\Omega_n\| \to 0$. We use the relations $H(t) = (1 - \psi_\nu(t))H^\nu(u)$, $\nu = 1, 2$ and conclude that the right-hand side of (1) equals $ts_n(t)s_n(t')\mathbf{E}\,H(t)\Sigma H(t') + \Omega_n$, where the new remainder term is such that $\|\Omega_n\| \to 0$ as $n \to \infty$. Multiplying these matrix relations by $\vec{\mu}$ from the left and from the right we obtain the first lemma statement.

To evaluate the left-hand side of (9), we deal with sample covariance matrices C of the form (5). For normal distributions, sample means $\bar{\mathbf{x}}$ do not depend on C. Therefore, we obtain

$$\mathbf{E}\,\Psi(t, t') = \mathbf{E}\,\Phi(t, t') + \mathbf{E}\,N_0^{-1}\mathrm{tr}(\Sigma H(t)\Sigma H(t')).$$

Using (10), we obtain the second lemma statement. The proof of Lemma 5.6 is complete. $\square$

Limit Probabilities of Errors

Now we pass to integrals with respect to $d\eta(t)$. Let us perform the limit transition and find the limiting expressions for $\widehat{G} = 1/2\,\vec{\mu}^T\Gamma(C)\bar{\mathbf{x}}$ and $\widehat{D} = \bar{\mathbf{x}}^T\Gamma(C)\Sigma\Gamma(C)\bar{\mathbf{x}}$.

THEOREM 5.10. *Under Assumptions A–E, the following limits exist*

$$\operatorname*{l.i.m.}_{n\to\infty} \theta_n^{\mathrm{opt}} = \theta^{\mathrm{opt}},$$

$$\operatorname*{l.i.m.}_{n\to\infty} \vec{\mu}^T \Gamma(C)\bar{\mathbf{x}} = 2G, \quad \operatorname*{l.i.m.}_{n\to\infty} \bar{\mathbf{x}}^T \Gamma(C)\Sigma\Gamma(C)\bar{\mathbf{x}} = D, \quad (11)$$

where

$$\theta^{\mathrm{opt}} = 1/2(y_2 - y_1) \int \frac{1 - h(t)}{ts(t)}\, d\eta(t),$$

$$G = 1/2 \int b(t)\, d\eta(t),$$

$$D = \int\int \frac{k(t) - k(t')}{s(t)s(t')\,(t' - t)}\, d\eta(t)d\eta(t'), \quad (12)$$

and the latter integrand is extended by continuity to $t = t'$.

Proof. The functionals θ_n^{opt} equal $1/2\,(\overset{\mathrm{o}}{\mathbf{x}}{}_2^T\Gamma\,\overset{\mathrm{o}}{\mathbf{x}}{}_2 - \overset{\mathrm{o}}{\mathbf{x}}{}_1^T\Gamma\,\overset{\mathrm{o}}{\mathbf{x}}{}_1)$, where $\overset{\mathrm{o}}{\mathbf{x}}_\nu = \mathbf{x}_\nu - \vec{\mu}_\nu$, $\nu = 1, 2$, and $\Gamma(C) = (I + tC)^{-1}$ belongs to the class $\mathcal{L}_3$ of functionals allowing ε-normal evaluation with the polynomial $\varepsilon_3(t) \to 0$. By Lemma 5.4, its variance vanish as $n \to \infty$. For normal variables, we have

$$\mathbf{E}\,\theta_n^{\mathrm{opt}} = 1/2(N_2^{-1} - N_1^{-1})\mathbf{E}\,\mathrm{tr}\Sigma\Gamma(C),$$

and by Lemma 5.5, this quantity tends to the first expression in (12). This is the first lemma statement.

Next, for normal distributions, we have

$$\mathbf{E}\,\widehat{G} = \mathbf{E}\int \vec{\mu}^T(I + tC)^{-1}\vec{\mu}\, d\eta(t) = \mathbf{E}\int b_n(t)\, d\eta(t) \to \int b(t)\, d\eta(t),$$

and we obtain the second relation in (11).

For the third functional, we apply Lemmas 5.5 and 5.6. As $n \to \infty$ we have $s_n(t) \to s(t) > 1 - y > 0$. Dividing by $s(t)$ and $s(t')$, we obtain the third limit in (11). Theorem 5.10 is proved. $\square$

THEOREM 5.11. *Suppose Assumptions A–E hold, $D > 0$, the discriminant function (3) is used, and the threshold θ is chosen for each n so that it minimizes $(\alpha_1 + \alpha_2)/2$.*
Then,

$$\operatorname*{plim}_{n\to\infty} (\alpha_1 + \alpha_2)/2 = \Phi(-\sqrt{J^{\mathrm{eff}}}/2),$$
$$\text{where} \quad J^{\mathrm{eff}} = J^{\mathrm{eff}}(\eta) = 4G^2/D,$$

and $G = G(\eta)$ and $D = D(\eta)$ are defined by (12).

Proof. The minimum of $(\alpha_1 + \alpha_2)/2$ is attained for $\theta = \theta_n^{\mathrm{opt}}$, where θ_n^{opt} is defined by (11). The statement of Theorem 5.11 immediately follows from Theorem 5.10. $\square$

Example 1. Let us choose a special form of $\eta(t)$: $\eta(t') = 0$ for $t' \leq t$, and $\eta(t') = t$ for $t' \geq t$. Then, $\Gamma(C) = t\,(I + tC)^{-1}$ is a ridge estimator of the matrix Σ^{-1}. In this case, $\theta^{\mathrm{opt}} = 1/2\,(y_2 - y_1)$ $(1 - h(t))/s(t)$, $G = 1/2\,b(t)$, and $D = -t^2 s(t)^{-2} k'(t) > 0$.

Example 2. Let $\Sigma = \sigma^2 I$, $\sigma > 0$ for all $n = 1, 2, \ldots$, and let $\eta(t) = 1$ for all arguments $t > 0$. Then, we have $\Gamma(C) = I$, $G = J/2$, $D = J + y_1 + y_2$, and $\theta^{\mathrm{opt}} = (y_2 - y_1)/2$ in agreement with limit formula $J^{\mathrm{eff}} = J^{\mathrm{eff}}(1) = J/2\sqrt{J + y_1 + y_2}$ for independent components of normal $\mathbf{x}$ (see Introduction).

Example 3. Let $\eta(t') = \operatorname{ind} t' \leq t)$, where $t \to \infty$. This corresponds to the transition to the standard discriminant function. As $t \to \infty$ we find that $h(t) \to 0$, $s(t) \to 1 - y$, and $tb(t) \to J/(1 - y)$, where $J = \lim_{n\to\infty} \vec{\mu}^T \Sigma^{-1} \vec{\mu}$. We obtain

$$G \to J/2(1 - y)^{-1}, \quad D \to (J + y_1 + y_2)(1 - y)^{-3},$$
$$\text{and} \quad J^{\mathrm{eff}} \to J^2(1 - y)/(J + y_1 + y_2),$$

in agreement with the well-known Deev formula (see Introduction).

Example 4. Let $\eta(t') = \operatorname{ind}(t' \leq t)$, and consider a special case when $\mu_i^2/\lambda_i = J/n$ for all components μ_i of the vector $\vec{\mu}$ in the system of coordinates where Σ is diagonal, where λ_i are the corresponding eigenvalues of Σ, $i = 1, 2, \ldots, n$ (the case of equal contributions to the Mahalanobis distance). Suppose that the limit spectrum of sample covariance matrices is given by the "ρ-model" that was described in Section 3.1. Then, the dependence of J^{eff} on t can be found in an explicit form as follows:

$$J^{\text{eff}} = J^{\text{eff}}(h) = J^2 \frac{J - y + 2yh - (\rho + y)h^2}{J + y_1 + y_2},$$

where

$$2h = \left[1 + \rho + d(1 - y)t \right.$$
$$\left. + \sqrt{(1 + \rho + d(1 - y)\ t)^2 + 4(dy\ t - \rho)}\right]^{-1},$$

and $d = \sigma^2(1 - \rho)^2$, $\sigma > 0$, and $\rho < 1$ are parameters of the model. We note that $J^{\text{eff}}(h)$ is a monotone function of h, and the maximum is attained for $h = h^{\text{opt}} = y\ (\rho + y)^{-1}$, when $t = t^{\text{opt}} = \rho/(dy)$ for $\rho > 0$ and $y > 0$. The maximum value is

$$J^{\text{eff}}(t^{\text{opt}}) = \frac{J^2}{J + y_1 + y_2} \left(1 - \frac{\rho y}{\rho + y}\right). \tag{13}$$

This limit formula presents a version of Deev's formula to this special model and asymptotically optimized discriminant function. The ratio $1 + y^2/(\rho + y(1 - \rho) - y^2)$ characterizes the gain in the limit provided by the improved discriminant procedure as compared with the standard one. In the case when $\rho = 0$, the value $J^{\text{eff}} = J^2/(J + y_1 + y_2)$; this corresponds to the discrimination in the system of coordinates where $\Sigma = I$ for independent variables (although this fact is unknown to the observer). In the case when $y = 1$, we have $J^{\text{eff}} = J^2/[(1 + \rho)/(J + y_1 + y_2)]$ in contrast to the standard discriminant procedure for which $J^{\text{opt}} = 0$.

THEOREM 5.12. *Suppose that some function $\eta(t) = \eta^{\text{opt}}(t)$ of finite variation exists such that*

$$\int \frac{k(t) - k(t')}{s(t')\ (t - t')}\ d\eta^{\text{opt}}(t') = b(t)s(t), \quad t \geq 0. \tag{14}$$

Then, for any function $\eta(t)$ of finite variation on $[0, \infty]$, we have $J^{\text{eff}}(\eta) \leq J^{\text{eff}}(\eta^{\text{opt}})$.

This assertion immediately follows from Theorems 5.10 and 5.11.

Best-in-the-Limit Discriminant Procedure

Theorem 5.11 shows that minimal half-sum of limit probabilities of discrimination errors is achieved when the ratio G^2/D is maximum. The numerator and denominator of this ratio are quadratic in $\eta(t)$ and this allows the minimization in a general form. It is convenient to apply methods of complex analysis. Let us perform the analytical extension of functions $h(t)$, $s(t)$, $b(t)$, and $k(t)$ to the plane of complex z.

Consider the contour $\mathfrak{L} = (\sigma - i\varepsilon, \sigma + i\varepsilon)$ and functions

$$\rho(t) = \int_{-\infty}^{t} \frac{1}{s(x)}\, d\eta(x) \quad \text{and} \quad \alpha(z) = \int \frac{1}{z+t}\, d\rho(t).$$

LEMMA 5.7. *Under Assumptions A–E, there exists $\sigma > 0$ such that*

$$G = \frac{1}{4\pi i}\int_{\mathfrak{L}} b(z)s(z)\alpha(z)dz, \quad D = \frac{1}{4\pi i}\int_{\mathfrak{L}} k(z)\alpha^2(z)\, dz. \quad (15)$$

Proof. From Lemma 5.5 and Remark 4, it follows that the functions $h(z)$, $b(z)$, and $k(z)$ are analytic and regular everywhere except the segment $[v_1, v_2]$, with $v_{1,2}$ defined by Remark 4, $v_1 > 0$. For $|z| \to \infty$, we have $h(z) = O(|z|^{-1})$, $s(z) = O(|z|^{-1})$, $b(z) = O(|z|^{-1})$, and $\alpha(|z|) = O(|z|^{-1})$. Therefore, we can deform the contour $\mathfrak{L}$ and close it from the above by half-circumference with the radius $R \to \infty$. We substitute the expression for $\alpha(z)$ to right-hand sides of (15), change the order of integration, and use the residue theorem. It follows that

$$\frac{1}{2\pi i}\oint_{\mathfrak{L}} b(z)s(z)\alpha(z)dz = \frac{1}{2\pi i}\int \left(\int_{\mathfrak{L}} \frac{b(z)s(z)}{z+t}\, dz\right) d\rho(t) =$$

$$= \int (\text{Res}[b(z)s(z)]_{z=-t})\, d\rho(t) = \int b(t)\, d\eta(t) = G.$$

Similarly, we obtain

$$\frac{1}{2\pi i} \oint k(z)\alpha^2(z)\, dz =$$

$$= \frac{1}{2\pi i} \int \left[\oint \frac{1}{s(z)s(z')(z+t)(z+t')} k(z)\, dz \right] d\rho(t)d\rho(t') =$$

$$= \int\int \frac{k(t)-k(t')}{s(t)s(t')\,(t'-t)}\, d\rho(t)\, d\rho(t') = D.$$

Our lemma is proved. □

Let us use the Hölder inequality and extend functions $h(z)$, $b(z)$, $s(z)$, and $k(z)$ to the real axis. Denote

$$g(v) = \mathrm{Im}(b(-v)s(-v)),$$

$$d(v) = \mathrm{Im}\ k(-v) = \mathrm{Im}\ b(-v) + (y_1 + y_2)\frac{\mathrm{Im}\ h(-v)}{|vs(-v)|^2}.$$

LEMMA 5.8. *Under Assumptions A–E, we have*

$$G = \frac{1}{2\pi} \int_{v_1}^{v_2} g(v)\alpha(v)\, dv, \quad D = \frac{1}{\pi} \int_{v_1}^{v_2} d(v)\alpha^2(v)\, dv,$$

where the segment $\mathfrak{V} = \{v:\ 0 < v_1 \le v \le v_2, \text{and}\ d(v) > 0\}$.

Proof. We use Remark 4 and close the contour $\mathfrak{L}$ by a half-circumference from the above. The functions $h(z), b(z), s(z)$, and $k(z)$ are analytical and have no singularities for $\mathrm{Re}\ z < -\sigma$ every-where, except the cut on the half-axis located at $v \in [v_1, v_2]$. When $|z| \to \infty$, the functions $\alpha(z), b(z), k(z)$ decrease as $|z|^{-1}$ while $s(z)$ is bounded. It follows that the integral over the infinitely remote region can be made arbitrarily small. The function $\alpha(z)$ has no singularities for $z > 0$. Contracting the contour to the real axis, we see that the real parts of the integrands are mutually canceled while the imaginary parts double. We obtain G and D from (12). The lemma is proved. □

Note that $0 \leq g(v) \leq d(v)$, $v > 0$. Define

$$\alpha^{\mathrm{opt}}(v) = \frac{g(v)}{d(v)}, \quad v > 0, \qquad J^{\mathrm{opt}} = J^{\mathrm{eff}}(\eta^o) = \frac{1}{\pi} \int \frac{g^2(v)}{d(v)} \, dv.$$

THEOREM 5.13. *Let conditions A–E be fulfilled and there exists a function of bounded variation $\eta^{\mathrm{opt}}(t)$ such that*

$$\int \frac{1}{s(t)(v+t)} \, d\eta^{\mathrm{opt}}(t) = \alpha^{\mathrm{opt}}(v), \quad v \in \mathfrak{V}, \tag{16}$$

and $J^{\mathrm{opt}} > 0$.
   Then,

   1. *using the classification rule $w^{\mathrm{opt}}(\mathbf{x}) > \theta_n^{\mathrm{opt}}$ against $w^{\mathrm{opt}}(\mathbf{x}) \leq \theta_n^{\mathrm{opt}}$ and the discriminant function*

$$w^{\mathrm{opt}}(\mathbf{x}) = (\bar{\mathbf{x}}_1 - \bar{\mathbf{x}}_2)^T \Gamma^{\mathrm{opt}}(C)(\mathbf{x} - (\bar{\mathbf{x}}_1 + \bar{\mathbf{x}}_2)/2)$$

$$\text{and } \Gamma^{\mathrm{opt}}(C) = \int (I + tC)^{-1} \, d\eta^{\mathrm{opt}}(t)$$

we have

$$\operatorname*{plim}_{n \to \infty} (\alpha_1 + \alpha_2)/2 = \Phi(-\sqrt{J^{\mathrm{opt}}}/2),$$

where $J^{\mathrm{eff}} = J^{\mathrm{eff}}(\eta^{\mathrm{opt}})$;
   2. *for the classification with any other $w(\mathbf{x})$ of the form (3), we have $J^{\mathrm{opt}} \leq J^{\mathrm{eff}}(\eta)$ for any function $\eta(t)$ of bounded variation.*

Example 5. Let $\mu_i^2 = \bar{\mu}^2/n$ for each $i = 1, 2, \ldots, n$. Then,

$$g(v) = J/v \operatorname{Im} h(v), \quad d(v) = (J + y_1 + y_2)/v \, |s(v)|^{-1} \operatorname{Im} h(v),$$

where $J = \lim_{n \to \infty} \bar{\mu}^T \Sigma^{-1} \bar{\mu}$ and $\alpha^{\mathrm{opt}}(v) = \mathrm{const}|s(v)|^{-2}$. For the ρ-model considered in Section 3.1 with $\rho > 0$, we have $|s(v)|^2 = \mathrm{const} \, (v + t^{\mathrm{opt}})^{-1}$, where $t^{\mathrm{opt}} = \rho d^{-1} y^{-1}$ and there exists the solution $\eta^{\mathrm{opt}}(t)$ of equations (14) and (16) in the form of a unit jump at the point $t = t^{\mathrm{opt}}$. The corresponding asymptotically unimprovable discriminant function is

$$w^{\mathrm{opt}}(\mathbf{x}) = (\bar{\mathbf{x}}_1 - \bar{\mathbf{x}}_2)^T (C + \lambda I)^{-1} (\mathbf{x} - (\bar{\mathbf{x}}_1 + \bar{\mathbf{x}}_2)/2),$$

where $\lambda = dy/\rho$. This discriminant function includes a ridge estimator of the inverse covariance matrix Σ with the regularization $\lambda > 0$. In this case, the maximum J^{opt} of $J^{\mathrm{eff}}(\eta)$ is achieved within the class of discriminant functions (3) with $\Gamma^{\mathrm{opt}}(C) = (I + tC)^{-1}$ and the maximum value

$$J^{\mathrm{opt}} = \frac{J^2}{J + y_1 + y_2} \left(1 - \frac{\rho y}{\rho + y} \right)$$

that coincides with (13). In case of $\rho = 0$, we obtain asymptotically unimprovable discriminant function $w(\mathbf{x}) = (\bar{\mathbf{x}}_1 - \bar{\mathbf{x}}_2)^T (\mathbf{x} - (\bar{\mathbf{x}}_1 + \bar{\mathbf{x}}_2)/2)$ (the spherical classificator) with $J^{\mathrm{opt}} = J^2/(J + y_1 + y_2)$ that corresponds to the discriminant function constructed for the case of a priori known covariance matrix Σ.

Note that the solution of (14) and (16) may not exist. If it exists, however, serious difficulties are still to be expected when trying to replace $k(t)$ and $s(t)$ by their natural estimators. Relation (14) is the Fredholm integral equation of the first kind and its solution may be ill conditioned. For applications, some more detailed theoretical investigations are necessary.

However, some researches show that these difficulties can be overcome. In [64] and [98], the asymptotic expressions for this solution were used to construct a practical improved discriminant procedure (see Introduction). This procedure was examined numerically in academic tests and successfully applied to a practical problem.

The Extension to a Wide Class of Distributions

Let us use the Normal Evaluation Principle proposed in Section 3.4. Note that under Assumptions A–E, the functionals θ_n^{opt}, $\widehat{G}$, and $\widehat{D}$ belong to the class $\mathfrak{L}_3$ of functionals allowing ε-normal evaluation in the square mean with $\varepsilon \to 0$. We may conclude that Theorem 5.10 remains valid for a wide class of distributions described in Section 3.4. However, the assertion of Theorem 5.11 is not valid for the extended class of distributions since the discriminant function (3) may be not normally distributed. Let us return to the first definition of quality of discrimination introduced by R. Fisher. He proposed the empirical quality function defined as $\widehat{F}_n = \widehat{G}^2/\widehat{D}$, that is the argument in (4).

We consider an extended class of populations restricted by the following Assumptions A1–A3. For $\mathbf{x}$ from $\mathfrak{X}_\nu$, denote

$$\vec{\mu}_\nu = \mathbf{E}\,\mathbf{x}, \qquad \overset{o}{\mathbf{x}} = \mathbf{x} - \vec{\mu}_\nu, \qquad \nu = 1, 2,$$

$$M_\nu = \sup_{|\mathbf{e}|=1} \mathbf{E}\,(\mathbf{e}^T \overset{o}{\mathbf{x}})^4 > 0, \qquad M = \max(M_1, M_2),$$

$$\gamma_\nu = \sup_{\|\Omega\|=1} \mathrm{var}\,(\overset{o}{\mathbf{x}}^T \Omega \overset{o}{\mathbf{x}}/n)/M, \qquad \nu = 1, 2, \qquad \gamma = \max(\gamma_1, \gamma_2),$$

where Ω are nonrandom, symmetric, positive, semidefinite matrices of unit spectral norm.

A1. All variables in the both populations in $\mathfrak{P}$ have the fourth moments.

A2. The parameters M_ν and $\vec{\mu}_\nu^2$ are uniformly bounded in $\mathfrak{P}$, $\nu = 1, 2$.

A3. The parameters $\gamma_\nu \to 0$, $\nu = 1, 2$.

Remark 5. Under Assumptions A1–A3 and B–E, if the discriminant function (3) is used and $D(\eta) \geq 0$, then for $\widehat{F}_n = \widehat{F}_n(\eta)$ as $n \to \infty$, we have

$$\plim_{n\to\infty} \widehat{F}_n(\eta) = J^{\mathrm{eff}}(\eta),$$

where $J^{\mathrm{eff}}(\eta)$ is defined in Theorem 5.11.

Theorem 5.12 remains valid for distributions described in conditions A1–A3. The asymptotically unimprovable discriminant function $w^{\mathrm{opt}}(\mathbf{x})$ remains as in Theorem 5.3. The following statement is a refinement of Theorem 5.13.

THEOREM 5.14. *Let conditions A1–A3 and B–E be fulfilled, the discriminant function* (3) *used, and there exist a function of* *bounded variation $\eta^{\mathrm{opt}}(t)$ such that*

$$\int \frac{1}{s(t)(v+t)}\, d\eta^{\mathrm{opt}}(t) = \alpha^{\mathrm{opt}}(v), \qquad v \in \mathfrak{V},$$

where $\mathfrak{V}$ is defined in Lemma 5.8.
   Then, if $J^{\mathrm{opt}} > 0$, we have

   *1. the classification rule $w^{\mathrm{opt}}(\mathbf{x}) > \theta_n^{\mathrm{opt}}$ against $w^{\mathrm{opt}}(\mathbf{x}) \leq \theta_n^{\mathrm{opt}}$
with the discriminant function*

$$w^{\mathrm{opt}}(\mathbf{x}) = (\bar{\mathbf{x}}_1 - \bar{\mathbf{x}}_2)^T \Gamma^{\mathrm{opt}}(C)(\mathbf{x} - (\bar{\mathbf{x}}_1 + \bar{\mathbf{x}}_2)/2)$$

$$\text{and } \ \Gamma^{\mathrm{opt}}(C) = \int (I + tC)^{-1}\, d\eta^{\mathrm{opt}}(t)$$

leads to the quality function $\widehat{F}_n(\eta^o)$ such that

$$\operatorname*{plim}_{n\to\infty} \widehat{F}_n(\eta^{\mathrm{opt}}) = J^{\mathrm{eff}}(\eta^{\mathrm{opt}}) = J^{\mathrm{opt}};$$

   *2. for the classification with any other $w(\mathbf{x})$ of the form (3) we
have $J^{\mathrm{opt}} \geq J^{\mathrm{eff}}(\eta)$ for any function $\eta(t)$ of bounded variation.*

Proof is reduced to citing Theorems 5.11, 5,12, 5.14, and
Remark 5.

Estimating the Error Probability

For the usage in practical problems, we should propose consistent estimators for the functions $s(t)$, $b(t)$, $k(t)$, $g(v)$, and $d(v)$
defining the optimum discriminant function and suggest an estimator of the limit error probability.
   First, we estimate the parameters G and D and then the limit
probability of error $\alpha = \Phi(-G^2/D)$.
   Consider the statistics

$$\widehat{h}(t) = N^{-1}\mathrm{tr}\, H(t), \quad \widehat{s}(t) = 1 - t\, \mathrm{tr}\,(CH(t)), \quad \widehat{k}(t) = \bar{\mathbf{x}}^T H(t)\bar{\mathbf{x}},$$

$$\widehat{g}(t) = \widehat{k}(t)/2 - (1 - \widehat{s}(t))/\widehat{s}(t), \quad \widehat{d}(t, u) = t u \bar{\mathbf{x}}^T H(t) C H(u)\bar{\mathbf{x}}.$$

Denote $c = c(\|\Sigma\|, \vec{\mu}^2)$, $N = N_1 + N_2$, $N_0 = N_1 N_2/N$, where N_ν is sample size for $\mathfrak{X}_\nu$, $y_\nu = n/N_\nu$, $\nu = 1, 2$, and $\varepsilon = 1/n + 1/N$.

LEMMA 5.9. *If* $0 \le y < 1$, *then*

$$\mathbf{E}(\widehat{h}(t) = h(t))^2 \le c/N, \quad \mathbf{E}(\widehat{s}(t) - s(t))^2 \le c/N,$$

$$\mathbf{E}\widehat{k}(t) = \vec{\mu}^T (I + t\Sigma)^{-1}\vec{\mu} + (y_1 + y_2)\frac{1 - h(t)}{s(t)} + \omega,$$

$$(1 - y)^2 \, \mathbf{E}(\widehat{g}(t) - g(t))^2 \le c\varepsilon,$$

$$(1 - y)^2 \, \mathbf{E}\left|\frac{\widehat{d}(t, u)}{\widehat{s}(t)\widehat{s}(u)} - d(t, u)\right|^2 \le c\varepsilon,$$

where $|\omega| \le c\varepsilon$.

Define

$$\widehat{G} = \int \widehat{g}(t)d\eta(t) = \int [\widehat{k}(t) - \frac{n}{N_0}\frac{1 - \widehat{h}(t)}{\widehat{s}(t)}] d\eta(t),$$

$$\widehat{D} = \int \int \frac{u\widehat{k}(t) - t\widehat{k}(u)}{\widehat{s}(t)\widehat{s}(u)} d\eta(t) \, d\eta(u).$$

THEOREM 5.15. *If* $D > 0$ *and* $0 \le y < 1$, *then as* $n \to \infty$, *the estimator*

$$\widehat{\alpha} = \Phi(-\widehat{G}/\sqrt{\widehat{D}}) \overset{\mathbf{P}}{\to} \alpha.$$

To construct estimators of limit functions $g(v)$ and $d(v)$ consistent in the double-limit transition as $\varepsilon \to +0$ and $n \to \infty$ with the optimal dependence $\varepsilon = \varepsilon(n)$, further investigations are necessary.

CHAPTER 6

THEORY OF SOLUTION TO HIGH-ORDER SYSTEMS OF EMPIRICAL LINEAR ALGEBRAIC EQUATIONS

In this chapter, we develop *the statistical approach* to solving empirical systems of linear algebraic equations (SLAE). We consider the unknown solution vector as the vector of parameters, consider pseudosolutions as estimators, and seek estimators from wide classes minimizing the quadratic risk. First, we establish a general form of an unimprovable solution found in [54] under the Bayes approach by averaging over possible equations and over their solution vectors. Then, we use the asymptotic theory developed in [55]–[58] and isolate principal part of the quadratic risk function under the assumption that the number of unknown $n \to \infty$ along with number of equations $N \to \infty$ so that $N/n - 1 \to \kappa > 0$. We apply the techniques of spectral theory of Gram matrices developed in Chapter 3 and derive the dispersion equation expressing limit spectral functions of unknown matrices of coefficients in terms of limit spectral functions of the observable empirical matrices. We consider a wide class of regularized pseudosolutions depending on an arbitrary function of finite variation and find the limit value of their quadratic risk. Then, we solve the extremum problem and find equations defining asymptotically best pseudosolution. Using dispersion equations, we remove the unknown parameters and express this pseudosolution in terms of limit functions of empirical matrix of coefficients and empirical right-hand side vectors only and thus obtain always stable pseudosolutions of minimum-limit quadratic risk.

6.1. THE BEST BAYES SOLUTION

Suppose a consistent system of linear equations is given $A\mathbf{x} = \mathbf{b}$, where A is the coefficient matrix of N rows and n columns, $N \geq n$, $\mathbf{x}$ is the unknown vector of the dimension n, and $\mathbf{b}$ is the right-hand side vector of the dimension N. The observer only knows the matrix $R = A + \delta A$, and vector $\mathbf{y} = \mathbf{b} + \delta\mathbf{b}$, where δA are matrices with random entries $\delta A_{ij} \sim \mathbf{N}(0, p/n)$ and $\delta\mathbf{b}$ are vectors with components $\delta b_i \sim \mathbf{N}(0, q/n)$. Assume that all these random values are independent, $i = 1, \ldots, N$, $j = 1, \ldots, n$.

We construct a solution procedure that is unimprovable on the average for a set of problems with different matrices A and vectors $\mathbf{b}$, and consider Bayes distributions $A_{ij} \sim \mathbf{N}(0, a/n)$ and $\mathbf{x}_j \sim \mathbf{N}(0, 1/n)$ for all i and j, assuming all random values are independent. It corresponds to a uniform distribution of entries of A, given the quadratic norm distributed as $\bar{\nu}^2$ and a uniform distribution of directions of vectors $\mathbf{x}$. We minimize the quadratic risk of the estimator $\widehat{\mathbf{x}}$

$$D = D(\widehat{\mathbf{x}}) = \mathbf{E}(\mathbf{x} - \widehat{\mathbf{x}})^2. \tag{1}$$

Here (and in the following) the square of a vector denotes the square of its length, and the expectation is calculated over the joint distribution of $A, \delta A, \mathbf{x}$, and $\delta\mathbf{b}$.

Let us restrict ourselves with a class $\mathfrak{K}$ of regularized pseudosolutions of the form

$$\widehat{\mathbf{x}} = \Gamma R^T \mathbf{y}, \tag{2}$$

where the matrix function $\Gamma = \Gamma(R^T R)$ is diagonalized together with $R^T R$ and has eigenvalues $\gamma(\lambda)$ corresponding to the eigenvalues λ of $R^T R$. Let us restrict scalar function $\gamma(u)$ by the requirement that it is a non-negative measurable function such that $u\gamma(u)$ is bounded by a constant.

For the standard solution, $\gamma(u) = 1/u$.

Define a nonrandom distribution function for the set of eigenvalues of the empirical matrices $R^T R$ as the expectation

$$F(u) = \mathbf{E} n^{-1} \sum_{i=1}^{n} \text{ind}(\lambda_i \le u),$$

where λ_i are eigenvalues of $R^T R$. Note that $W = R^T R$ is the Wishart matrix and $F(u)$ can be derived from the well-known distribution of eigenvalues.

Using this function we can write, for example,

$$\mathbf{E} n^{-1} \text{tr}(R\Gamma^2 R^T) = n^{-1} \mathbf{E} \sum_{i=1}^{n} \lambda_i \gamma^2(\lambda_i) = \int u \gamma^2(u) \, dF(u).$$

PROPOSITION 1. *Let an estimator* $\widehat{\mathbf{x}} \in \mathfrak{K}$ *of the solution to the system* $A\mathbf{x} = \mathbf{b}$ *be calculated using the observed empiric matrix* $R = A + \delta A$ *and vector* $\mathbf{y} = \mathbf{b} + \delta\mathbf{b}$, *where entries of the matrices* $A = \{A_{ij}\}$, $\delta A = \{\delta A_{ij}\}$, *and components of the vectors* $\mathbf{x} = \{x_j\}$ *and* $\delta\mathbf{b} = \{\delta b_i\}$ *are independent random values*

$$A_{ij} \sim \mathbf{N}(0, a/n), \quad x_j \sim \mathbf{N}(0, 1/n), \quad \delta A_{ij} \sim \mathbf{N}(0, p/n),$$

and $\delta b_i \sim \mathbf{N}(0, q/n)$, $i = 1, \ldots, N$, $j = 1, \ldots, n$, *where* $a > 0$. *Then, the functional* (1) *equals*

$$D = \int [1 - 2a(a + p)^{-1} u\gamma(u) + a^2(a + p)^{-2} u^2 \gamma^2(u)$$
$$+ su\gamma^2(u)] \, dF(u), \tag{3}$$

where $s = ap/(a + p) + q$.

Proof. The quantity D is a sum of three addends

$$D = \mathbf{E}\mathbf{x}^T \mathbf{x} - 2\mathbf{E}\mathbf{x}^T \Gamma R^T \mathbf{y} + \mathbf{E}\mathbf{y}^T R\Gamma^2 R^T \mathbf{y}.$$

Let angular brackets denote the normed trace of a matrix: $\langle M \rangle = n^{-1} \text{tr} M$.

First, we average over the distribution of $\delta \mathbf{b}$. Let us use the obvious property of normal distributions

$$\mathbf{E}(\delta \mathbf{b}^T M \delta \mathbf{b}) = qn^{-1} \mathrm{tr} M,$$

where M is an $N \times N$ matrix of constants. We find that

$$D = \mathbf{E} \mathbf{x}^T \mathbf{x} - 2 \mathbf{E} \mathbf{x}^T \Gamma R^T \mathbf{b} + \mathbf{E} \mathbf{b}^T R \Gamma^2 R^T \mathbf{b} + q \mathbf{E} \langle R \Gamma^2 R^T \rangle.$$

Then, we average with respect to the distribution of unknowns $\mathbf{x}$. In view of the equality $\mathbf{b} = A\mathbf{x}$, we have

$$D = 1 - 2 \mathbf{E} \langle \Gamma R^T A \rangle + \mathbf{E} \langle A^T R \Gamma^2 R^T A \rangle + q \mathbf{E} \langle R \Gamma^2 R^T \rangle. \quad (4)$$

In further transformations, we use a simple property of normal distributions: if a random value r is normally distributed with zero average and the variance σ^2, then for any differentiable function $f(\cdot)$, we have

$$\mathbf{E} r f(r) = \sigma^2 \mathbf{E} \frac{\partial f}{\partial r}. \quad (5)$$

For example, consider the second term in (3). Apply (7) to the normal entries of A. We obtain

$$\mathbf{E} \langle \Gamma R^T A \rangle = n^{-1} \mathbf{E} A_{ij} (\Gamma R^T)_{ji} = \frac{a}{n^2} \mathbf{E} \frac{\partial (\Gamma R^T)_{ji}}{\partial A_{ij}},$$

where (and in the following) the summation over repeated indexes is implied.

All entries of random matrix R have the identical variance $(a+p)/n$. Therefore,

$$\mathbf{E} \langle \Gamma R^T R \rangle = n^{-1} \mathbf{E} R_{ij} (\Gamma R^T)_{ji} = \frac{a+p}{n^2} \mathbf{E} \frac{\partial (\Gamma R^T)_{ji}}{\partial R_{ij}}.$$

The matrix ΓR^T depends only on $R = A + \delta A$, and the partial derivatives with respect to elements A and R coincide. Comparing these two latter expressions, we find that

$$\mathbf{E} \langle \Gamma R^T A \rangle = a(a+p)^{-1} \mathbf{E} \langle \Gamma R^T R \rangle.$$

Similarly, we derive the relation

$$\mathbf{E}\langle R^T R \Gamma^2 R^T A \rangle = a(a+p)^{-1} \mathbf{E}\langle R^T R \Gamma^2 R^T R \rangle.$$

Once again, we use (5) with respect to random A and obtain

$$\mathbf{E}\langle A^T R \Gamma^2 R^T A \rangle = \frac{a}{n^2} \mathbf{E} A_{ij} \frac{\partial (R \Gamma^2 R^T)_{jk}}{\partial A_{kj}} + a \mathbf{E}\langle R \Gamma^2 R^T \rangle,$$

where (and in the following) $k = 1, \ldots, N$. Further, we use (5) with respect to entries of δA. It follows that

$$\mathbf{E}\langle A^T R \Gamma^2 R^T \delta A \rangle = \frac{p}{n^2} \mathbf{E} A_{ji} \frac{\partial (R \Gamma^2 R^T)_{jk}}{\partial (\delta A)_{ki}}.$$

Adding two latter equalities we obtain

$$\mathbf{E}\langle A^T R \Gamma^2 R^T R \rangle = \frac{a+p}{n^2} \mathbf{E} A_{ji} \frac{\partial (R \Gamma^2 R^T)_{jk}}{\partial A_{ki}} + a \mathbf{E}\langle R \Gamma^2 R^T \rangle.$$

Now we combine the five latter equalities and get the equation

$$\mathbf{E}\langle A^T R \Gamma^2 R^T A \rangle = \frac{a^2}{(a+p)^2} \mathbf{E}\langle R^T R \Gamma^2 R^T R \rangle + \frac{ap}{a+p} \mathbf{E}\langle R \Gamma^2 R^T \rangle. \tag{6}$$

These relations are sufficient to express D in terms of the observed variables only. Substitute (6) to (4). We obtain

$$D = 1 - 2 \frac{a}{a+p} \mathbf{E}\langle \Gamma R^T R \rangle + \frac{a^2}{(a+p)^2} \mathbf{E}\langle R^T R \Gamma^2 R^T R \rangle + s \mathbf{E}\langle R \Gamma^2 R^T \rangle, \tag{7}$$

where $s = ap/(a+p) + q$. Passing to the (random) system of coordinates in which $R^T R$ is diagonal, we can replace formally

$$\mathbf{E} \, n^{-1} \mathrm{tr} \, \varphi(R^T R) = \mathbf{E} \, n^{-1} \sum_{i=1}^{n} \varphi(\lambda_i) = \int \varphi(u) \, dF(u),$$

where λ_i are eigenvalues of matrices $R^T R$. The expression (3) follows from (7). The proof of Proposition 1 is complete. $\square$

Note that the integrand in (3) is quadratic with respect to $\gamma(u)$ and allows the standard minimization for each $u > 0$ independently on $F(u)$.

Denote $\theta = a/(a + p)$.

COROLLARY. *The expression* (3) *reaches the minimum in the class $\mathfrak{K}$ for* $\gamma(u) = \gamma_{\text{opt}}(u) = \theta/(\theta^2 u + s)$ *with*

$$\widehat{\mathbf{x}} = \widehat{\mathbf{x}}_{\text{opt}} = \frac{\theta}{\theta^2 R^T R + sI} R^T \mathbf{y},$$

and the minimum value of (1) *is*

$$D = D_{\text{opt}} = D(\widehat{\mathbf{x}}_{\text{opt}}) = \int \frac{s}{s + \theta^2 u} \, dF(u).$$

Note that functions $\gamma(u)$ minimizing (3) do not depend on the unknown function $F(u)$. Therefore, the optimal solution $\widehat{\mathbf{x}}_{\text{opt}}$ is expressed in terms of only observable variables in spite of the fact that the minimum risk remains unknown. In Section 6.2, it will be shown that the function $F(u)$ allows a simple limit expression as $n \to \infty$, $N \to \infty$, and $n/N \to c < 1$.

Conclusions

Thus, we can draw a conclusion that the solution of empirical linear algebraic equations with normal errors can be stabilized and made more accurate by replacing standard minimum square solution by an optimal on the average regular estimator

$$\widehat{\mathbf{x}} = \alpha(R^T R + tI)^{-1} R^T \mathbf{y}, \tag{8}$$

with the "ridge" parameter $t = s/\theta^2$ and a scalar scaling coefficient $\alpha = 1/\theta$. Empirical ridge parameters are often used for stabilizing solutions of the regression and discriminant problems (see Introduction). But the problem of finding optimum ridge parameters is solved only for special cases.

The multiple $\alpha \geq 1$ presents an *extension coefficient* in contrast to shrinkage multiples for well-known Stein estimators (see Chapter 2). The matrix $\alpha(R^T R + tI)^{-1}$ in (8) is, actually, a scaled and regularized estimator of $(A^T A)^{-1}$ optimal with respect to the minimization of D. Such combined "shrinkage-ridge" solutions are characteristic of a number of extremum problems solved in the theory of high-dimensional, multivariate statistical analysis (see Chapter 4).

We may compare the optimum value D_{opt} with the square risk of the standard solution D_{std} if we set $\gamma(u) = 1/u$ in (3). Then,

$$D_{\text{std}} = \int [(1 - \theta)^2 + \frac{s}{u}] \, dF(u). \tag{9}$$

It is easy to prove that the difference $D_{\text{std}} - D_{\text{opt}} \geq 0$, and it is equal to 0 for $q = 0$ and $p = 0$.

For $p = 0$ and $q > 0$, we have $\theta = 1$, the matrix R is nonrandom, $\alpha = 1$, $t = q$, and D_{opt} is less D_{std} due to the factor $u/(u + q)$ in the integrals with respect to $dF(u)$.

If $p > 0$ or $q > 0$, the expression (9) contains a term proportional to

$$\int u^{-1} \, dF(u) = \mathbf{E}n^{-1}\text{tr}W^{-1} = \text{const}\,\frac{1}{N - n - 1},$$

where W is the well-known Wishart matrix calculated for variables distributed as $\mathbf{N}(0, N(a + p)/n)$ over a sample of size N. For $N = n + 1$, this integral is singular, and the quadratic risk of the standard solution is infinitely large, while $D_{\text{opt}} \leq 1$ for our optimal solution.

6.2.   ASYMPTOTICALLY UNIMPROVABLE SOLUTION

V. L. Girko first used the increasing-dimension asymptotics (1973) for the investigation of random SLAE, in which the number of unknowns n increases along with the number of equations N so that $n/N \to c > 0$ and found limit spectral distributions arising in this asymptotics. He found that, under uniformly small variance of coefficients, these limit spectra do not depend on distribution of coefficients. In monographs [24] and [25], V. L. Girko applied the increasing-dimension asymptotics for studying ridge solutions of SLAE and obtained some limit equations for expected solution vectors in terms of parameters: matrices A and vectors $\mathbf{b}$. His main result was to derive asymptotically unbiased estimators of the solution vectors. He also found unbiased estimators for the vectors $(A^T A + tI)^{-1} A\mathbf{b}$, $t > 0$.

We consider a class $\mathfrak{K}$ of pseudosolutions that present linear combinations of ridge solutions $\mathbf{x} = (R^T R + tI)^{-1} R^T \mathbf{y}$, $t > 0$. Our goal is to calculate the principal part of the quadratic risk of such pseudosolutions in the increasing-dimension asymptotics, and then to minimize it, obtaining algorithms that are asymptotically not worse than any pseudosolutions from $\mathfrak{K}$. The main results were obtained in [55]–[58]. The theory is developed as follows.

1. We consider a sequence of problems of random SLAE solution

$$\mathfrak{P} = (N, n, A, \mathbf{x}, \mathbf{b}, R, \mathbf{y}, \widehat{\mathbf{x}})_n, \quad n = 1, 2, \ldots \tag{1}$$

(we drop the indexes n for arguments of $\mathfrak{P}$), in which the system $A\mathbf{x} = \mathbf{b}$ of N equations with n unknowns is solved, $\mathbf{x} \in \mathbb{R}^n$, $\mathbf{b} \in \mathbb{R}^N$, using the observed matrix $R = A + Xi$, where Ξ is a nuisance matrix, and $\mathbf{y} = \mathbf{b} + \vec{\nu}$ is an observed right-hand vector with a nuisance vector $\vec{\nu}$, $\widehat{\mathbf{x}}$ being the pseudosolution.

2. We consider the parametrically defined family $\mathfrak{K}$ of estimators $\widehat{\mathbf{x}}$ of $\mathbf{x}$ (the class of estimation algorithms) depending on n, N, and on some vector of parameters or on some function that we call an "estimation function."

3. In the asymptotics $N/n \to 1 + \kappa$, $\kappa > 0$, we calculate the limit quadratic risk of estimators from $\mathfrak{K}$.

4. We solve an extremum problem and find an optimal nonrandom estimation function guaranteeing the limit quadratic risk not greater than the quadratic risk of any algorithm from $\mathfrak{K}$. Thus, we find an estimator *asymptotically dominating* $\mathfrak{K}$.

5. Then, we construct a statistics (a consistent estimator) approximating the nonrandom optimal function (or parameters) that defines the pseudosolution algorithm.

6. It remains to prove that thus constructed algorithm leads to a quadratic risk not greater than any algorithm from $\mathfrak{K}$.

In this chapter, we carry out this program completely only for two-parametric "shrinkage-ridge" estimators of unknown vector that have the form $\widehat{x} = \alpha \, (R^T R + tI)^{-1} R^T \mathbf{y}$. For arbitrary linear combinations of ridge estimators, we carry out only items 1–5 of the program.

In the first section, we develop the asymptotic theory of spectral properties of random Gram matrices $R^T R$. In the increasing-dimension asymptotics, the leading parts of spectral functions and of the resolvent are isolated, and it is proved that the variance of the corresponding functionals is small (Lemmas 6.1–6.4). We isolate the principal parts of expectations (Theorem 6.1).

Then, we pass to the limit and deduce the basic "dispersion" equations relating limit spectra of matrices $R^T R$ to spectra of $A^T A$ (Theorem 6.2). We study the analytic continuation of the limit normed trace of the resolvent and find spectra bounds and the form of spectral density (Theorem 6.3).

We investigate the quadratic risk of pseudosolutions in the form of linear combinations of ridge solutions with different ridge parameters. The leading part of the quadratic risk is isolated (Theorem 6.4), and its consistent estimator is suggested (Theorem 6.5). Then, we perform the limit transition (Theorem 6.6).

We pass to the analytical continuation of spectral functions and solve the limit extremum problem (Theorem 6.7). Then, we construct an ε-consistent estimator of the best limit estimation function (Theorem 6.8).

As a special case, we study two-parametric pseudosolutions of the shrinkage-ridge form, characterised by a regularizing "ridge parameter" and the "shrinkage coefficient." Theorem 6.9 provides the limit quadratic risk of these solutions under a double-limit transition: as the order of equations increases and the regularization parameter decreases.

We separate proofs from the main text and place them at the end of Chapter 6.

Spectral Functions of Large Gram Matrices

We consider a sequence (1) in which the matrices A and R are of size $N \times n$, $\Xi = R - A$, and the square matrices $\Sigma = A^T A$, $S = R^T R$ are of size $n \times n$ as $n \to \infty$, and $N \to \infty$. The random matrices Ξ have independent components (we do not write the indexes n and N). Let A_{mj}, R_{mj}, Ξ_{mj} denote entries of matrices A, R, Ξ, $m = 1, \dots, N$, $j = 1, \dots, n$.

Assume that $\mathfrak{P}$ satisfies the following requirements (let constants c and c with subscripts do not depend on n).

1. For each n, the inequality holds $n \leq c_0 N$ (c_1 is reserved for the lowest boundary of limit spectra).

2. For each n, the norm (only spectral norms of matrices are used) is bounded by a constant: $\|A\| \leq c_2$.

3. As $n \to \infty$, we have $\kappa_n = N/n - 1 \to \kappa \geq 0$, $\kappa < c_3$.

4. For each n, the entries Ξ_{mj} of Ξ are independent and normally distributed as $\mathbf{N}(0, d/n)$, $m = 1 \dots, N$, $j = 1, \dots, n$. Let $d = c_4 > 0$ do not depend on n.

Under these conditions, the entries of A have the order of magnitude $1/\sqrt{n}$ on the average and are comparable with standard deviations for entries of the matrix R.

Let us apply methods developed in spectral theory of random matrices of increasing dimension (Chapter 3). We consider the resolvent

$$H = H(t) = (S + tI)^{-1}, \quad t \geq 0$$

and spectral function depending on H. First, we establish the fact of small variance for some spectral functions. For this purpose, we use the method of one-by-one exclusion of independent variables.

Let $\mathrm{var}_m(X)$, $X = (X_1,\ldots,X_N)$, denote the conditional variance calculated by integration over the distribution of only one variable X_m, $m = 1,\ldots,N$.

Remark 1. Let $X = (X_1,\ldots,X_N)$ be a vector with independent components, and let a function $f(X)$ be of the form $f(X) = f^m(X) + \delta_m(X)$, where $f^m(X)$ does not depend on X_m, $m = 1,\ldots,N$. Suppose that two first moments of $f(X)$ and $f^m(X)$ exist, $m = 1,\ldots,N$. Then,

$$\mathrm{var}\, f(X) \le \sum_{m=1}^{N} \mathbf{E}\mathrm{var}_m(\delta_m(X)).$$

This assertion is a consequence of the Burkholder inequality. Denote

$$\Sigma = A^T A, \quad S = R^T R, \quad S^m = S - \mathbf{r}_m \mathbf{r}_m^T, \quad H^m = (S^m + tI)^{-1},$$

and $\Phi_m = \mathbf{r}_m^T H^m \mathbf{r}_m$, $m = 1,\ldots,N$. Consider n-dimensional vectors $\mathbf{a}_m = \{A_{mj}\}$, $\mathbf{r}_m = \{R_{mj}\}$, $j = 1,\ldots,n$, $m = 1,\ldots,N$. Note that $\mathbf{a}_m^2 \le c_2^2$, $\mathbf{E}\mathbf{r}_m^2 \le c_2^2 + d$. We exclude the independent variables $\mathbf{r}_m$ one-by-one.

Remark 2. Let n and N be fixed, $n \le N$, $\|A\| \le c_2$ and $t \ge c > 0$. Under Assumption 4, the variance

$$\mathrm{var}_m \Phi_m \le ad(c_2^2 + d)\, c^{-2} n^{-1}, \quad m = 1,\ldots,N.$$

LEMMA 6.1. *Under Assumptions 1–4 for $t \ge c > 0$, the variance*

$$\mathrm{var}\big(n^{-1}\mathrm{tr}H(t)\big) \le kN/n^3, \quad \text{where} \quad k = ad(c_2^2 + d)[c^2 + (c_2^2 + d)^2]/c^6$$

and a is a numeric coefficient.

LEMMA 6.2. *Under Assumptions 1–4 for any nonrandom unity vectors* **e** *as* $n \to \infty$ *uniformly in* $t \geq c > 0$, *we have*

$$\mathrm{var}(\mathbf{e}^T H(t)\mathbf{e}) = O(n^{-1}).$$

LEMMA 6.3. *Under Assumptions 1–4 for any nonrandom unity vectors* $\mathbf{e}, \mathbf{f} \in \mathbb{R}^N$ *as* $n \to \infty$ *uniformly with respect to* $t \geq c > 0$, *we have*

$$\mathrm{var}\big(\mathbf{e}^T RH(t)R^T\mathbf{f}\big) = O(n^{-1}).$$

LEMMA 6.4. *Under Assumptions 1–4 for* $t, t' \geq c > 0$ *and any nonrandom unity vectors* $\mathbf{e} \in \mathbb{R}^N$ *as* $n \to \infty$ *for* $t, t' \geq c > 0$ *uniformly, we have*

$$\mathrm{var}[\mathbf{e}^T R(S + tI)^{-1}(S + t'I)^{-1}R^T\mathbf{e}] = O(n^{-1}).$$

Denote $h_n = \mathbf{E}n^{-1}\mathrm{tr}H$, $s_n = 1 + h_n d$.

LEMMA 6.5. *Under Assumptions 1–4 as* $n \to \infty$ *for* $t \geq c > 0$ *uniformly, we have*

$$\mathbf{E}HR^T = \mathbf{E}HA^T/s_n + \Omega,$$

where $\|\Omega\| = O(n^{-1})$.

LEMMA 6.6. *Under Assumptions 1–4 as* $n \to \infty$ *uniformly for* $t \geq c > 0$, *we have*

$$\mathbf{E}HR^T R = \mathbf{E}HR^T A + d(\kappa_n + h_n t)\mathbf{E}H + O(n^{-1}).$$

Now we derive the equation connecting spectra of random matrices $S = R^T R$ of large dimension with spectra of nonrandom matrices $\Sigma = A^T A$.

THEOREM 6.1. *Under Assumptions 1–4 as $n \to \infty$ uniformly with respect to $t \geq c > 0$, we have*

1. $h_n = (1 + h_n d) \, n^{-1} \mathrm{tr}(A^T A + r_n s_n I)^{-1} + O(n^{-1})$,
2. $\mathbf{E}H = s_n (A^T A + r_n s_n I)^{-1} + \Omega_n$,

where $h_n = \mathbf{E}n^{-1}\mathrm{tr}H$, $s_n = 1 + h_n d$, $r_n = ts_n + \kappa_n d$, and $\|\Omega_n\| = O(n^{-1})$ (in spectral norm).

Limit Spectral Functions of Gram Matrices

Assume additionally that the weak convergence holds

$$F_{0n}(u) \stackrel{def}{=} n^{-1} \sum_{j=1}^{n} \mathrm{ind}(\lambda_{0j} \leq u) \to F_0(u), \quad u \geq 0, \qquad (2)$$

where λ_{0j} are eigenvalues of the matrices $A^T A$, $j = 1, \ldots, n$.

THEOREM 6.2. *Under Assumptions 1–4 and (2), the following statements are valid.*

1. *For $t \geq c > 0$, the convergence in the square mean*

$$h_n = \mathbf{E}n^{-1}\mathrm{tr}H(t) \to h(t), \quad s_n = 1 + h_n(t)d \to s(t),$$
$$r_n = ts_n(t) + \kappa_n d \to r(t)$$

holds uniformly in $t \geq c > 0$, where $s(t) = 1 + dh(t)$ and $r(t) = ts(t) + \kappa d$.

2. *For each $t > 0$, the equation holds*

$$h(t) = s(t) \int (u + r(t)s(t))^{-1} \, dF_0(u). \qquad (3)$$

3. *As $n \to \infty$, the weak convergence in probability holds*

$$F_n(u) \stackrel{def}{=} n^{-1} \sum_{j=1}^{n} \mathrm{ind}(\lambda_j \leq u) \stackrel{P}{\to} F(u), \quad u > 0,$$

where λ_j are eigenvalues of the matrices $R^T R$, $j = 1, \ldots, n$.

4. *For each $t > 0$, the equation holds*

$$h(t) = \int (u + t)^{-1} \, dF(u). \tag{4}$$

5. *As $n \to \infty$, we have*

$$\mathbf{E}H(t) = s(t)\left(A^T A + r(t)s(t)\right)^{-1} + \Omega_n,$$

where $\|\Omega_n\| = O(n^{-1})$ uniformly in $t \geq c > 0$.

The limit equation (3) is called *the dispersion equation* for random matrices $R^T R$ of increasing dimension.

Consider the support $\mathfrak{S} = \{v > 0 : \ F'(v) > 0\}$. A set of $z = -v$, where $v \in \mathfrak{S}$, may be called the *limit spectrum region* of the matrices $R^T R$.

THEOREM 6.3. *Under Assumptions 1–4 and (2), the following is true.*

1. *The function $h(t)$ defined by (3) for $t > 0$ allows the analytic continuation to complex arguments z. The function $h(z)$ is regular on the plane of complex z, and inside of any compact not containing points $z = -v$ such that $v \in \mathfrak{S}$ is uniformly bounded and has a uniformly bounded derivative.*

2. *For any $v = -\mathrm{Re}\, z > 0$ and $\varepsilon = -\mathrm{Im}\, z \to +0$, there exists the limit $\lim \pi^{-1}\mathrm{Im}\, h(z) = F'(v)$.*

3. *For $v > 0$, $F'(v) \leq \pi^{-1}(vd)^{-1/2}$.*

4. *For $\kappa > 0$, the set of points $\mathfrak{S}$ is located on the segment $[v_1, v_2]$, where*

$$v_1 = \frac{\kappa^2 d}{4(\sqrt{1 + \kappa} + 1)^2}, \quad v_2 = 2d(2 + \kappa) + 2.25 \, c_2,$$

while the diameter of $\mathfrak{S}$ is not less than $\pi d\kappa/[4 \, (1 + \sqrt{1 + \kappa})]$.

5. *For $\kappa > 0$ for all $v \geq 0$, the function $F'(v)$ satisfies the Lipschitz condition with the exponent $1/3$ and is differentiable*

everywhere except, maybe, of the boundary points of the set $\mathfrak{S}$.

6. *For $\kappa > 0$, the function $h(z)$ satisfies the Hölder condition on the whole plane of complex z with the exponent $1/3$.*

Special cases

Now we study characteristic features of limit spectra of the matrices $S = R^T R$ for a special case when $F_0(v) = \mathrm{ind}(v \geq a^2)$. In particular, it is true if for each n the matrix $\Sigma = A^T A = a^2 I$. From (3), we obtain the equation

$$ths^2 + h(a^2 - d) + \kappa hds = 1, \qquad (5)$$

where $h = h(t)$, $s = s(t) = 1 + hd$. Introduce the parameter of the "signal-to-noise ratio" $T = a^2/d$.

1. If $d = 1$ and the ratio $T = 0$, then matrices $R^T R$ are the Wishart matrices, $(s-1)(\kappa - vs) = 1$, and one can find the density

$$F'(v) = (2\pi v)^{-1} \sqrt{(v_2 - v)\,(v - v_1)},$$

where $v_1 = (\sqrt{1+\kappa} - 1)^2$, $v_2 = (\sqrt{1+\kappa} + 1)^2$, $v_1 \leq v \leq v_2$. This is the well-known Marchenko–Pastur distribution (it holds, in particular, for limit spectra of increasing sample covariance matrices). For $\kappa > 0$, the limit spectrum is separated from zero, and for $\kappa = 0$, it is located on the segment $[0, 4]$.

2. If $\kappa = 0$ and $T > 0$, then at the spectrum points as $\varepsilon \to +0$, we find from (3) that $2vx = |s - 1|^{-2}d$, where $x = \mathrm{Re}\ s$. Using equation (4), one can calculate the limit density $F'(v) = \pi^{-1} \mathrm{Im}\ h(v)$. It is of the form

$$F'(v) = \frac{1}{\pi}\left(x\,\frac{3x - 2 - 2(T-1)\,(1-x)^2}{1 + 2x\,(T-1)} \right)^{1/2},$$

where the parameter $x = \mathrm{Re}\ s \geq 1/2$ can be determined from the equation

$$v = \frac{1 + 2x\,(T-1)}{2x\,(1-2x)^2}.$$

Let $T \leq 1$. Then, the left boundary of the limit spectrum defined by the distribution function $F(v)$ is fixed by the point $v = 0$.

Let $T = 1$. Then, $F'(v) = \pi^{-1}\sqrt{x\ (3x - 2)}$, where x is defined by the equation $2x\ (1 - 2x)^2\ v = 1$. In this case, the limit spectrum is located on the segment $[0, 27/4]$. For small v, the function $F'(v) \approx 2\pi^{-1}\sqrt{3}\ v^{-1/3}$, and $F''(27/4) = -\infty$.

Let $T = 1 + \varepsilon$, where $\varepsilon > 0$ is small. Then, the lower spectrum boundary is located approximately at the point $4/27\ \varepsilon^3$; the spectrum is located within the boundaries increasing with T.

Quadratic Risk of Pseudosolutions

We study estimators of the solutions to SLAE $A\mathbf{x} = \mathbf{b}$ in the form of linear combinations of regularized minimum square solutions such that vectors $\mathbf{x} = (x_1, \ldots, x_n)$ are estimated from the known random matrices of coefficients $R = A + \Xi$ of size $N \times n$ and random right-hand side vectors $\mathbf{y} = \mathbf{b} + \vec{\nu}$. We consider a class $\mathfrak{K}$ of estimators of the form

$$\hat{\mathbf{x}} = \Gamma R^T \mathbf{y}, \quad \text{where } \Gamma = \Gamma(R^T R) = \int_{t \geq c} (R^T R + tI)^{-1}\, d\eta(t), \quad (6)$$

and $\eta(t)$ is a function having the variation on $[0, \infty)$ not greater than $V = V(\eta)$, and such that $\eta(t) = 0$ for $t < c$.

Consider a sequence of the solution problems (1), in which N equations are solved with respect to n unknown variables as $n \to \infty$ and $N = N(n) \to \infty$. The function $\eta(t)$ and the quantity $c > 0$ do not depend on n. Assume that the following conditions are fulfilled (constants c with indexes do not depend on n).

A. As $n = 1, \ldots, \infty$ $\kappa_n = N/n - 1 \to \kappa \geq 0$.

B. For each n, the system of equations $A\mathbf{x} = \mathbf{b}$ is soluble; the vector $\mathbf{b}$ lengths are not greater than c_5, spectral norms of

$\|A\| \leq c_2$, and the vectors $\mathbf{x}$ are normed so that (the scalar square) $\mathbf{x}^2 = 1$.

C. For each n, the entries Ξ_{mj} of matrices Ξ and components $\vec{v}_m$ of vectors $\vec{v}$ are independent, mutually independent, and normally distributed as $\Xi_{mj} \sim \mathbf{N}(0, d/n)$, $\vec{v}_m \sim \mathbf{N}(0, q/n)$, $j = 1, \ldots, n$, $m = 1, \ldots, N$, where $d = c_4 > 0$ and $q = c_6$ do not depend on n.

Under these conditions for fixed constants $c_1 - c_6$, entries of the matrices A and components of $\mathbf{b}$ on the average are of the order of magnitude $n^{-1/2}$ and are comparable with standard deviations for entries of R and components of $\mathbf{y}$.

We set the minimization problem for the quantity

$$\Delta_n(\eta) \stackrel{def}{=} \mathbf{E}(\mathbf{x} - \widehat{\mathbf{x}})^2 = \mathbf{E}[\mathbf{x}^2 - 2\mathbf{x}^T\Gamma R^T \mathbf{y} + \mathbf{y}^T R\Gamma^2 R^T \mathbf{y}]$$

with the accuracy up to terms small as $n \to \infty$.

Denote $H = H(t) = (R^T R + tI)^{-1}$, $h_n = h_n(t) = \mathbf{E}n^{-1}\mathrm{tr}H(t)$.

LEMMA 6.7. *Under Assumptions A–C for $t \geq c > 0$, we have*

$$\mathbf{E}RHR^T R = \mathbf{E}RHR^T A + d(\kappa_n + h_n t)\mathbf{E}RH + \omega, \qquad (7)$$

where $|\omega| \leq adc^{-1/2}n^{-1}$, and a is a numeric coefficient.

Denote

$$\widehat{h}_n(t) = n^{-1}\mathrm{tr}H(t), \quad \widehat{s}_n(t) = 1 + d\widehat{h}_n(t), \quad \widehat{r}_n(t) = t\,\widehat{s}_n(t) + \kappa_n d,$$

$$\widehat{\varphi}_n(t) = (\mathbf{y}^2 - \mathbf{y}^T RH(t)R^T \mathbf{y}), \quad \widehat{g}_n(t) = \frac{\widehat{\varphi}_n(t) - (t\,\widehat{h}_n(t) + \kappa_n)q}{\widehat{r}_n(t)},$$

$$r_n = r_n(t) = \mathbf{E}\widehat{r}_n(t), \quad s_n = s_n(t) = \mathbf{E}\widehat{s}_n(t),$$

$$\varphi_n = \varphi_n(t) = \mathbf{E}\widehat{\varphi}_n(t), \quad g_n = g_n(t) = \mathbf{E}\widehat{g}_n(t). \qquad (8)$$

Remark 3. Under Assumptions A–C for fixed $c_1 - c_4$ and $t \geq c > 0$, the uniform convergence as $n \to \infty$ holds

$$\mathrm{var}\,\widehat{\varphi}_n(t) = O(n^{-1}), \quad \mathrm{var}\,\widehat{g}_n(t) = O(n^{-1}).$$

Remark 4. Under Assumptions A–C as $n \to \infty$,

$$g_n(t) = \frac{\varphi_n(t) - (th_n(t) + \kappa_n)\, q}{t\,(1 + h_n(t)d) + \kappa_n d} + O(n^{-1}).$$

LEMMA 6.8. *Under Assumptions A–C as $n \to \infty$ uniformly in $t \geq c > 0$, we have*

$$g_n(t) = \mathbf{E}\,\frac{\mathbf{x}^T H(t)\mathbf{x}}{s_n(t)} + O(n^{-1}) = \mathbf{x}^T \Sigma (\Sigma + r_n(t)s_n(t)I)^{-1}\mathbf{x} + O(n^{-1}),$$

$$\varphi_n(t) = \frac{r_n(t)}{s_n(t)}\,\mathbf{E}\mathbf{x}^T \Sigma H(t)\mathbf{x} + q(th_n(t) + \kappa_n) + O(n^{-1}) =$$

$$= r_n(t)\mathbf{x}^T \Sigma (\Sigma + r_n(t)s_n(t)I)^{-1}\mathbf{x} + q(th_n(t) + \kappa_n) + O(n^{-1}),$$

where $r_n(t) = ts_n(t) + \kappa_n d$.

Denote

$$K_n(t, t') = \frac{\varphi_n(t) - \varphi_n(t')}{t - t'} = \mathbf{E}\mathbf{y}^T R H(t)H(t')R^T \mathbf{y}.$$

THEOREM 6.4. *Under Assumptions A–C, the quantity*

$$D_n(\eta) \overset{\text{def}}{=} 1 - 2\int g_n(t)\,d\eta(t) + \iint K_n(t, t')\,d\eta(t)\,d\eta(t') \qquad (9)$$

is such that for $\widehat{\mathbf{x}}$ of the form (6) as $n \to \infty$, the quadratic risk is

$$\Delta_n(\eta) = \mathbf{E}(\mathbf{x} - \widehat{\mathbf{x}})^2 = D_n(\eta) + O(n^{-1}).$$

Theorem 6.4 allows to calculate the quadratic risk of regularized pseudosolutions and to search for the minimum in subclasses of functions $\eta(t)$. For example, let $\eta(t) = \alpha\,\text{ind}(t' > t)$, $t \geq c > 0$ (shrinkage-ridge estimator). Then, the minimum of (9) is reached when $\alpha = g_n(t)/\varphi_n(t)$ and $2g_n'(t)\varphi_n'(t) = \varphi_n''(t)$. This last equation may be used to determine the t providing the minimum of $\Delta_n(t) = 1 - g_n^2(t)/\varphi_n(t)$. In another class, we consider linear combination of such estimators with a priori magnitudes of t and find an unimprovable set of coefficients by solution of a system of a small number of linear equations.

Let us construct an unbiased estimator of the leading part of the quadratic risk that was found in Theorem 6.4. Define

$$\widehat{K}_n(t,t') = \frac{\varphi_n(t) - \varphi_n(t')}{t - t'} = \mathbf{y}^T R H(t) H(t') R^T \mathbf{y},$$

$$\widehat{D}_n(\eta) = 1 - 2 \int \widehat{g}_n(t) \, d\eta(t) + \int \int \widehat{K}_n(t,t') \, d\eta(t) \, d\eta(t').$$

Now we prove the unbiasedness of this estimator.

THEOREM 6.5. *Under Assumptions A–C for $\widehat{\mathbf{x}}$ of the form (6), we have*

$$\widehat{D}_n(\eta) = D(\eta) + \widehat{\omega}_n, \quad \text{where} \quad \mathbf{E}\widehat{\omega}_n^2 = O(n^{-1}).$$

Thus, under Assumptions A–C for $\widehat{\mathbf{x}}$ of the form (6) as $n \to \infty$ in the square mean, we have $\widehat{D}_n(\eta) \to (\mathbf{x} - \widehat{\mathbf{x}})^2$ in the square mean.

Theorems 6.4 and 6.5 establish upper boundaries of bias and variance of the quadratic risk estimator $\widehat{D}_n(\eta)$. These theorems are remarkable by that they allow to compare the efficiency of different versions of solution algorithms (defined by the function $\eta(\cdot)$) using a *single realization* of the random system.

Limit relations

Assume that in $\mathfrak{P}$ the convergence (2) holds.

Let us pass to the limit. To provide the convergence of functions of R and $\mathbf{x}$, we introduce an additional condition. In addition to Assumptions 1–4, suppose that for all v the limit exists

$$G(v) = \lim_{n \to \infty} \sum_{i=1}^{n} x_j^2 \, \text{ind}(\lambda_j^0 \leq v), \tag{10}$$

where λ_j^0 are eigenvalues of the matrices $\Sigma = A^T A$, and x_j are components of the vector $\mathbf{x}$ in a system of coordinates where Σ is diagonal. By condition B, we have $G(c_2) = \mathbf{x}^2 = 1$.

LEMMA 6.9. *Under Assumptions A–C, (2), and (10) for $t, t' \geq c > 0$ as $n \to \infty$, the uniform convergence holds*

$$h_n(t) \to h(t) \overset{def}{=} s(t) \int (v + r(t)s(t))^{-1} \, dF_0(v),$$

$$g_n(t) \to g(t) \overset{def}{=} \int v(v + r(t)s(t))^{-1} \, dG(v),$$

$$\varphi_n(t) \to \varphi(t) \overset{def}{=} r(t)g(t) + q(th(t) + \kappa),$$

$$K_n(t, t') \to K(t, t') \overset{def}{=} \frac{\varphi(t) - \varphi(t')}{t - t'},$$

where the last expression is continuously extended to $t = t'$.

THEOREM 6.6. *Under Assumptions A–C, (2), and (10) for the pseudosolutions $\widehat{\mathbf{x}} = \Gamma R^T \mathbf{y}$, the quadratic risk $\Delta_n(\eta)$ converges to the limit*

$$\lim_{n \to \infty} \Delta_n(\eta) = \lim_{n \to \infty} D_n(\eta) =$$
$$= D(\eta) \overset{def}{=} 1 - 2 \int g(t) \, d\eta(t) + \iint K(t, t') \, d\eta(t) \, d\eta(t'). \quad (11)$$

This statement is a direct consequence of the uniform convergence in Lemma 6.9 and the variance decrease in Remark 3.

The formula (11) expresses the limit quadratic risk of all pseudosolutions $\widehat{\mathbf{x}} = \Gamma(R^T R) R^T \mathbf{y}$ from $\mathfrak{K}$. The right-hand side of (11) presents a quadratic function of $\eta(t)$ and allows an obvious minimization. For example, one may search for the minimum in a class of stepwise functions $\eta(t)$ that present a sum of a small number of jumps of different size.

Minimization of the Limit Risk

Let us minimize (11) in the most general class of all functions $\eta(t)$ of bounded variation. To obtain an explicit form of the extremal solution, we pass to complex parameters and present the limit risk (11) in the form of function of only one argument.

LEMMA 6.10. *Under Assumptions A–C, (2), (10) for $\kappa > 0$ the functions $g(z)$ and $\varphi(z)$ are regular for all z on the plane of complex z except, perhaps, points of the half-axis $z < 0$.*

Consider the scalar function corresponding to the matrix expression in (6)

$$\Gamma(z) = \int_{t \geq c > 0} (t + z)^{-1} \, d\eta(t).$$

Let $\mathfrak{L} = (-i\infty, +i\infty)$ be the integration contour.

LEMMA 6.11. *Under Assumptions A–C, (2), and (10) for $\kappa > 0$, the limit expression in (11) equals*

$$D(\eta) = 1 - \frac{1}{\pi i} \int_{\mathfrak{L}} g(z)\, \Gamma(-z) dz - \frac{1}{2\pi i} \int_{\mathfrak{L}} \varphi(z)\, \Gamma^2(-z) dz. \quad (12)$$

The minimization of (12) remains difficult due to the fact that functions $\Gamma(-z)$ are complex. We pass to real $\Gamma(\cdot)$. We notice that the function $g(z)$ may be singular in the left half-plane of complex z for negative z. Actually, for example, let $\mathbf{x}$ be an eigenvector of the matrix $\Sigma = A^T A$ having only one eigenvalue λ for all n. Then, the function $g(z) = (\lambda + w)^{-1}\lambda$ has a pole at $w = r(z)s(z) = -\lambda$. To provide the smoothness of $g(z)$ and $\varphi(z)$, we introduce an additional assumption.

Suppose that for all z and z' with $\operatorname{Im} z \neq 0$ and $\operatorname{Im} z' \neq 0$, the Hölder condition holds

$$|g(z) - g(z')| \leq c_5 |z - z'|^{\varsigma}, \quad (13)$$

where c_5 and $0 < \varsigma < 1$ do not depend on z and z'.

Under this assumption for each $\operatorname{Re} z = -v < 0$ and $\varepsilon = -\operatorname{Im} z \to +0$, the limits exist $g(-v) = \lim g(-v - i\varepsilon)$ and $\varphi(-v) = \lim \varphi(-v - i\varepsilon)$.

Remark 5. Let $\kappa > 0$ and for each n in the system of coordinates where the matrix $\Sigma = A^T A$ is diagonal, components of $\mathbf{x} = (x_1, \ldots, x_n)$ are uniformly bounded: $x_j^2 = O(n^{-1})$, $j = 1, \ldots, n$. Then, function $g(z)$ satisfies the Hölder condition (13).

LEMMA 6.12. *Let conditions A–C, (2), (10), and (13) be fulfilled and $\kappa > 0$. Then,*

$$D(\eta) = 1 - \frac{2}{\pi} \int_{\mathfrak{S}} \operatorname{Im} g(-v)\,\Gamma(v)dv -$$

$$\frac{1}{\pi} \int_{\mathfrak{S}} \operatorname{Im} \varphi(-v)\,\Gamma^2(v)dv + O(\varepsilon),$$

where $\mathfrak{S} = \{v > 0 : F'(v) > 0\}$, $\operatorname{Im} g(-v) \geq 0$, $\operatorname{Im} \varphi(-v) \leq 0$.

Denote by $\lambda_0 \geq 0$ the smallest eigenvalue of the matrices $\Sigma = A^T A$, for $n = 1, 2, \ldots$.

LEMMA 6.13. *Let conditions A–C, (2), (10), and (13) be fulfilled, $\kappa > 0$, and $\lambda_0 > 0$. Then, for $z = -v < 0$, where $v \in \mathfrak{S}$, we have*

$$|\operatorname{Im} \varphi(z)| \geq c_6 > 0, \qquad \left| \frac{\operatorname{Im} g(z)}{\operatorname{Im} \varphi(z)} \right| \leq B \overset{def}{=} \frac{2}{\sqrt{\lambda_0 v_1}} + \frac{\kappa d}{\lambda_0 v_1}. \quad (14)$$

Denote

$$\rho(v) = -\pi^{-1}\operatorname{Im} \varphi(-v) \geq 0, \quad \Gamma^0(v) = -\operatorname{Im} g(-v)/\operatorname{Im} \varphi(-v) \geq 0. \quad (15)$$

For $\kappa > 0$ and $\lambda_0 \geq c_1 > 0$, the function $\Gamma^0(v)$ is defined and uniformly bounded everywhere for $v \in \mathfrak{S}$. It is easy to examine that

$$\int_{\mathfrak{S}} \rho(v)dv = \lim_{n \to \infty} \frac{1}{\pi} \int_{\mathfrak{S}} \operatorname{Im} \varphi_n(-v)\,dv =$$

$$= \lim_{n \to \infty} \frac{1}{\pi} \mathbf{E}\mathbf{y}^T RR^T \mathbf{y} \leq (c_3 + c_1 c_5)(c_2 + d).$$

As a consequence, from Lemmas 6.11–6.13, we obtain the following statement.

THEOREM 6.7. *Let conditions A–C, (2), (10), and (13) be fulfilled, $c_1 > 0$, and $\kappa > 0$. Then,*

$$D(\eta) = \lim_{n \to \infty} \mathbf{E}(\mathbf{x} - \Gamma(R^T R)R^T \mathbf{y})^2 = D^0 + \int_{\mathfrak{S}} [\Gamma(v) - \Gamma^0(v)]^2 \rho(v)dv,$$

where

$$D^0 = 1 - \int_{\mathbb{S}} [\Gamma^0(v)]^2 \, \rho(v) dv.$$

Thus, under these conditions, the estimator $\widehat{\mathbf{x}} = \Gamma^0(R^T R)R^T \mathbf{y}$ as $n \to \infty$ asymptotically dominates the class of estimators $\mathfrak{K}$.

However, the function $\Gamma^0(v)$ is determined by parameters and is unknown to the observer. As an estimator $\Gamma^0(v)$, we consider the ε-regularized statistics

$$\widehat{\Gamma}^0_{n\varepsilon}(v) = \min\left[\frac{\operatorname{Im} \widehat{g}_n(z)}{\operatorname{Im} \widehat{\varphi}_n(z)}, \quad B\right],$$

where $B \geq \sup_{v \in \mathbb{S}} \Gamma^0(v)$ is from (14), and $z = -v - i\varepsilon$, $v, \varepsilon > 0$.

We define the variance for complex variables Z: $\operatorname{Var} Z = \mathbf{E}\,|Z - \mathbf{E}Z|^2$.

LEMMA 6.14. *Under Assumptions A–C for any fixed z with $\operatorname{Im} z < 0$, as $n \to \infty$, we have*

$$\operatorname{Var} \widehat{h}_n(z) = \operatorname{Var}[n^{-1}\operatorname{tr}(R^T R + zI)^{-1}] = O(n^{-2})$$
$$\operatorname{Var} \widehat{r}_n(z) = O(n^{-2}), \quad \operatorname{Var} \widehat{\varphi}_n(z) = O(n^{-1}),$$
$$\operatorname{Var} \widehat{g}_n(z) = O(n^{-1}).$$

THEOREM 6.8. *Let conditions A–C, (2), (10), (13), $c_1 > 0$, and $\kappa > 0$ be fulfilled. Then,*

$$\lim_{\varepsilon \to +0} \lim_{n \to \infty} \int_{\mathbb{S}} \mathbf{E}[\Gamma^0(v) - \widehat{\Gamma}^0_{n\varepsilon}(v)]^2 \rho(v) dv = 0. \qquad (16)$$

Thus, we proved that the statistics $\widehat{\Gamma}^0_{n\varepsilon}(\cdot)$ involved in the algorithm $\widehat{\mathbf{x}} = \widehat{\Gamma}^0_{n\varepsilon}(R^T R) \, R^T \mathbf{y}$ approximates the unknown estimation function $\Gamma^0(\cdot)$ of the asymptotically dominating algorithm $\widehat{\mathbf{x}} = \Gamma^0(R^T R)R^T \mathbf{y}$. The question of the limit quadratic risk provided by the function $\widehat{\Gamma}^0_{n\varepsilon}(v)$ requires further investigation.

Shrinkage-Ridge Pseudosolution

Let us study a subclass of the class of pseudosolutions $\mathfrak{K}$ that have the form

$$\widehat{\mathbf{x}} = \alpha (R^T R + tI)^{-1} R^T \mathbf{y}, \quad t, \alpha \geq 0, \tag{17}$$

where $\alpha > 0$ and $t > 0$ do not depend on n. The coefficients $\alpha > 0$ may be called "shrinkage coefficients" similarly to the Stein estimators, and the parameters t may be called "ridge parameters" similarly to the well-known regularized estimators of the inverse covariance matrices.

For pseudosolutions (17), we have

$$(\mathbf{x} - \widehat{\mathbf{x}})^2 = 1 - 2\alpha\, \widehat{g}_n(t) + \alpha^2 \widehat{\varphi}_n'(t) + w_n(t), \quad \text{where } \mathbf{E}|w_n(t)|^2 = O(n^{-1}).$$

Remark 6. For any $t > 0$ and $\alpha > 0$, the pseudosolutions (17) have the limit quadratic risk

$$D(\eta) = D(\alpha, t) = \lim_{n \to \infty} (\mathbf{x} - \widehat{\mathbf{x}})^2 = 1 - 2\alpha\, g(t) + \alpha^2\, \varphi'(t),$$

$$D(\alpha^0, t) = 1 - g^2(t)/\varphi'(t), \quad \text{where } \alpha^0 = g^2(t)/\varphi'(t). \tag{18}$$

The case $\alpha = 1$ and $t = 0$ corresponds to the standard minimum square solution. For $\alpha = \alpha^0 = g(t)/\varphi'(t)$, we have the minimum $D(\alpha^0, t) = 1 - g^2(t)/\varphi'(t)$.

Now we find the limit quadratic risk of regularized solutions (17) under the decreasing regularization: let $\kappa > 0$ and $t \to +0$.

LEMMA 6.15. *Under Assumptions A–C, conditions* (2), (10), *and* $\kappa > 0$, *the following is true.*

1. *For* $t > 0$, *the equation* (3) *has a unique solution* $h(t)$, *and there exists the limit* $h(0) = \lim h(t)$ *for* $t \to +0$ *such that*

$$\frac{1}{\kappa d} \leq h(0) \leq \frac{1}{(1 + \kappa)\, d + c_2}.$$

Denote

$$s(0) = 1 + h(0)d, \quad b_\nu = \int \frac{v}{[v + \kappa d s(0)]^\nu}\, dG(v), \quad \nu = 1, 2.$$

2. *The limits exist*

$$g(0) = \lim_{t \to +0} g(t) = b_1 \le 1, \quad \varphi(0) = \lim_{t \to +0} \varphi(t) = \kappa(g(0)d + q).$$

3. *For* $t \to +0$, *the derivative exists*

$$s'(0) = \frac{ds(t)}{dt}\bigg|_{t=0} = -\frac{dh^2(0)}{1 + \kappa d^2} \frac{s^2(0)}{h^2(0)}.$$

4. *For* $t \to +0$, *there exist the derivatives*

$$g'(0) = -b_2[s^2(0) + \kappa d \ s'(0)];$$

$$\varphi'(0) = \left[b_1 - \frac{\kappa d \ s(0)}{1 + \kappa d^2 \ h^2(0)} b_2 \right] s(0) + q h(0),$$

and $\varphi'(0) \ge K(\kappa) \ c_1$, *where* $K = K(\kappa) > 0$ *with the right-hand side derivative* $K'(0) > 0$.

Denote by $\widehat{a}^0 = \widehat{a}^0(t) = \widehat{g}_n(t)/\widehat{\varphi}'_n(t)$ an estimator of the function $a^0 = a^0(t) = g(t)/\varphi'(t)$, where $\widehat{g}_n(t)$ is defined by (8). The derivative $\widehat{\varphi}'_n(t) = \mathbf{y}^T R H^2(t) R^T \mathbf{y}, \ t > 0$.

Consider the ridge pseudosolutions (estimators of $\mathbf{x}$) of the form

$$\widehat{\mathbf{x}}_t = (R^T R + tI)^{-1} R^T \mathbf{y}, \quad \widehat{\mathbf{x}}_t^0 = \widehat{a}^0 \ (R^T R + tI)^{-1} R^T \mathbf{y},$$
$$\text{as} \ t \to +0.$$

THEOREM 6.9. *Under Assumptions of Theorem 6.8, we have*

$$\lim_{t \to +0} \lim_{n \to \infty} \mathbf{E}(\mathbf{x} - \widehat{\mathbf{x}}_t)^2 = 1 - 2g(0) + \varphi'(0) \overset{def}{=} D(1,0),$$

$$\lim_{t \to +0} \operatorname*{plim}_{n \to \infty} (\mathbf{x} - \widehat{\mathbf{x}}_t^0)^2 = 1 - g^2(0)/\varphi'(0) \overset{def}{=} D(a^0, 0). \quad (19)$$

Here the first relation presents the limit quadratic risk of the minimum square solution $\widehat{\mathbf{x}}_t$ with decreasing regularization. The pseudosolution $\widehat{\mathbf{x}}_t^0$ may be called asymptotically optimal shrinkage minimum square solution with decreasing regularization. Denote

$$\Lambda_\nu = \int v^{-\nu} dF_0(v), \quad \nu = 1, 2.$$

Remark 7. Under assumptions of Theorem 6.8 as $\kappa \to 0$, we have $k_1(d - c_2)/\kappa \leq D(1,0) \leq k_2/\kappa$, where k_1 and k_2 are constants.

Thus, for sufficiently large "noise" d, the limit quadratic risk of the regularized minimum square solution can be indefinitely large. Finite risk may be guaranteed either by a ridge regularization (fixed $t > 0$) or by an excessive number of equations (for sufficiently large portion of additional equations κ).

Special form of SLAE

Now we investigate in more detail a special form of equations when for all n the matrices $\Sigma = I$, but the fact that the equations have such trivial form is unknown to the observer. For this case, we have $F_0(v) = \text{ind}(v \geq 1)$, and the functions $g(t)$, $\varphi(t)$, as well as the limit quadratic risk do not depend on $\mathbf{x}$ for all n since $\mathbf{x}^2 = 1$ and $G(v) = F_0(v)$. Let us write out the limit relations. To be concise, denote $h = h(t)$, $s = s(t)$, and $r = r(t)$. The dispersion equation (3) takes on the form

$$1 + rs = d + 1/h. \tag{20}$$

The functions $g = g(t) = h/s$ and $\varphi = \varphi(t) = hr/s + q(ts + \kappa)$.

Let us compare the quadratic risks (18).

First, suppose the "noise level" is low: $0 < d < 1$. Let $t \to +0$. Then, it is easy to find that $h \to (1 - d)^{-1}$, $s \to (1 - d)^{-1}$, $g \to 1$, and $\varphi'(0) = (1 + q)/(1 - d)$. By Remark 6, the limit quadratic risk of standard regularized solution with decreasing regularization is $\lim_{n \to \infty} D(1, t) = (d + q)/(1 - d)$. Under the optimum shrinkage, we have $D(\alpha^0, t) \to (d + q)/(1 + q)$. One can see that as $t \to +0$, the decrease of regularization leads to the degeneration, while the shrinkage leads to a finite risk $D(\alpha^0, 0) < 1$.

Suppose $d \geq 1$. Let $t \to +0$. Then, solving equation (20) we obtain infinite $h = h(t)$ and $t^{1/3}h(t) \to d^{-2/3}$. The functions $g = g(t) \to 1$ and $\varphi = \varphi(t) \approx (d + q)\, d^{-2/3}t^{2/3}$. The quadratic risk without shrinkage is $D(1, t) \approx 2/3\,(d+q)d^{-2/3}\,t^{-1/3}$ increases infinity, while the optimum shrinkage risk does not exceed 1.

Now we investigate the dependence of the quadratic risk on the excessiveness of SLAE, that is, on the parameter $\kappa > 0$. Let $d = 1$. Then, equation (20) is simply $rhs = 1$. If $\kappa > 0$ and $t \to +0$ the function $h = h(t)$ remains finite and tends to $2(\sqrt{4 + \kappa} + \sqrt{\kappa})^{-1/2}\,\kappa^{-1/2}$. Set $t = 0$. Then, as $\kappa \to 0$ it is easy to see that the quadratic risk $D(1, 0) \approx q\kappa^{-1/2}$ while $D(\alpha^0, 0) \approx 1$.

Setting $d = 1$ and $q = 0$, we study the effect of regularization. Then, we have $rhs = 1$ that is the cubic equation with respect to $h(t)$. Functions $g = g(t) = r/s$, and $\varphi'(t) = 2h^2/[1 + 3h - \kappa h^2 s]$. The regularized solution $\hat{\mathbf{x}}_t = (R^T R + tI)^{-1}R^T\mathbf{y}$ leads to the limit risk

$$D(1, t) = D(h) \overset{def}{=} \frac{1 - h}{s} + \frac{2h^2}{1 + 3h - \kappa sh^2}.$$

The limit risk of the regularized solution $\hat{\mathbf{x}}_t^0 = \alpha\,\hat{\mathbf{x}}_t$ with the best limit shrinkage coefficient $\alpha = \alpha^0 = g(t)/\varphi'(t)$ is equal to

$$D(\alpha^0, t) = D^0(h) \overset{def}{=} 1 - \frac{1 + 3h - \kappa sh^2}{2s^2} < 1.$$

This last expression presents the function of h that has a unique minimum for some $h = h^0 \leq 1$. In view of the monotone dependence of $h = h(t)$, there exists a unique minimum of the function $D(\alpha^0, t)$ in t, that defines the *asymptotically unimprovable shrinkage-ridge solution*.

In the case when $\Sigma = I$, $d = 1$, and $\kappa > 0$, we can calculate the best limit function (15) analytically. It is

$$\Gamma^0(v) = \frac{\text{Im } g(-v)}{\text{Im } \varphi(-v)} = \frac{1}{1 + q}\,\frac{1}{v\,|s|^2 - \kappa},$$

where complex $s = s(-v)$. The asymptotically unimprovable solution $\widehat{\mathbf{x}}^0 = \Gamma^0(R^T R)\, R^T\mathbf{y}$ leads to the limit risk

$$D^0 = 1 - \frac{1}{1+q}\int \frac{1}{|s|^2(v\,|s|^2 - \kappa)}\; F'(v)dv,$$

where $F'(v)$ is the limit spectrum density for $R^T R$. If $\kappa = 0$, then s is a (complex) root of the equation $vs^2(s-1)+1 = 0$. Denote $x = \mathrm{Re}\; s$. Then, $2x(2x-1)^2 v = 1$, $2x = |s|^4$. The spectral density $F'(v) = \pi^{-1}\sqrt{3x^2 - 2x}$, $x \geq 2/3$, and the spectrum is located on the segment $[0, 27/4]$. For $q = 0$, the limit quadratic risk is $D^0 \approx 0.426$.

Proofs for Section 6.2

We preserve all notations and the numeration of formulas, lemmas, and theorems from the above.

Remark 2 (proof).
Indeed, denote $\vec{\xi}_m = \mathbf{r}_m - \mathbf{a}_m$. It is obvious that

$$\mathrm{var}_m \Phi_m = \mathrm{var}_m(2\mathbf{a}_m^T\Omega\, \vec{\xi}_m^T + \vec{\xi}_m^T\Omega\, \vec{\xi}_m), \quad \text{where } \Omega = \mathbf{E}H^m.$$

Here the variance of the first addend is $4d\, \mathbf{a}_m^T\Omega^2\mathbf{a}_m/n \leq 4d\, c_2/c^2 n$. The variance of the second addend is $2d^2\mathrm{tr}\,\Omega^2/n \leq 2d/c^2 n$. Thus,

$$\mathrm{var}_m \Phi_m \leq ad(c_2 + d)\, c^{-2}n^{-1}.$$

Remark 2 holds true.

LEMMA 6.1 (proof).
We have the identity $H\mathbf{r}_m = (1 + \Phi_m)^{-1}H^m\mathbf{r}_m$. By Remark 1

$$\mathrm{var}(n^{-1}\mathrm{tr}H) \leq \frac{1}{n^2}\sum_{m=1}^{N}\mathbf{E}\mathrm{var}_m\left(\frac{\mathbf{r}_m^T\Omega^2\mathbf{r}_m}{1 + \Phi_m}\right),$$

where $\Omega = H^m$ does not depend on $\mathbf{r}_m$ and $\|\Omega\| \leq 1/c$. The right-hand side of this inequality equals $Nn^{-2}\mathbf{E}\mathrm{var}_m(\psi\varphi)$, where

$\psi = \mathbf{r}_m^T \Omega^2 \mathbf{r}_m$, $\varphi = (1 + \Phi_m)^{-1}$. We use the fact that if any two random values ψ and φ have appropriate moments and $|\varphi| \leq 1$, then

$$\text{var}(\psi\varphi) \leq 2 \text{ var } \psi + 4\mathbf{E}\psi^2\text{var } \varphi.$$

First, we estimate $\text{var}_m\psi = \text{var}_m(2\mathbf{a}_m^T\Omega^2 \vec{\xi}_m + \vec{\xi}_m^T\Omega^2 \vec{\xi}_m)$, where the matrix $\Omega = H^m$ is fixed. The variance of the first addend is not greater than $4d \; n^{-1}\mathbf{a}_m^T\Omega^4\mathbf{a}_m$. The variance of the second addend is not greater than $2d^2 \; n^{-1}\text{tr } \Omega^4$, and $\mathbf{a}_m^2 \leq \|A\| \leq c_2$. We may conclude that $\text{var}_m\psi \leq \leq ad(c_2 + d) \; c^{-4}n^{-1}$, where a is a numeric coefficient.

Further, it is obvious that $\text{var}_m\varphi \leq \text{var}_m\Phi_m$. Let us calculate the expectation integrating only over the distribution $\vec{\xi}_m \sim \mathbf{N}(0, d/n)$. We obtain

$$\mathbf{E}\psi^2 \leq \mathbf{E}(\mathbf{r}_m^2)^2/c^4 = \left[(\mathbf{a}_m^2)^2 + 2\mathbf{a}_m^2d + 2d^2\right]/c^4 \leq 2 \; (c_2 + d)^2/c^4.$$

Consequently,

$$\text{var}(\psi\varphi) \leq ad(c_2 + d)(c^2 + (c_2 + d)^2)/nc^6.$$

The Lemma 6.1 statement follows. $\square$

LEMMA 6.2 (proof).

Denote additionally $v_m = \mathbf{e}^T H^m\mathbf{r}_m$, $u_m = \mathbf{e}^T H\mathbf{r}_m$. Using the identity $v_m = (1 + \Phi_m)u_m$ and Remark 1, we find that

$$\text{var}(\mathbf{e}^T H(t)\mathbf{e}) \leq \mathbf{E} \sum_{m=1}^{N} \text{var}_m\left(\frac{v_m^2}{1 + \Phi_m}\right). \tag{21}$$

Denote $\varphi = (1 + \Phi_m)^{-1}$, $\mathbf{f}_m = H^m\mathbf{e}$, $p_m = \mathbf{e}^T H^m\mathbf{a}_m$. We have

$$\text{var}_m(v_m^2\varphi) \leq 2 \text{ var}_m(v_m^2) + 2\mathbf{E}_m v_m^4\text{var}_m\varphi$$

where the expectation $\mathbf{E}_m$ is calculated by integrating only over the distribution of $\vec{\xi}_m$. We obtain that $|\mathbf{f}_m| = O(1)$, $p_m = O(1)$, and

$$\text{var}_m(v_m^2) = \text{var}_m\left(2p_m\ \mathbf{f}_m^T\vec{\xi}_m + (\mathbf{f}_m^T\vec{\xi}_m)^2\right) \le$$
$$\le a|\mathbf{f}|^2(p_m^2 d/n + |\mathbf{f}|^2 d^2/n^2) = p_m^2 O(1) + O(n^{-2});$$
$$\mathbf{E}_m v_m^4 \le a\ [p_m^4 + \mathbf{E}(\mathbf{f}^T\vec{\xi}_m)^4] \le O(n^{-1})\ p_m^2 + O(n^{-2}),$$

where a is a number. In the former lemma, we have shown that $\text{var}_m\varphi = O(n^{-1})$. Consequently, the left-hand side of (21) is not greater than

$$\sum_m [O(n^{-1})\ p_m^2 + O(n^{-2})].$$

To prove Lemma 6.2, it suffices to show that the sum of p_m^2 is bounded. Since $H\mathbf{r}_m = (1 + \Phi_m)^{-1}H^m\mathbf{r}_m$, we have

$$p_m^2 \le \mathbf{E}_m(e^T H^m \mathbf{r}_m)^2 = \mathbf{E}_m v_m^2 = \mathbf{E}_m(1 + \Phi_m)^2\ u_m^2.$$

Notice that

$$\sum_m u_m^2 = e^T H R^T R H e \le 1/c.$$

Denote

$$\delta_m = \Phi_m - \mathbf{E}\Phi_m = 2\mathbf{a}_m^T\Omega\ \vec{\xi}_m + (\vec{\xi}_m^T\Omega\ \vec{\xi} - \mathbf{E}\vec{\xi}_m^T\Omega\ \vec{\xi}_m),$$
$$\Delta_m = (1 + \Phi_m)^2 - \mathbf{E}(1 + \Phi_m)^2 = 2(1 + \mathbf{E}\Phi_m)\ \delta_m + \delta_m^2,$$

where $\Omega = \mathbf{E}H^m$. The expectation of Φ_m^2 is bounded, and the other moments of Φ_m, u_m and $\vec{\xi}_m^T\Omega\ \vec{\xi}_m$ exist. We substitute $(1 + \Phi_m)^2 = O(1) + \Delta_m$ and estimate the contribution of the addend Δ_m. Note that $u_m^2\Delta_m \le 1/2(u_m^2 + \Delta_m^2 u_m^2)$. In the right-hand side of this equality, the sum of first summands is bounded. It remains to show that $\mathbf{E}\Delta_m^2 u_m^2 = O(n^{-1})$, or that $\mathbf{E}\Delta_m^4 = O(n^{-2})$. Let us substitute $\Delta_m = O(1)\delta_m + \delta_m^2$. Passing to the coordinate system in which Ω is diagonal and using the normality of $\vec{\xi}_m$, one can easily check that all moments of $\sqrt{n}\delta_m$ are bounded by constants not depending on n. Consequently, $\mathbf{E}\Delta_m^4 = O(n^{-2})$. It follows that the sum of p_m^2 is bounded. We arrive at Lemma 6.2 statement. $\square$

LEMMA 6.3 (proof).

Using the rotation invariance of the problem setting, let us rotate the axes so that the vectors $\mathbf{e}$ become parallel to one of the axis. Then, for $l, m = 1, \ldots, N$ as $n \to \infty$, it is required to prove that

$$\operatorname{var}\left(\mathbf{r}_l^T H(t)\mathbf{r}_m\right) = O(n^{-1}).$$

First, consider the case when $l = m$. Denote $\Psi_m = \mathbf{r}_m^T H \mathbf{r}_m$, $\Phi_m = \mathbf{r}_m^T H^m \mathbf{r}_m$. For $N = 1$, $\Phi_m = 1/t$. The identity $H = H^m - H^m \mathbf{r}_m \mathbf{r}_m^T H$ implies var $\Psi_m \leq$ var Φ_m. In view of the independence H^m of $\mathbf{r}_m$, we have

$$\operatorname{var} \Phi_m = \operatorname{var}(\mathbf{r}_m^T \Omega\ \mathbf{r}_m) + \mathbf{E}[\mathbf{r}_m^T (H^m - \Omega)\ \mathbf{r}_m]^2,$$

where $\Omega = \mathbf{E}H^m$. Here the first variance in the right-hand side is estimated similarly to $\operatorname{var}_m \psi$ in Lemma 6.1 and is not greater than $ad(c_2 + d)\ c^{-2}n^{-1}$. We estimate the second term using Lemma 6.2 (for fixed $\mathbf{r}_m$), since the matrix H^m is of the form of the matrix H with smaller number of variables. We find that it is $O(n^{-1})$ with the coefficient $\mathbf{E}(\mathbf{r}_m^2)^2 = O(1)$. Thus, we proved the lemma statement for $l = m$.

Now, let $l \neq m$, $N > 1$. We isolate the variables $\mathbf{r}_l$ and $\mathbf{r}_m$. Denote

$$\widetilde{S} = S - \mathbf{r}_l\mathbf{r}_l^T - \mathbf{r}_m\mathbf{r}_m^T, \quad \widetilde{H} = (\widetilde{S} + tI)^{-1} = H - \widetilde{H}\mathbf{r}_l\mathbf{r}_l^T H - \\ -\widetilde{H}\mathbf{r}_m\mathbf{r}_m^T H, \quad \Phi_{lm} = \mathbf{r}_l^T \widetilde{H}\mathbf{r}_m^T, \quad \Psi_{lm} = \mathbf{r}_l H \mathbf{r}_m,$$

where $\widetilde{H}$ does not depend on $\mathbf{r}_l$ and $\mathbf{r}_m$. We obtain the identity

$$\Psi_{lm} = \Phi_{lm} - \Phi_{ll}\Psi_{lm} - \Phi_{lm}\Psi_{mm}. \tag{22}$$

As in the proof of Lemma 6.2, we obtain (with N less by 1) that var $\Phi_{ll} = O(n^{-1})$, var $\Psi_{mm} = O(n^{-1})$, while $\Psi_{mm} \leq 1$,

$\Psi_{lm} \leq \Psi_{ll}\Psi_{mm} \leq 1$, and $\Phi_{lm} = O(1)$, $l, m = 1, \ldots, N$. From (22) it follows that

$$(1 + \mathbf{E}\Phi_{ll})\Psi_{lm} = (1 - \mathbf{E}\Psi_{mm})\Phi_{lm} + \omega,$$

where $\mathbf{E}\omega^2 = O(n^{-1})$. Estimating the expected squares of deviations Ψ_{lm} and Φ_{lm}, we find that var $\Psi_{lm} = $ var $\Phi_{lm} + O(n^{-1})$. In view of the independence of $\widetilde{H}$ from $\mathbf{r}_l$ and $\mathbf{r}_m$, we have

$$\text{var } \Phi_{lm} = \mathbf{E}\mathbf{r}_l^2\mathbf{r}_m^2 \text{ var}(\mathbf{e}^T\widetilde{H}\mathbf{e}) + \text{var}(\mathbf{r}_l^T\Omega\mathbf{r}_m),$$

where $\mathbf{e}$ does not depend on $\mathbf{r}_l$ and $\mathbf{r}_m$, and $\Omega = \mathbf{E}\widetilde{H}$. Here the first summand is $O(n^{-1})$ by Lemma 6.2. The second summand is

$$\text{var}(\mathbf{a}_l^T\Omega\,\vec{\xi}_m + \vec{\xi}_l^T\Omega\,\mathbf{r}_m + \xi_l^T\Omega\,\vec{\xi}_m),$$

where the first two summand expectations are $O(n^{-1})$, and the variance of the third summand is not greater than $2d^2\text{tr }\Omega^2/n = O(n^{-1})$. We conclude that var $\Phi_{lm} = O(n^{-1})$ and, consequently, var $\Psi_{lm} = O(n^{-1})$. Lemma 6.3 is proved. $\square$

LEMMA 6.4 (proof). The proof is similar to that of Lemma 6.3.

LEMMA 6.5 (proof). We use the relation

$$\mathbf{E}\psi(x)\, x = \mathbf{E}\psi(x)\mu + \sigma^2\mathbf{E}\,\frac{\partial\psi(x)}{\partial x} \tag{23}$$

that is valid for $x \sim \mathbf{N}(\mu, \sigma^2)$ for any differentiable function $\psi(x)$ if there exist $\mathbf{E}|\psi(x)|$ and $\mathbf{E}|x\psi(x)|$. For $x = R_{mi}$, we have

$$\mathbf{E}(HR^T)_{im} = \mathbf{E}H_{ij}A_{mj} + \frac{d}{n}\mathbf{E}\,\frac{\partial H_{ij}}{\partial R_{mj}} \tag{24}$$

(we mean the summation over repeated indexes). Differentiating we obtain

$$\mathbf{E}\,\frac{\partial H_{ij}}{\partial R_{mj}} = -\mathbf{E}(H^2R^T)_{im} - \mathbf{E}\text{tr}H(HR^T)_{im}. \tag{25}$$

Here (spectral) norms

$$\|HR^T\| \le \sqrt{\lambda_{\max}(HR^TRH)} \le c^{-1/2},$$
$$\|H^2R^T\| \le \sqrt{\lambda_{\max}(H^2R^TRH^2)} \le c^{-3/2}.$$

It follows that

$$\mathbf{E}(1 + \frac{d}{n} \, \mathrm{tr}H)HR^T = \mathbf{E}HA^T + \Omega,$$

where $\|\Omega\| = O(n^{-1})$. But $\|HR^T\| = O(1)$, and in view of Lemma 6.1, we have $\mathrm{var}(n^{-1}\mathrm{tr}H) = O(n^{-2})$. It follows that the expectation in the left-hand side can be replaced by the product of expectations with the accuracy up to $O(n^{-1})$. The statement of Lemma 6.5 follows. □

LEMMA 6.6 (proof).
Let us use (23) with $x = R_{mk}$. We find that

$$\mathbf{E}(HR^TR)_{ik} = \mathbf{E}(HR^TA)_{ik} + \frac{d}{n}\mathbf{E}\frac{\partial H_{ij}R_{mj}}{\partial R_{mk}}$$

(we mean the summation over j and m). Here the derivative

$$\mathbf{E}\frac{\partial(H_{ij}R_{mj})}{\partial R_{mk}} = N\mathbf{E}H_{ik} - \mathbf{E}(HR^TRH)_{ik} - \mathbf{E}H_{ik}\mathrm{tr}(R^TRH). \quad (26)$$

Substituting $R^TRH = I - tH$ and differentiating with respect to R_{mj} in (26), we obtain the principal part with the additional terms $\mathbf{E}dn^{-1}H(I-tH)$, and $\Omega = \mathbf{E}(Hn^{-1}\mathrm{tr}H) - \mathbf{E}H \cdot \mathbf{E}n^{-1}\mathrm{tr}H$. The first of these is $O(n^{-1})$ in norm. Now we estimate the matrix in the subtrahend. Denote by $\mathbf{e}$ the eigenvector corresponding to the largest eigenvalue of this symmetric matrix. Let us apply the Cauchy–Bunyakovskii inequality. We find that the second term in norm is not greater than the square root of the product $\mathrm{var}(n^{-1}\mathrm{tr}H)$. $\mathrm{var}(\mathbf{e}^T H\mathbf{e})$. Here the first multiple is $O(n^{-2})$ by Lemma 6.1, while the second one is not greater than $1/c$. We obtain the Lemma 6.6 statement. □

THEOREM 6.1 (proof).

We start from the expression established in Lemma 6.6. Substitute $HR^T R = I - tH$ and use (24) in the left-hand side. Rearranging the summands we find that

$$I = \mathbf{E}H[tI + (1 + h_n d)^{-1} A^T A + \kappa_n dI + h_n dt I] + \Omega,$$

where $\|\Omega\| = O(n^{-1})$. Inverting the matrix we obtain the both statements of Theorem 6.1. $\square$

THEOREM 6.2 (proof).

The convergence of $F_{0n}(u)$ implies the convergence

$$h_{0n}(t) \stackrel{def}{=} \int (u + t)^{-1} \, dF_{0n}(u) \to h_0(t), \quad t \geq c > 0.$$

We rewrite Statement 2 of Theorem 6.1 as

$$h_n / s_n = n^{-1} \text{tr}(A^T A + uI)^{-1} + O(n^{-1}) \text{ with } u = r_n s_n, \quad (27)$$

where the remainder term is small uniformly in $t \geq c > 0$ as $n \to \infty$. We have $u = r_n(t) s_n(t) \geq c > 0$. It is not difficult to obtain the estimate $a_{nm}(s_n - s_m) = o(1)$ as $n, m \to \infty$, where all the coefficients $a_{nm} \geq (ds_n s_m)^{-1} \geq 1/d$. It follows that we have the uniform convergence $s_n(t) \to s(t)$, $h_n(t) \to h(t) = (s(t)-1)/d$, and $r_n(t) \to r(t) = ts(t)+\kappa d$. By Lemma 6.1 uniformly $\text{var}(n^{-1} \text{tr} H(t)) \to 0$. It follows that all these functions converge in the square mean (Statements 1 and 2). Relation (3) follows.

Further, we use V. L. Girko's theorems on the convergence of spectral functions of random matrices. By Theorem 3.2.3 from [21], the convergence $h_n(t) \to h(t)$ for all $t > 0$ implies the weak convergence $F_n(u)$ almost surely (Statement 3) and the relation between $h(t)$ and $F(u)$ (Statement 4). Also from Theorem 6.1, Statement 5 follows easily. The proof of Theorem 6.2 is complete. $\square$

THEOREM 6.3 (proof).

Starting from (4) one can easily establish the possibility of analytical continuation of the limit function $h(z)$, its uniform boundedness, and boundedness of its derivative. Let $-\text{Re } z = v > \delta > 0$

and assume the contrary: there exists some sequence of z such that $|h(z)| \to \infty$. Then, the ratio $h(z)/s(z)$ tends to a positive constant, and the integral in the right-hand side of (3) vanishes, which is not true. It follows that the values $h(z)$ are uniformly bounded. The first statement is proved.

Statement 2 immediately follows from (4).

Next, to be concise, denote $h = h(z)$, $s = s(z) = 1 + h(z)d$, $r = r(z) = zs_{(z)} + \kappa d$, $s_0 = \operatorname{Re} s(z)$, and $s_1 = \operatorname{Im} s(z)$, and consider the function of $\varepsilon > 0$

$$\mu_\nu = \mu_\nu(z) = \int |v + rs|^{-2} v^\nu \, dF_0(v), \quad \nu = 1, 2. \qquad (28)$$

We fix some $v = -\operatorname{Re} z > 0$ and tend $\varepsilon = -\operatorname{Im} z \to +0$.

Let $v \in \mathfrak{S}$ (at the spectrum). Then, $\operatorname{Im} h(z) \to \pi F'(v) > 0$. Divide both parts of (3) by s, take imaginary parts, and divide by $\operatorname{Im} h > 0$. We find that

$$1/\mu_0 = p|s|^2 d, \quad \text{where} \quad p = 2vs_0 - \kappa d, \quad v \in \mathfrak{S}, \qquad (29)$$

where $|s| \geq s_1 \to d\pi F'(v) > 0$. Now let us equate imaginary parts of the left- and right-hand sides of (3) and divide by $\operatorname{Im} h$. We find that

$$1/\mu_0 = d\mu_1/\mu_0 + d|s|^2 v + s_0|s|^2 \varepsilon/\operatorname{Im} h.$$

For $v \in \mathfrak{S}$ and $\varepsilon \to +0$, the last summand presents $O(\varepsilon)$. From these two equations, it follows that

$$|s|^2 (p - v) = \mu_1/\mu_0 + O(\varepsilon), \quad v \in \mathfrak{S}. \qquad (30)$$

Since $|s| \geq |s_1| \to d\pi F'(v) > 0$ and $v > 0$, we infer that $p > v + O(\varepsilon)$.

Further, applying the Cauchy–Bunyakovskii inequality to (3), we derive the inequality $|h|^2 \leq \mu_0|s|^2$. Substituting μ_0 from (28), we obtain $|h|^2 pd \leq 1 + O(\varepsilon)$ and as $\varepsilon \to +0$ we have

$$|s - 1|^2 \leq d/p \leq d/v, \quad v \in \mathfrak{S}. \qquad (31)$$

In particular, it follows that for $v \in \mathfrak{S}$, the value $\mathrm{Im}\, h \leq 1/\sqrt{vd}$, and $F'(v) \leq \pi^{-1}(vd)^{-1/2}$. This is Statement 3 of our theorem.

Let us find spectrum bounds defined by the function $F(v)$. From (30) and (31), the inequality follows $(2s_0 - 1)\, v \geq \kappa d + O(\varepsilon)$ and $v\,(1 + 2\sqrt{d/v}) \geq \kappa d + O(\varepsilon)$. Solving the quadratic equation with respect to $\sqrt{v}$, we find that $v \geq v_1 + O(\varepsilon)$, where

$$v_1 \overset{def}{=} \frac{\kappa^2 d}{4(\sqrt{1+\kappa}+1)^2}, \quad v \in \mathfrak{S}. \tag{32}$$

Now note that $\mu_1 \leq \mu_0 c_2$, where c_2 is the upper bound of the matrices $A^T A$ spectrum. Equation (30) implies

$$(2s_0 - 1)v - \kappa d = p - v \leq c_2\, |s|^{-2} + O(\varepsilon), \quad v \in \mathfrak{S}. \tag{33}$$

We have $s_0 \geq 1 - \sqrt{d/v} + O(\varepsilon)$. Let $v > 9d$. Then, $s_0 \geq 2/3 + O(\varepsilon)$ and $|s|^2 \geq 4/9 + O(\varepsilon)$. From (33), it follows that $(2s_0 - 1)\, v \leq 9/4\, c_2 + \kappa d + O(\varepsilon)$. Finding the lower boundary of s_0 from (31), we obtain the inequality $v - 2\sqrt{dv} \leq \kappa d + 9/4\, c_2 + O(\varepsilon)$. Let us solve this quadratic inequality with respect to $\sqrt{v}$ as $\varepsilon \to +0$. We find that $v \leq 2d\,(2 + \kappa) + 2.25\, c_2 + O(\varepsilon)$. This relation provides the upper spectrum bound in the theorem formulation. Spectrum bounds v_1 and v_2 are established.

Now we integrate both parts of the inequality over $v \in \mathfrak{S}$, where $F'(v) \leq \pi^{-1}(vd)^{-1/2}$, and conclude that $\pi\sqrt{d}/2 \leq \sqrt{v_{20}} - \sqrt{v_{10}}$, where $v_{20} \geq v_{10} \geq 0$ are (unknown) extreme points of the set $\mathfrak{S}$. Thus, we find the lower estimate of the set $\mathfrak{S}$ diameter shown in Statement 4.

Let us establish the Hölder inequality for $s(z)$. In view of (4) for $\varepsilon = \mathrm{Im}\, z \neq 0$, the function $h(z)$ is differentiable. Let $v = -\mathrm{Re}\, z \in \mathfrak{S}$. Differentiating the left- and right-hand sides of (3), we find the following equation for the derivative $s' = s'(z)$:

$$s'/(ds^2) = -m_0 s'(\kappa d - 2vs) - m_0 s^2, \quad \text{where}$$

$$m_0 \overset{def}{=} \int (u + rs)^{-2}\, dF_0(u).$$

We rewrite this equation in the form $s'(B - p) = -s^2$, where $B = 1/(dm_0 s^2)$. Now estimate $|s'|$ from below. In view of (3) for z

from any compact within $\mathfrak{S}$, the values s are bounded from below and from above. By (29) for sufficiently small $\varepsilon > 0$, we have $p > \delta/2$ and, consequently, $|m_0 s^2| \leq \mu_0|s|^2 = O(1)$. It follows that $|s'| \leq \text{const } /|\rho|$, where

$$\rho = \mu_0|s|^2 - m_0 s^2 = \int |u/s + r|^{-2}\, dF_0(u) - \int (u/s + r)^{-2}\, dF_0(u).$$

Denote $w = u/s + r$. It is easy to verify that

$$\text{Re } \rho = 2 \int |w|^{-4}(\text{Im } w)^2\, dF_0(u).$$

The value $\text{Im } w = -|s|^{-2}s_1 - vs_1 + O(\varepsilon)$. For sufficiently small $\varepsilon \to +0$, we have $|\text{Im } w| > vs_1/\sqrt{2}$. Applying the Cauchy–Bunyakovskii inequality and using (27) for sufficiently small $\varepsilon > 0$, we obtain

$$\text{Re } \rho \geq v^2 s_1^2 \int |w|^{-4}\, dF_0(u) \geq v^2 s_1^2\, |s|^4 \mu_0^2 \geq (vd)^2\, p^{-2}s_1^2.$$

Here $v \geq \delta > 0$, and the values p are bounded from above. We conclude that $|s'| \leq \text{const } /s_1^2$. Thus, the function s_1^3 has the derivative uniformly bounded in $\mathfrak{S}$. As $z \to -v - i\varepsilon$, where $v < 0$ and $\varepsilon \to +0$, the limit function exists $s_1(-v) = \lim s_1(z)$. The difference $s_1^3(z) - s_1^3(-v) = O(\varepsilon)$ and therefore $s_1(z) \leq s_1(-v) + O(\varepsilon^{1/3})$. By (4) we can infer that s_1^3 has the uniformly bounded derivative on any compact beyond $\mathfrak{S}$. Thus, the function $s_1(z)$ satisfies the Hölder inequality with the exponent $1/3$ for all z. In view of (4) for $v > 0$, we have $F'(v) = \pi^{-1}\text{Im } h(-v)$. One can see that $F'(v)$ also satisfies the Lipschitz condition with the exponent $1/3$. For $\kappa > 0$, we have $v_1 > 0$. It follows that for $0 < v < v_1$, we have $F'(v) = 0$. Assume that $F(v)$ has a jump at the point $v = 0$. Then by (4), the function $h(z)$ has a pole at the point $z = 0$. It is easy to see that this contradicts (3). We conclude that $F'(v)$ satisfies the Lipschitz condition for all $v \geq 0$. Set $F'(v) = 0$ for $v < 0$. Then, $F'(v)$ satisfies the Lipschitz condition for all real v. Consider the integral in equation (4) as the Cauchy-type integral. The integration contour can be extended to the negative half-axis and be closed by an infinitely remote circumference. We apply well-known theorems on the Cauchy-type integrals. It follows that the function $h(z)$ defined by (4) satisfies the Hölder equation with the exponent $1/3$ on the whole plane of complex z. The proof of Theorem 6.3 is complete. □

LEMMA 6.7 (proof).
Let us use relation (23) with $x = R_{mj}$. We find

$$\mathbf{E}(RHR^T R)_{lk} = \mathbf{E}RHR^T A_{lk} + \frac{d}{n}\, \mathbf{E}\, \frac{\partial R_{li}H_{ij}R_{mj}}{\partial R_{mk}}, \qquad (34)$$

$l = 1, \ldots, N$, $k = 1, \ldots, n$ (here the summation over i, j, and m is implied). The derivative

$$\mathbf{E}\, \frac{\partial R_{li}H_{ij}R_{mj}}{\partial R_{mk}} = (N+1)\mathbf{E}(RH)_{lk} + R_{ij}\frac{\partial H_{ij}}{\partial R_{mk}}R_{mj}, \qquad (35)$$

where the second summand equals

$$-(RH)_{lk}\mathrm{tr}(R^T RH) - (RHR^T H)_{lk}$$

and the spectral norms $\|RH\| \leq 1/\sqrt{c}$, $\|RHR^T H\| \leq 1/c$. For $t \geq c > 0$, we have

$$n^{-1}\mathrm{tr}(R^T RH) = 1 - tn^{-1}\mathrm{tr}H,$$
$$\|\mathbf{E}RHn^{-1}\mathrm{tr}H - \mathbf{E}RH\mathbf{E}n^{-1}\mathrm{tr}H\| \leq t^{-1}\sqrt{\mathrm{var}(n^{-1}\mathrm{tr}H)} = O(n^{-1}).$$

By Lemma 6.1 we find that (34) are entries of $[N - n(1 - h_n)]\mathbf{E}RH + \Omega$, where the matrix Ω is uniformly bounded in spectral norm. The lemma is proved. □

Remark 3 (proof).
Indeed, in view of the independence of H of $\mathbf{y}$, the variance

$$\mathrm{var}\, \widehat{\varphi}_n(t) = \mathrm{var}(\mathbf{y}^T\Omega\mathbf{y}) + \mathbf{E}\mathbf{E}'|\mathbf{y}^T\Delta_n\mathbf{y}|^2 \leq$$
$$\leq \|\Omega\|^2\, \mathrm{var}\, \mathbf{y}^2 + \mathbf{E}|\mathbf{y}|^2\, (\mathbf{e}^T\Delta_n\mathbf{e})^2 = O(n^{-1}),$$

where $\Omega = \mathbf{E}RH(t)H(t')R^T$, $\Delta_n = RH(t)H(t')R^T - \Omega$, the expectation $\mathbf{E}'$ is calculated for fixed $\mathbf{y}$, and $\mathbf{e}$ depends only on $\mathbf{y}$ but not on R. Here the variance var $\mathbf{y}^2 \leq 2(2q\mathbf{b}^2 + q^2/n)/n = O(n^{-1})$, $\mathbf{E}\mathbf{y}^2 = O(1)$, and by Lemma 6.3, we have $\mathbf{E}(\mathbf{e}^T \Delta_n \mathbf{e})^2 = O(n^{-1})$. This is the first statement of Lemma 6.3. It is easy to verify the second half of Remark 3 using Lemma 6.1 and the fact that $\widehat{r}_n(t) \geq c > 0$. $\square$

Remark 4 (proof).
Indeed, note that the denominator of $\widehat{g}_n(t)$ contains the function $\widehat{r}_n(t) \geq c > 0$. Using Remark 3, it is easy to justify Remark 4. $\square$

LEMMA 6.8 (proof).
First, let us transform

$$\varphi_n = \mathbf{E}(\mathbf{y}^2 - \mathbf{y}^T RHR^T \mathbf{y}) = \mathbf{E}[\mathbf{b}^2 - \mathbf{b}^T RHR^T \mathbf{b} + qn^{-1}$$
$$\mathrm{tr}(I - RHR^T)].$$

We substitute here the right-hand side of the equation $\mathbf{b} = A\mathbf{x}$ and use Lemma 6.7 and the identities $HR^T R = I - tH$ and $\mathrm{tr}(RHR^T) = n - t\,\mathrm{tr}H$. It follows that

$$\varphi_n = r_n\, \mathbf{E}\mathbf{b}^T RH\mathbf{x} + q(N/n - 1 + th_n) + O(n^{-1}).$$

By Lemma 6.5, we have

$$\varphi_n = r_n s_n^{-1} \mathbf{x}^T \Sigma\, \mathbf{E}H\mathbf{x} + q(th_n + \kappa_n) + O(n^{-1}).$$

Using Theorem 6.1, we obtain the second lemma statement.
Further, we apply Remark 4 and obtain the first lemma statement. The proof of Lemma 6.8 is complete. $\square$

THEOREM 6.4 (proof).
First, using the identity $Ht + HR^T R = I$, we rewrite the expression in Lemma 6.7 formulation in the form

$$A = \mathbf{E}RHR^T A + (t + h_n td + \kappa_n d)\mathbf{E}RH + \omega,$$

with the same upper estimate of ω as in Lemma 6.7.

Let us multiply the left- and right-hand sides of this matrix by $\mathbf{x}$ from the right and by $\mathbf{b}^T$ from the left. We find that

$$\mathbf{b}^2 = \mathbf{E}\mathbf{b}^T R H R^T \mathbf{b} + (t + th_n d + \kappa_n d)\mathbf{E}\mathbf{b}^T R H \mathbf{x} + \omega',$$

where $|\omega'| \leq$ is greater than $|\omega|$ by no more than $\sqrt{c_2}$ times. It is obvious that $\mathbf{E}\mathbf{y}^2 = \mathbf{b}^2 + Nn^{-1}q$. Note that for $\mathbf{y} \sim N(0, q/n)$ for any constant matrix M, we have the relation $\mathbf{E}\mathbf{y}^T M \mathbf{y} = \mathbf{b}^T M \mathbf{b} + qn^{-1}\mathbf{E}\mathrm{tr}M$. It follows that

$$\mathbf{E}\mathbf{y}^2 - q(\kappa_n + 1) =$$
$$= \mathbf{E}\mathbf{y}^T R H R^T \mathbf{y} - q\,(1 - h_n t) + (t + th_n d + \kappa_n d)\mathbf{E}\mathbf{y}^T R H \mathbf{x} + \omega'$$

with the same remainder term. Thus, we find that for $t \geq c > 0$, the expectation

$$\mathbf{E}\mathbf{y}^T R H \mathbf{x} = \frac{\varphi_n - (th_n + \kappa_n)\, q}{t(1 + h_n d) + \kappa_n d} + \omega'', \qquad (36)$$

where $|\omega''| \leq ad\, c^{-3/2}c_2^{1/2}$ and by Remark 4, the left-hand side of (36) equals $g_n(t) + O(n^{-1})$. Integrating over $d\eta(t)$, we obtain the second term of the right-hand side of (9) with the accuracy up to $O(n^{-1})$.

Let us transform the third term of the right-hand side using the identity $H(t)H(t') = (t' - t)(H(t) - H(t'))$. We obtain that

$$\mathbf{E}\mathbf{y}^T R\Gamma^2 R \mathbf{y} = \int\int \frac{\mathbf{E}\mathbf{y}^T R H(t)R^T \mathbf{y} - \mathbf{E}\mathbf{y}^T R H(t')R^T \mathbf{y}}{t' - t}\, d\eta(t)\, d\eta(t') =$$
$$= \int\int \frac{\mathbf{E}\varphi_n(t) - \mathbf{E}\varphi_n(t')}{t - t'}\, d\eta(t)\, d\eta(t') = \int\int K_n(t, t')\, d\eta(t)\, d\eta(t').$$
$$(37)$$

Thus, we obtain the third term in the theorem formulation. The proof of Theorem 6.4 is complete. $\square$

THEOREM 6.5 (proof).

For $t \geq c > 0$, the functions $\widehat{h}_n(t) \leq c^{-1}$, $\widehat{\varphi}_n(t) \leq \mathbf{y}^2$, and $\widehat{g}_n(t) \leq \mathbf{y}^2$. By Remark 3, the variances of these functions are

$O(n^{-1})$ uniformly in $t \geq c > 0$. It follows that the integral of $\widehat{g}(t)$ with respect to the measure $\eta(t)$ converges in the square mean to the integral of $g(t)$ with respect to the same measure.

Further, for any $\varepsilon > 0$, the variance $\widehat{K}_n(t, t') = O(n^{-1})$ uniformly as $|t - t'| > \varepsilon$. For $|t - t'| \leq \varepsilon$, we have

$$\widehat{K}_n(t, t') = \widehat{\varphi}'_n(t) + \frac{1}{2} \widehat{\varphi}''_n(\theta) \, |t - t'|,$$

where θ is an intermediate value of the argument, and the second derivative is uniformly bounded. Choosing $\varepsilon = n^{-1/2}$, we find that $\widehat{K}_n(t, t')$ is uniformly approximating $K_n(t, t')$ in the square mean. By Lemma 6.4, the variance $\widehat{K}_n(t, t')$ also is $O(n^{-1})$. It follows that the double integral in (9) converges in the square mean to the double integral in (11). We obtain the theorem statement. $\square$

LEMMA 6.9 (proof).

The first two statements of Lemma 6.9 follow from Theorem 6.2, Lemma 6.8, and Remark 3. By (8) for $t \geq c > 0$, the derivatives $\varphi'_n(t)$ and $\varphi'(t)$ exist and are uniformly bounded. The uniform convergence

$$K_n(t, t') \to K(t, t') \stackrel{def}{=} \frac{\varphi(t) - \varphi(t')}{t - t'}$$

follows from the uniform convergence $\varphi_n(t) \to \varphi(t)$ and uniform convergence of derivatives $\varphi'_n(t)$. Thus, the lemma is proved. $\square$

THEOREM 6.6 (proof).

This statement is a direct consequence of uniform convergence in Lemma 6.9 and the decrease of variances in Remark 3.

LEMMA 6.10 (proof).

By Theorem 6.3, the analytical continuations of functions $s(z)$ and $r(z)$ are regular. To prove the existence of the derivative $g'(z)$, it suffices to show that the denominator in the integrand in (37) does not vanish. By (4) for $\kappa > 0$ and z around $z = 0$, the function $s(z)$ is bounded, $r(0) = \kappa d$, and it follows that $s(z)$ is regular. We conclude that $g'(0)$ exists. For $\mathrm{Im}\, z \neq 0$ by (4), we have $\mathrm{Im}\, h(z) \neq 0$, and

(3) implies that $\operatorname{Im} h(z) = |s(z)|^2 \mu_0 \operatorname{Im}(r(z)s(z))$, where μ_0 is the integral defined in the proof of Theorem 6.3. Consequently, for these z, the derivative $g'(z)$ exists, and it follows that the derivative $\varphi'(z)$ also exists. The lemma is proved. $\square$

LEMMA 6.11 (proof).

We start from the right-hand side of (12). To the right from $\mathfrak{L}$ the functions $g(z)$ and $\varphi(z)$ have no singularities, and for $|z| \to \infty$, we have $|\Gamma(-z)| = O(|z|^{-1})$. For $\kappa > 0$, the value $v_1 > 0$ and as $|z| \to \infty$ by (4), we have $zh(z) \to \int v^{-1} dF(v) \le 1/v_1$. Thus, $s(z) = O(1)$, $zs^2(z) \approx z$, and one can see that $g(z) = O(|z|^{-1})$ and $\varphi(z) = O(1)$. Consequently, we can close the contour $\mathfrak{L}$ in the integrals (12) to the right side by an infinitely remote semicircle. The closed contours are formed. We find that

$$\int g(t)\ d\eta(t) = -\frac{1}{2\pi i}\int \left[\oint \frac{g(z)}{z-t}\,dz\right] d\eta(t) =$$

$$= \frac{1}{2\pi i}\int_{\mathfrak{L}} g(z)\Gamma(-z)\ dz,$$

$$\int \frac{\varphi(t) - \varphi(t')}{t - t'}\ d\eta(t) =$$

$$= \frac{1}{2\pi i}\int\int \left[\oint \varphi(z)\frac{1}{t'-t}\left(\frac{1}{z-t} - \frac{1}{z-t'}\right)dz\right] d\eta(t)\,d\eta(t') =$$

$$= -\frac{1}{2\pi i}\int_{\mathfrak{L}} \varphi(z)\ \Gamma^2(-z)\,dz.$$

The expression (12) passes to the right-hand side of (11). Our lemma is proved. $\square$

Remark 5 (proof).

Indeed, we note that in this case, the Radon–Nikodym derivatives $dG(v)/dF_0(v)$ are uniformly bounded, and the integral in (37) is $O(1)\mu_0$, where μ_0 is defined in Theorem 6.3. Therefore, $|g'(z)| \le \mu_0|(r(z)s(z))|$. For $z = -v - i\varepsilon$, $v, \varepsilon > 0$, the value $1/\mu_0 = p|s|^2 d + O(\varepsilon)$, where $p \ge v \ge v_1$, $|s| \ge 1/2$, and $\varepsilon = |\operatorname{Im} z|$.

Therefore, for sufficiently small ε, the values $\mu_0 \leq 8/v_1 d$ are uniformly bounded. By Theorem 6.3, the functions $r(z)$ and $s(z)$ satisfy the Hölder condition. We conclude that for $v \in \mathfrak{S}$, the functions $g(z)$ and $\varphi(z)$ also satisfy the Hölder condition and allow the continuous extension to $z = -v$. Remark 5 justified.

LEMMA 6.12 (proof).
We start from the right-hand side of (12). From (4), one can see that as $|z| \to \infty$, the functions $s(z)$ and $\varphi(z)$ are uniformly bounded and $g(z)$ decreases as $O(|z|^{-1})$. We can deform the contour $\mathfrak{L}$ by moving its remote parts to the region of negative Re z. Let us move these parts of $\mathfrak{L}$ from above and from below to the half-lines $(0, \pm i\varepsilon; -\infty \pm i\varepsilon)$, where some $\varepsilon > 0$. Since all functions in the integrands are bounded for Re $z = 0$, the contribution of the vertical segment of the closed integration path is $O(\varepsilon)$. Let $\varepsilon \to +0$ tending the upper half-line to the lower one. The functions $g(z)$ and $\varphi(z)$ satisfy the Hölder condition (13) at the half-line $z \leq 0$. Therefore, the real parts of these functions mutually cancel, whereas the imaginary parts double. The right-hand side of (12) takes the form

$$1 - \frac{2}{\pi} \int_0^\infty \text{Im } g(-v)\Gamma(v)\, dv - \frac{1}{\pi} \int_0^\infty \text{Im } \varphi(-v)\Gamma^2(v)\, dv.$$

For $v \notin \mathfrak{S}$, Im $h(-v) = 0$ and imaginary parts of the functions $r(-v), g(-v), \varphi(-v)$ vanish. We obtain the statement of Lemma 6.12. $\square$

LEMMA 6.13 (proof).
Denote $v = -\text{Re } z$, $s_0 = \text{Re } s$, $s_1 = \text{Im } s$, $r = r(z) = -vs + \kappa d$, $r_0 = \text{Re } r$, $g_1 = \text{Im } g(z)$, and $\varphi_1 = \text{Im } \varphi(z)$,

$$A_\nu = \int \frac{u^\nu}{|u + r(z)s(z)|^2}\, dG(u), \quad \nu = 0, 1.$$

From Lemma 6.9, one can find that the imaginary parts $g_1 = -A_0$ $\text{Im}(r(z)s(z)) = A_0(2vs_0 - \kappa d) \geq 0$, $\varphi_1 = -(A_1 v + A_0|r|^2 + qv/d)$.

It is clear that $A_0 \lambda_0 \le A_1$. Therefore,

$$|\varphi_1| \ge A_0 \, \lambda_0 v_1 \ge \frac{\lambda_0 v_1}{2 \, (c_2^2 + |rs|^2)},$$

where $|s|^2 \le 1 + d/v$, $v \ge v_1 > 0$, and $|r|^2 \le \kappa^2 d^2 + c_2^2 |s|^2$. Statement 1 is proved. The ratio

$$\left| \frac{g_1}{\varphi_1} \right| \le \frac{2|r_0| + \kappa d}{\lambda_0 v + |r|^2} \le \frac{2}{\sqrt{\lambda_0 v}} + \frac{\kappa d}{\lambda_0 v},$$

$v \ge v_1 > 0$. Our lemma is proved. $\square$

LEMMA 6.14 (proof).

First, we reconsider the proof of Lemma 6.1. We note that for any complex vectors $\mathbf{q}$ and for any symmetric real positive semidefinite matrices A, the inequalities hold

$$\|A + zI\| \le |z|^2/|z_1|, \quad |1 + \mathbf{q}^H z(A + zI)^{-1}\mathbf{q}|^{-1} \le |z/z_1|, \quad (38)$$

where the upper H is the Hermitian conjugation sign. In Remark 2, we have now $\|\Omega\| \le |z/z_1|$; we obtain (as z is fixed) the estimate $\mathrm{Var}_m \Phi_m = O(n^{-1})$. In the proof of Lemma 6.1, the upper estimates change: now we have $\|\Omega\| = O(1)$, $|\varphi| = O(1)$. It follows that $\mathrm{Var}_m \psi = O(n^{-1})$. The variance

$$\mathrm{Var}_m \varphi \le \sqrt{\mathbf{E}|1 + \Phi_m|^{-2} \, |1 + \mathbf{E}\Phi_m|^{-2} \, \mathrm{Var}_m \Phi_m},$$

where the multiples under the radical sign are $O(1)$ by the second of the inequalities (38) in the above. We find that $\mathrm{Var}_m \varphi \le O(1)$ $\mathrm{Var}_m \Phi_m$, $\mathbf{E}\psi^2 = O(1)$, and $\mathrm{Var}_m(\psi \varphi) = O(n^{-1})$. Thus, we proved the first lemma statement.

We now reconsider the proof of Lemma 6.3. Let $\mathbf{e} = \mathbf{f}$ be a complex vector of unit length. For fixed $z_1 \ne 0$ by (38), we have $\|\Omega\| = O(1)$ and, as before, $\mathbf{E}(\mathbf{r}_m^2)^2 = O(1)$. We obtain a refinement of Lemma 6.3: for any $z_1 \ne 0$, we have $\mathrm{Var}[\mathbf{e}^H RH(z) R^T \mathbf{e}] = O(n^{-1})$. Thus, the first and the second statements of our lemma are proved. Next, we reconsider Remark 3. Now $\Omega = \mathbf{E}RH(z)H(z')$

R^T, $\|\Omega\| \le |z/z_1|$, and $|e^H \Delta_n e|^2 = O(n^{-1})$. We obtain the third statement of Lemma 6.14. The value $r_n = r_n(z) = z s_n(z) + \kappa d$ and $|\mathrm{Im}\, r_n| \ge |z_1|$. From (8) it follows that for fixed $z_1 \ne 0$, we have $\mathrm{Var}\, \widehat{\varphi}_n(z) = O(n^{-1})$. The proof of Lemma 6.14 is complete. $\square$

THEOREM 6.8 (proof).

Let $z = -v - i\varepsilon$, $v, \varepsilon > 0$. Denote $s_1 = \mathrm{Im}\, s(z)$, $g_1 = \mathrm{Im}\, g(z)$, $\varphi_1 = \mathrm{Im}\, \varphi(z)$, $\widehat{g}_1 = \mathrm{Im}\, \widehat{g}_n(z)$, $\widehat{\varphi}_1 = \mathrm{Im}\, \widehat{\varphi}_n(z)$, and $\Gamma^0(v) = g_1/\varphi_1$.

The values of v for which $|\widehat{g}_1/\widehat{\varphi}_1| > B$ lay outside $\mathfrak{S}$, where $\rho(v) = 0$, and the contribution of these v to (16) equals 0. For $|\widehat{g}_1/\widehat{\varphi}_1| \le B$, we subtract

$$\frac{g_1}{\varphi_1} - \frac{\widehat{g}_1}{\widehat{\varphi}_1} = \frac{\widehat{g}_1 - g_1}{\varphi_1} + \frac{\widehat{g}_1}{\widehat{\varphi}_1}(\varphi_1 - \widehat{\varphi}_1).$$

Denote by $\mathfrak{S}'$ a subset of $\mathfrak{S}$ for which $\rho(v) \ge \varepsilon > 0$. From Lemmas 6.13 and 6.14, it follows that for fixed $\varepsilon > 0$, the contribution of $v \in \mathfrak{S}'$ tends to zero as $n \to \infty$. The ratios g_1/φ_1 and $\widehat{g}_1/\widehat{\varphi}_1$ are bounded. It follows that the contribution of the region $\mathfrak{S} - \mathfrak{S}'$ tends to zero as $\varepsilon \to 0$ independently of n. Thus, the theorem is proved. $\square$

Remark 6 (proof).

Indeed, the first statement follows from Theorem 6.6. The existence of the derivative $\varphi'(t)$ and the expressions for $(\mathbf{x} - \widehat{\mathbf{x}})^2$ follows from Lemma 6.14 and (29).

LEMMA 6.15 (proof).

From (3), we find that for $t > 0$, this equation has a unique solution $h(t)$ that tends to $h(0)$ as $t \to +0$, and

$$\frac{1}{s(0)} + \int \frac{d}{(v + \kappa d s(0))}\, dF_0(v) = 1, \quad s(0) = 1 + d h(0).$$

The left-hand side of this equation depends on $s(0)$ monotonously, and this equation is uniquely solvable with respect to $s(0)$. One can see that $1/s(0) + 1/(\kappa s(0)) \ge 1$ and $s(0) \le 1 + 1/\kappa$. The second statement of our lemma follows from Lemma 6.9. Next, we differentiate both parts of the equation (3) with respect to $t > 0$. Isolating the coefficient of the derivative $s'(t)$ we obtain

$$s'(t) = -\frac{dh^2(t)\, s^2(t)}{1 + dh^2(t)\, (r(t) + ts(t))}.$$

For $t \to +0$, we obtain statement 3. Let us calculate the derivatives $g'(t)$ and $\varphi'(t)$ from Lemma 6.9 for $t > 0$. Then, we tend $t \to +0$. From Lemma 6.9 by differentiating, we obtain the expression for $\varphi'(0)$. Further, it is obvious that $b_1 \geq c_1/(c_2 + d\,(1 + \kappa))$, $b_2 \leq b_1/(\kappa d\, s(0))$. For $\varphi'(0)$, we find that

$$\varphi'(0) \geq \frac{\lambda_0}{c_2 + (1 + \kappa)d}\, \frac{\kappa d^2\, h^2(0)}{1 + \kappa d^2\, h^2(0)}.$$

We derive the last inequality in the lemma statement. Lemma 6.15 is proved. $\square$

THEOREM 6.9 (proof).

The first statement immediately follows from Remark 6. To be concise, we denote $g = g(t)$, $\widehat{g} = \widehat{g}_n(t)$, $p = \varphi_n(t)$, $\widehat{p} = \widehat{\varphi}'_n(t)$. We obtain

$$\mathbf{E}(\mathbf{x}_t^0 - \widehat{\mathbf{x}}_t^0)^2 = \mathbf{E}\widehat{p}\big(g^2/p - \widehat{g}^2/\widehat{p}^2\big)^2. \tag{39}$$

By Remark 6, we have $0 \leq g^2/p \leq 1$, and $0 \leq \widehat{g}^2/\widehat{p} \leq 1$ with probability tending to 1.

By Lemma 6.9 for fixed $t > 0$ as $n \to \infty$, the convergence in the square mean $\widehat{g} \to g$, and $\widehat{p} \to p$ holds. We conclude that the expression (39) tends to zero as $n \to \infty$. Theorem 6.9 is proved. $\square$

EXPERIMENTAL INVESTIGATION OF SPECTRAL FUNCTIONS OF LARGE SAMPLE COVARIANCE MATRICES

The development of asymptotical theory of large sample covariance matrices made it possible to begin a systematic construction of improved and unimprovable statistical procedures of multivariate analysis (see Chapters 4, 5, and [71]). New procedures have a number of substantial advantages over traditional ones: they do not degenerate for any data and are applicable and approximately unimprovable independently of distributions.

Most of theoretical results in multiparametric statistics were obtained in the "increasing-dimension asymptotics," in which the observation dimension $n \to \infty$ along with sample sizes N so that $n/N \to y > 0$. This asymptotics serves as a tool for isolating principal parts of quality functions, for constructing their estimators, and for the solution of extremum problems. However, the problem of practical applicability of the improved procedures requires further investigations. Theoretically, it is reduced to obtaining upper estimations for the remainder terms. But weak upper estimates of the remainder terms derived until now seem to be too restrictive for applications. The question of practical advantages of asymptotically improved and unimprovable procedures remains open.

In this appendix, we present data of the experimental (numerical) investigation of spectral functions of large sample covariance matrices presented in [73] and discuss results of their comparison with asymptotical relations of spectral theory.

1. THEORETICAL RELATIONS

A remarkable feature of the spectral theory of random matrices of increasing dimension is its independence of distributions. In Chapter 3, it is shown that applicability of asymptotic formulas of this theory is determined by magnitudes of two special parameters. Let $\mathbf{x}$ be an observation vector, and $\mathfrak{S}$ be a population for which $\mathbf{E}\mathbf{x} = 0$. Denote

$$M = \sup_{|\mathbf{e}|=1} \mathbf{E}(\mathbf{e}^T\mathbf{x})^4 > 0, \tag{1}$$

where $\mathbf{e}$ are nonrandom unity vectors;

$$\gamma = \sup_{\|\Omega\|\leq 1} \operatorname{var}(\mathbf{x}^T\Omega\mathbf{x}/n)/M, \tag{2}$$

where Ω are nonrandom, symmetric, positive semidefinite matrices with spectral norms not greater than 1. In Chapter 3, it was shown that the applicability of asymptotical theory is guaranteed in the situation when the moment M is bounded and the parameter γ is vanishing.

Denote $\Sigma = \operatorname{cov}(\mathbf{x}, \mathbf{x})$. For a nondegenerate normal distribution $\mathbf{N}(0, \Sigma)$, the values $M = 3\|\Sigma\|^2$, $\gamma = 2/3\ n^{-2}\operatorname{tr}\Sigma^2$. If $\Sigma = I$ then $M = 3$ and $\gamma = 2/3n$.

Let $\mathfrak{X} = \{\mathbf{x}_m\}$ be a sample of size N from population $\mathfrak{S}$. Let

$$\bar{\mathbf{x}} = N^{-1}\sum_{m=1}^{N}\mathbf{x}_m, \quad C = N^{-1}\sum_{m=1}^{N}(\mathbf{x}_m - \bar{\mathbf{x}})(\mathbf{x}_m - \bar{\mathbf{x}})^T.$$

Here C is sample covariance matrix (a biased estimator of Σ).
Denote $y = n/N$,

$$H = H(t) = (I + tC)^{-1}, \quad h_n(t) = n^{-1}\operatorname{tr}H(t),$$
$$s_n(t) = 1 - y + yh_n(t).$$

We cite the basic limit spectral relation (dispersion equation) for functions of the matrices Σ and matrices C derived in Theorem 3.1 and in [25].

THEOREM 1 (corollary of Theorem 3.1). *For any populations with $M > 0$, for $t \geq 0$,*

$$h_n(t) = n^{-1}\,\mathrm{tr}(I + ts_n(t)\Sigma)^{-1} + \omega(t), \tag{3}$$

$$\mathbf{E}H(t) = (I + ts_n(t)\Sigma)^{-1} + \Omega, \tag{4}$$

$$\mathrm{var}\left(n^{-1}H(t)\right) \leq a\tau^2/N,$$

where $\omega(t) = \|\Omega\| \leq a\tau \,\max(1, y)\,[\tau\sqrt{\delta} + (1 + \tau^2)/\sqrt{N}]$, a is a numeric coefficient, $\tau = \sqrt{M}t$, and $\delta = 2y^2\,(\gamma + \tau^2/N)$.

For normal distributions, the expectation of $n^{-1}\mathrm{tr}H(t)$ was investigated in more detail in [12], and it was found that for $\Sigma = I$

$$\mathbf{E}\,|\omega(t)| \leq \omega_1(t) \overset{def}{=} 2yt^2(1 + ty)/\sqrt{nN} + 3t/N,$$

while the variance $\mathrm{var}[n^{-1}\mathrm{tr}H(t)] = O(n^{-1}N^{-1})$.

The problem of recovering spectra of matrices Σ from the empirical matrices C may be solved in the form of limit relations (see Introduction). To pass to the limit, we consider a sequence of problems of the statistical analysis

$$\mathfrak{P} = \{(\mathfrak{G}, \Sigma, \mathrm{N}, \mathfrak{X}, \bar{\mathbf{x}}, C)_n\}, \quad n = 1, 2, \ldots, \tag{5}$$

where (indexes n are omitted) $\mathfrak{G}$ is n-dimensional population with $\Sigma = \mathrm{cov}(\mathbf{x}, \mathbf{x})$, and $\mathfrak{X}$ is a sample of size N from $\mathfrak{G}$.

THEOREM 2 (corollary of Theorem 3.1). *Let moments (1) exist and be uniformly bounded in $\mathfrak{P}$; let $n \to \infty$ so that $n/N \to \lambda > 0$, $\gamma \to 0$, and for almost all $u \geq 0$*

$$F_0(u) \overset{def}{=} \lim_{n\to\infty} \frac{1}{n} \sum_{i=1}^{n} \mathrm{ind}(\lambda_i^0 \leq u), \tag{6}$$

where $\lambda_1^0, \lambda_2^0, \ldots, \lambda_n^0$ are the eigenvalues of Σ.
  Then, for each $t \geq 0$
  1. *there exists the limit*

$$h(t) = \lim_{n\to\infty} \mathbf{E}n^{-1}\mathrm{tr}H(t) = \int (1 + ts(t)u)^{-1}\,dF_0(u), \tag{7}$$

where $s(t) = 1 - \lambda + \lambda h(t)$;

2. the variance $\operatorname{var}\left(n^{-1} \operatorname{tr} H(t)\right) \to 0$ as $n \to \infty$.

The limit dispersion equation (7) presents the main result of spectral theory of sample covariance matrices of increasing dimension.

Consider an empiric distribution function for eigenvalues λ_i of the matrices C

$$F_n(u) = n^{-1} \sum_{i=1}^{n} \operatorname{ind}(\lambda_i \le u), \quad u \ge 0. \tag{8}$$

THEOREM 3 (corrolary of Theorems 3.2 and 3.3). *Let assumptions of Theorem 2 hold and, in addition, $\lambda > 0$ and all eigenvalues of matrices Σ in $\mathfrak{P}$ exceed $c_1 > 0$, where c_1 does not depend on n.*

Then,

1. *the function $h(z)$ allows an analytical continuation to the region of complex z and satisfies the Hölder condition for all z;*

2. *the weak convergence holds $F_n(u) \overset{\mathbf{P}}{\to} F(u), \quad u \ge 0$;*

3. *and for all z, except $z < 0$,*

$$h(z) = \int (1 - zu)^{-1} \, dF(u).$$

In [9], it was shown that as $n \to \infty$, all eigenvalues of matrices C lay on the limit support almost surely.

Consider a special case: let $\Sigma = I$, $n = 1, 2, \ldots$ In this case, the function $h(z)$ satisfies the quadratic equation $h(z) - 1 = zh(z)s(z)$, and the limit spectral density of C is

$$F'(u) = (2\pi\lambda u)^{-1}\sqrt{(u_2 - u)(u - u_1)}, \tag{9}$$

where $u_1 = (1 - \sqrt{\lambda})^2$, $u_2 = (1 + \sqrt{\lambda})^2$ for $u_1 \le u \le u_2$. The solution of (7) is

$$h(t) = 2\left(\sqrt{(1 + (1 - \lambda)\,t)^2 + 4t\lambda} + 1 + (1 - \lambda)\,t\right)^{-1}. \tag{10}$$

Suppose, additionally, that there exists an $\varepsilon > 0$ such that the quantity $\sup\limits_{|\mathbf{e}|=1} \mathbf{E}(\mathbf{e}^T\mathbf{x})^{4+\varepsilon}$ is uniformly bounded in $\mathfrak{P}$. Then, Theorem 11.1 from [25] states that as $n \to \infty$, the minimum and maximum eigenvalues of the matrices C converge in probability to the magnitudes α_1 and α_2 such that

$$\alpha_i = \text{plim}\left(1 - \frac{y}{n} \sum_{k=1}^{n} \frac{\lambda_k}{\lambda_k - x_i}\right), \quad i = 1, 2, \qquad (11)$$

where $x = x_1$ and $x = x_2$ are the minimum and maximum real roots of the equation

$$\frac{y}{n} \sum_{k=1}^{n} \frac{\lambda_k^2}{(\lambda_k - x)^2} = 1.$$

2. NUMERICAL EXPERIMENTS

Numerical simulation was performed

- to investigate the relation between spectral functions of true covariance matrices Σ and sample covariance matrices C, given by equations (3) and (7);

- to study the convergence of the empirical distribution function of eigenvalues of C for large n;

- to investigate boundaries of spectra of matrices C and to compare these with theoretical relations (9) and (11);

- to study the dependence of the remainder terms $w(t)$ in (3) and (5) on the parameters t, M, $y = n/N$, and γ, and to compare with theoretical upper estimates;

- to estimate experimentally the boundaries of the applicability of the asymptotic equation (3) to distributions different from normal.

In the experiments, for fixed numbers n and N and given distribution law, samples were generated and used for calculation of sample means, sample covariance matrices, and spectral functions $h_n(t)$, $H(t)$, and $F_n(u)$.

In tables, random functions $h_n(t)$ and $F_n(u)$ are denoted by $\widehat{h}(t)$ and $\widehat{F}(u)$, and their averages (over s experiments) are denoted by $\langle \widehat{h}(t) \rangle$ and $\langle \widehat{F}(u) \rangle$.

The random inaccuracy $w(t)$ of the equation (3) was calculated in a series of s experiments. The average inaccuracy presents the estimator of $\mathbf{E} w(t)$, and its mean square deviation presents the estimator of the variance of $w(t)$. These characteristics are compared in tables with theoretical upper estimates of the asymptotic formula inaccuracy.

Empirical spectral function $\widehat{h}(t)$
Distribution $x \sim N(0,I)$

In Tables 1–4 the first two rows present the values of the function $\widehat{h}(t)$ for two independent experiments, the third row contains an estimator of the expectation along with the mean square deviation (in a series of $s = 100$ experiments). In the last row, the theoretical function $h(t)$ is shown. The analytical dependence of the deviations of the form $|\langle \widehat{h}(t) \rangle - h(t)| = kN^{-b}$ was fitted by the minimum square method: it is shown under the tables with the mean square error.

From Tables 1–4, one can see the systematic decrease of $\langle \widehat{h}(t) \rangle$ with growing N for fixed $y = n/N$ that can be explained by the increase of accuracy in the resolvent denominator.

Table 1: Spectral function $\widehat{h}(t)$: $\mathbf{x} \sim N(0, I)$, $t = 1$, $y = 1$, $s = 100$

	$N = 2$	$N = 10$	$N = 20$	$N = 50$
$\widehat{h}(t){:}1$	0.806	0.634	0.652	0.622
$\widehat{h}(t){:}2$	0.820	0.630	0.612	0.619
$\langle \widehat{h}(t) \rangle$	0.794 ± 0.107	0.657 ± 0.025	0.639 ± 0.014	0.626 ± 0.005
$h(t)$	0.618	0.618	0.618	0.618

Empirical law: $\langle \widehat{h}(t) \rangle - h(t) \approx 0.32 N^{-0.93} \pm 0.23 N^{-0.96}$

Table 2: Spectral function $\widehat{h}(t)$: $\mathbf{x} \sim N(0, I)$, $t = 2$, $y = 1$, $s = 100$

	$N = 2$	$N = 10$	$N = 20$	$N = 50$
$\widehat{h}(t){:}1$	0.750	0.591	0.504	0.510
$\widehat{h}(t){:}2$	0.874	0.515	0.537	0.508
$\langle \widehat{h}(t) \rangle$	0.700 ± 0.103	0.544 ± 0.028	0.521 ± 0.015	0.508 ± 0.006
$h(t)$	0.500	0.500	0.500	0.500

Empirical law: $\langle \widehat{h}(t) \rangle - h(t) \approx 0.46 N^{-1.0} \pm 0.21 N^{-0.90}$

Table 3: Spectral function $\widehat{h}(t)$: $\mathbf{x} \sim N(0, I)$, $t = 1$, $y = 1/2$, $s = 100$

	$N = 2$	$N = 10$	$N = 20$	$N = 50$
$\widehat{h}(t){:}1$	0.493	0.555	0.580	0.573
$\widehat{h}(t){:}2$	0.537	0.585	0.578	0.566
$\langle \widehat{h}(t) \rangle$	0.740 ± 0.201	0.588 ± 0.048	0.581 ± 0.022	0.568 ± 0.010
$h(t)$	0.562	0.562	0.562	0.562

Empirical law: $\langle \widehat{h}(t) \rangle - h(t) \approx 0.42 N^{-1.1} \pm 0.43 N^{-0.96}$

Table 4: Spectral function $\widehat{h}(t)$: $\mathbf{x} \sim N(0, I)$, $t = 2$, $y = 1/2$, $s = 100$

	$N = 2$	$N = 10$	$N = 20$	$N = 50$
$\widehat{h}(t){:}1$	0.392	0.504	0.446	0.417
$\widehat{h}(t){:}2$	0.170	0.369	0.422	0.422
$\langle\widehat{h}(t)\rangle$	0.708 ± 0.233	0.459 ± 0.059	0.435 ± 0.026	0.424 ± 0.010
$h(t)$	0.414	0.414	0.414	0.414

Empirical law: $\langle\widehat{h}(t)\rangle - h(t) \approx 0.52N^{-1.0} \pm 0.50N^{-0.98}$

Values of $\langle\widehat{h}(t)\rangle$ tend to theoretical functions $h(t)$ as N increases, and the difference decreases approximately by the law N^{-1}. The scatter of empiric $\langle\widehat{h}(t)\rangle$ in two experiments is covered by 2.5 σ, where σ^2 is the sample variance.

Accuracy of the empirical dispersion equation.
Distribution $x \sim N(0,I)$

In Tables 5–8, the experimental inaccuracy $w(t)$ of the equation (3) is presented. The function $s(t)$ was replaced by $1 - y + y\widehat{h}(t)$, where $\widehat{h}(t)$ was taken from Tables 1–4.

As in the former tables, the first two rows present two independent experimental values of $\widehat{w}(t)$, the next row shows sample mean and the mean square deviation in a series of $s = 100$ experiments. The fourth row contains theoretical upper inaccuracy $w(t)$ in (3) with the coefficient $a = 1$ (this estimator is distribution free). The fifth row presents the minimum value of a, for which (3) holds with $w(t)$ substituted by its expectation; in the sixth row, the refined theoretical upper estimate $w_1(t)$ is shown (valid for normal distribution).

Using $\widehat{w}(t)$, the analytical dependence $w(t) = kN^{-c}$ on N was constructed by the minimum square method for the expectation and the square scatter. It is shown under the tables.

From Tables 5–8, one may see that the average inaccuracy $\widehat{w}(t)$ (in the series of 100 experiments) of the equation (3) decreases approximately by the law N^{-1} as $n = N$ increases. The theoretical upper estimate of $w(t)$ in Section 3.1 decreases as $N^{-1/2}$, while

Table 5: The remainder term $w(t)$ in equation (3): $\mathbf{x} \sim N(0, I)$, $y = 1$, $t = 1$, $s = 100$

	$N = 4$	$N = 10$	$N = 20$	$N = 50$
$w(t):1$	0.120	0.0570	0.0303	0.0161
$w(t):2$	0.115	0.0637	0.0308	0.0110
$\langle w(t) \rangle$	0.117 ± 0.058	0.0526 ± 0.0284	0.0254 ± 0.0141	0.00959 ± 0.00525
$w(t)$	7.53	4.76	3.37	2.13
a	0.0155	0.0111	0.00755	0.0045
$w_1(t)$	1.75	0.70	0.350	0.14

Empirical law: $\langle w(t) \rangle \approx 0.62 N^{-1.2} \pm 0.22 N^{-0.93}$

Table 6: The remainder term $w(t)$ in equation (3); $\mathbf{x} \sim N(0, I)$, $y = 1$, $t = 2$, $s = 100$

	$N = 4$	$N = 10$	$N = 20$	$N = 50$
$w(t):1$	0.172	0.0347	0.0668	0.0128
$w(t):2$	0.193	0.0880	0.0469	0.0113
$\langle w(t) \rangle$	0.167 ± 0.064	0.0684 ± 0.0329	0.0309 ± 0.0162	0.0129 ± 0.0057
$w(t)$	52.7	33.3	23.6	14.9
a	0.00318	0.00205	0.00131	0.000862
$w_1(t)$	7.50	3.00	1.50	0.60

Empirical law: $\langle w(t) \rangle \approx 0.65 N^{-1.0} \pm 0.24 N^{-0.92}$

Table 7: The remainder term $w(t)$ in equation (3): $\mathbf{x} \sim N(0, I)$, $y = 1/2$, $t = 1$, $s = 100$

	$N = 4$	$N = 10$	$N = 20$	$N = 50$
$w(t):1$	-0.151	-0.0252	0.0561	0.0154
$w(t):2$	0.0928	0.0253	0.0102	0.0153
$\langle w(t) \rangle$	0.124 ± 0.097	0.0416 ± 0.0442	0.0216 ± 0.0237	0.00970 ± 0.00857
$w(t)$	2.84	1.79	1.27	0.802
a	0.0438	0.0232	0.017	0.0121
$w_1(t)$	1.28	0.51	0.26	0.10

Empirical law: $\langle w(t) \rangle \approx 0.45 N^{-1.0} \pm 0.44 N^{-1.0}$

the more refined estimator $w_1(t)$ (valid for normal distributions) decreases as N^{-1}. For $n = N$, the upper estimate $w(t)$ in all cases is comparable with 1 or more and has no sense as much the inaccuracy $w(t)$ is not greater than 1 by definition. The upper estimate $w_1(t)$

Table 8: The remainder term $w(t)$ in equation (3): $\mathbf{x} \sim N(0, I)$, $y = 1/2$, $t = 2$, $s = 100$

	$N = 4$	$N = 10$	$N = 20$	$N = 50$
$w(t){:}1$	0.285	0.0707	−0.0126	0.0182
$w(t){:}2$	0.0647	0.0906	0.0332	0.0175
$\langle w(t) \rangle$	0.169 ± 0.127	0.0558 ± 0.0456	0.0284 ± 0.0241	0.0121 ± 0.0109
$w(t)$	19.0	12.0	8.50	5.38
a	0.00891	0.00464	0.00334	0.00224
$w_1(t)$	4.33	1.73	0.87	0.35

Empirical law: $\langle w(t) \rangle \approx 0.58 N^{-1.0} \pm 0.49 N^{-1.0}$

proves to be ten times overstated as compared with the experimental value. For $n = N$, the mean square deviation of $w(t)$ decreases by the law N^{-1} that corresponds to the estimator of the function $\widehat{h}(t)$ variance for normal distribution found in [63] and [12], while (3) guarantees only the law $N^{-1/2}$. The upper estimate of the variance of $\widehat{h}(t)$ in [63] proves to be overstated by five times as compared with the empiric mean square deviation of $w(t)$ for $t = 1$ and $n = N = 50$. The disagreement with the experiment increases with the increase of t.

Spectral function $\widehat{h}(t)$ for binomial and normal populations

Tables 9–10, present the estimators $\langle \widehat{h}(t) \rangle$ of $h(t)$ calculated in a series of $s = 100$ experiments together with the mean square deviation $\sigma(t)$ of $\widehat{h}(t)$. Spectral functions are compared for normal distribution $N(0, I)$ and discrete binomial law $B(1, p)$ (in which each component of $\mathbf{x}$ takes on values $a = -\sqrt{p^{-1}(1 - p)}$ or $b = \sqrt{p(1 - p)^{-1}}$ $(p > 0)$ with the probabilities $\mathbf{P}(a) = p$, $\mathbf{P}(b) = 1 - p$ so that $\mathbf{E}\mathbf{x} = 0$ and $\mathbf{E}\mathbf{x}^2 = 1$. In the last row, limit values $h(t)$ are shown calculated by (7).

Tables 9 and 10 for sample size $N = 50$ show a good agreement of experimental results for these two distributions and a good fit to theoretical limits as well as a small scatter of estimators. Theoretical expected mean square scatter may be measured by the quantity

Table 9: Spectral function $\langle\widehat{h}(t)\rangle \pm \sigma(t)$: $y = 1$, $N = 50$, $s = 100$

	$t = 0.1$	$t = 0.5$	$t = 1$	$t = 2$
N	0.917 ± 0.002	0.736 ± 0.005	0.625 ± 0.006	0.509 ± 0.007
$B(0.3)$	0.918 ± 0.001	0.736 ± 0.003	0.624 ± 0.005	0.507 ± 0.005
$B(0.5)$	0.918 ± 0.000	0.736 ± 0.002	0.623 ± 0.003	0.505 ± 0.004
$B(0.7)$	0.918 ± 0.001	0.736 ± 0.003	0.624 ± 0.004	0.507 ± 0.006
$h(t)$	0.916	0.732	0.618	0.500

Table 10: Spectral function $\langle\widehat{h}(t)\rangle \pm \sigma(t)$: $y = 1/2$, $N = 50$, $s = 100$

	$t = 0.1$	$t = 0.5$	$t = 1$	$t = 2$
N	0.914 ± 0.003	0.709 ± 0.006	0.572 ± 0.009	0.423 ± 0.011
$B(0.3)$	0.914 ± 0.002	0.705 ± 0.005	0.566 ± 0.007	0.421 ± 0.007
$B(0.5)$	0.914 ± 0.000	0.705 ± 0.003	0.565 ± 0.004	0.418 ± 0.005
$B(0.7)$	0.914 ± 0.002	0.705 ± 0.005	0.567 ± 0.007	0.420 ± 0.007
$h(t)$	0.913	0.702	0.562	0.414

$(nNs)^{-1/2} = 0.002$. This fact demonstrates well the insensibility of spectral functions of sample covariance matrices to distributions when dimension is high ($n = 50$).

Function $\widehat{h}(t)$ for nonstandard normal distribution

Table 11 presents results of numerical modeling of the expectation of $\widehat{h}(t)$ and its mean square deviation calculated in a series of experiments for normal distribution with the covariance matrix of the form

$$
\Sigma = \begin{pmatrix}
2 & 1 & -1 & \cdots & 1 & -1 \\
-1 & 2 & -1 & \cdots & 1 & -1 \\
\vdots & \vdots & \vdots & \ddots & \vdots & \vdots \\
-1 & 1 & -1 & \cdots & 2 & -1 \\
-1 & 1 & -1 & \cdots & 1 & 2
\end{pmatrix}.
$$

This matrix corresponds to observations with the correlation coefficient $\pm 1/2$ and variance 2. Its eigenvalues include (for even n) $n/2$ units and $n/2$ of $\lambda = 3$. In the last row, the limit function

$h(t)$ is shown. This function was calculated by (7) as a root of the equation

$$2h = (1+ts)^{-1} + (1+3ts)^{-1}, \quad h = h(t), \quad s = s(t).$$

Table 11: Spectral function $\widehat{h}(t)$: $N(0, \Sigma)$, $\Sigma \neq I$, $N = 20$, $s = 100$

	$y = 1$ $t = 0.5$	$y = 1$ $t = 1$	$y = 1$ $t = 2$	$y = 1/2$ $t = 0.5$	$y = 1/2$ $t = 1$	$y = 1/2$ $t = 2$
$\langle\widehat{h}(t)\rangle$	0.65 ± 0.01	0.54 ± 0.01	0.44 ± 0.02	0.60 ± 0.02	0.47 ± 0.03	0.34 ± 0.03
$h(t)$	0.64	0.52	0.42	0.59	0.45	0.32

One may see a good agreement between the experimental and theoretical values of the function $\widehat{h}(t)$.

Function $\widehat{h}(t)$ and the inaccuracy $\widehat{\omega}(t)$ for different coupling of variables

Tables 12–14 present results of numeric simulation of the function $\widehat{h}(t)$ and the inaccuracy $\widehat{\omega}(t)$ in $s = 100$ experiments for normal population with the covariance matrix

$$\Sigma = \begin{pmatrix} 1 & r & \cdots & r & r \\ r & 1 & \cdots & r & r \\ \vdots & \vdots & \ddots & \vdots & \vdots \\ r & r & \cdots & 1 & r \\ r & r & \cdots & r & 1 \end{pmatrix} \tag{11}$$

for different $r < 1$. This matrix has a spectrum with a single eigenvalue $\lambda_1 = 1 - r + rn$ and $n - 1$ eigenvalues with $\lambda = 1 - r$. Theoretical limit function $h(t)$ can be calculated from (7) and equals

$$2\left(\sqrt{(1 + t(1-y)(1-r))^2 + 4ty(1-r)} + 1 + t(1-y)(1-r)\right)^{-1}.$$

Tables 12–14 show a good fit between experimental average values of $\widehat{h}(t)$ and limit values calculated from the dispersion equation (7). The good fit keeps on even for $r = 0.9$. Theoretical

Table 12: Function $\widehat{h}(0.5)$: $N(0, \Sigma)$, $N = 10$, $s = 100$

	$r = 1$			$y = 0.5$		
	$\langle \widehat{h}(t) \rangle$	$h(t)$	$\langle \widehat{\omega}(t) \rangle$	$\langle \widehat{h}(t) \rangle$	$h(t)$	$\langle \widehat{\omega}(t) \rangle$
$r = 0$	0.755 ± 0.023	0.732	0.03 ± 0.02	0.723 ± 0.039	0.702	0.03 ± 0.04
$r = 0.2$	0.769 ± 0.022	0.766	0.05 ± 0.02	0.766 ± 0.035	0.742	0.07 ± 0.04
$r = 0.4$	0.791 ± 0.019	0.805	0.07 ± 0.02	0.759 ± 0.033	0.788	0.07 ± 0.04
$r = 0.6$	0.820 ± 0.017	0.854	0.11 ± 0.02	0.777 ± 0.031	0.844	0.09 ± 0.03
$r = 0.8$	0.861 ± 0.012	0.916	0.16 ± 0.01	0.819 ± 0.023	0.913	0.13 ± 0.02
$r = 0.9$	0.888 ± 0.009	0.954	0.20 ± 0.01	0.838 ± 0.019	0.953	0.16 ± 0.02

Table 13: Function $\widehat{h}(1)$; $N(0, \Sigma)$, $N = 10$, $s = 100$

	$y = 1$			$y = 0.5$		
	$\langle \widehat{h}(t) \rangle$	$h(t)$	$\langle \widehat{\omega}(t) \rangle$	$\langle \widehat{h}(t) \rangle$	$h(t)$	$\langle \widehat{\omega}(t) \rangle$
$r = 0$	0.652 ± 0.030	0.618	0.05 ± 0.03	0.595 ± 0.044	0.562	0.05 ± 0.04
$r = 0.2$	0.667 ± 0.028	0.656	0.06 ± 0.03	0.654 ± 0.043	0.608	0.11 ± 0.04
$r = 0.4$	0.696 ± 0.023	0.703	0.11 ± 0.02	0.649 ± 0.037	0.667	0.10 ± 0.04
$r = 0.6$	0.743 ± 0.019	0.766	0.16 ± 0.02	0.677 ± 0.042	0.742	0.14 ± 0.04
$r = 0.8$	0.805 ± 0.016	0.854	0.25 ± 0.02	0.746 ± 0.026	0.844	0.21 ± 0.03
$r = 0.9$	0.850 ± 0.010	0.916	0.31 ± 0.01	0.786 ± 0.020	0.913	0.26 ± 0.02

Table 14: Function $\widehat{h}(2)$; $N(0, \Sigma)$, $N = 10$, $s = 100$

	$y = 1$			$y = 0.5$		
	$\langle \widehat{h}(t) \rangle$	$h(t)$	$\langle \widehat{\omega}(t) \rangle$	$\langle \widehat{h}(t) \rangle$	$h(t)$	$\langle \widehat{\omega}(t) \rangle$
$r = 0$	0.543 ± 0.026	0.500	0.07 ± 0.03	0.457 ± 0.044	0.414	0.06 ± 0.05
$r = 0.2$	0.558 ± 0.030	0.538	0.09 ± 0.03	0.541 ± 0.040	0.461	0.15 ± 0.04
$r = 0.4$	0.593 ± 0.030	0.587	0.14 ± 0.03	0.533 ± 0.039	0.523	0.14 ± 0.04
$r = 0.6$	0.652 ± 0.025	0.656	0.22 ± 0.03	0.559 ± 0.036	0.608	0.17 ± 0.04
$r = 0.8$	0.732 ± 0.019	0.766	0.33 ± 0.02	0.652 ± 0.034	0.742	0.27 ± 0.03
$r = 0.9$	0.801 ± 0.016	0.854	0.42 ± 0.02	0.719 ± 0.024	0.844	0.36 ± 0.03

upper estimates of $\omega(t)$ are strongly overstated. For large matrices of the form (11), the fourth maximum moment $M \approx 3r^2 n^2$, and theoretical upper estimate increases n^2 times. These estimates are proportional to powers of M, whereas Tables 12–14 show the linear increase of $\omega(t)$ with r or $\sqrt{r}$. Thus, the basic limit spectral equation (7) remains also valid for strongly coupled variables (up to $r = 0.9$).

Spectra of sample covariance matrices.
Distribution $x \sim \mathbf{N(0, I)}$

Sample covariance matrices C and their spectra were simulated. Table 15 shows maximum eigenvalues of C together with mean square deviations and the deviation of the averaged value from the theoretical limit value $\lambda_{max} = 4$. In the last row, the empirical regularity obtained by the minimum square method is shown.

In Table 16, the sample mean and the mean square deviation is presented for the uniform norm of the difference $\|\widehat{F} - F\|$ of functions $\widehat{F}(u)$ and $F(u)$, where $\widehat{F}(u)$ was calculated by (8) and $F'(u)$ is the theoretical density. Table 17 presents sample mean

Table 15: Maximum eigenvalues of C; $y = 1$, $N(0, I)$, $s = 100$

	$N = 2$	$N = 4$	$N = 6$	$N = 10$	$N = 20$	$N = 50$
$\langle \lambda max \rangle$	0.799 ± 0.806	2.32 ± 1.03	2.46 ± 0.66	3.05 ± 0.56	3.35 ± 0.35	3.70 ± 0.23
$4 - \langle \lambda max \rangle$	3.20	1.68	1.54	0.946	0.655	0.298

Empirical law: $4 - \langle \lambda_{max} \rangle \approx 5.08 N^{-0.713}$

Table 16: The difference of $\widehat{F}(u)$ and $F(u)$: $N(0, I)$, $y = 1$, $s = 100$

	$\|\widehat{F} - F\| : 1$	$\|\widehat{F} - F\| : 2$	$\langle \|\widehat{F} - F\| \rangle$
$N = 2$	0.607	0.500	0.596 ± 0.137
$N = 4$	0.438	0.250	0.356 ± 0.081
$N = 6$	0.254	0.223	0.251 ± 0.057
$N = 10$	0.213	0.218	0.174 ± 0.041
$N = 20$	0.0862	0.0640	0.0970 ± 0.0169
$N = 50$	0.0562	0.0539	0.0458 ± 0.0076

Empirical law: $\langle \|\widehat{F} - F\| \rangle \approx 1.060 N^{-0.799}$

Table 17: Local difference $\widehat{F}(1) - F(1)$; $N(0, I)$, $y = 1$, $s = 100$

	$N = 4$	$N = 10$	$N = 20$	$N = 50$
$\langle \widehat{F}(1) \rangle$	0.683 ± 0.169	0.643 ± 0.062	0.632 ± 0.034	0.616 ± 0.015
$F(1)$	0.609	0.609	0.609	0.609
$\langle \widehat{F}(1) \rangle - F(1)$	0.0735	0.0340	0.0230	0.00680

Empirical law: $\langle \widehat{F}(1) \rangle - F(t) \approx 0.370 N^{-0.996}$

Table 18: Girko's G estimator of the function $\eta(t)$: $N(0, I)$, $s = 100$

	$y = 1$ $t = 1$	$y = 1$ $t = 2$	$y = 1/2$ $t = 1$	$y = 1/2$ $t = 2$
$N = 2$	0.786 ± 0.114	0.683 ± 0.131	0.738 ± 0.265	0.571 ± 0.309
$N = 4$	0.630 ± 0.087	0.537 ± 0.095	0.608 ± 0.129	0.462 ± 0.136
$N = 6$	0.612 ± 0.053	0.470 ± 0.068	0.571 ± 0.095	0.430 ± 0.099
$N = 10$	0.564 ± 0.039	0.412 ± 0.051	0.541 ± 0.051	0.386 ± 0.071
$N = 20$	0.529 ± 0.022	0.379 ± 0.025	0.520 ± 0.026	0.359 ± 0.027
$N = 50$	0.514 ± 0.009	0.351 ± 0.011	0.508 ± 0.012	0.342 ± 0.012
$\eta(t)$	0.500	0.333	0.500	0.333

of the function $\widehat{F}(1)$ and its mean square deviation calculated in a series of experiments. In the last row, the empiric formulas are shown fitting the data from Tables 16 and 17.

One may see that the disagreement with the theoretical limit formula (7) decreases proportionally to $N^{-0.8}$ and $N^{-1.0}$ and is approximately 0.04 by the uniform norm of the difference, and 0.007 for the function $F(t)$ at the point $u = 1$. This fact indicates the lessening of the agreement at the endpoints of the spectra.

Girko's G_2 estimator

To estimate the spectral function $\eta(t) = n^{-1}\text{tr}(I + t\Sigma)^{-1}$ V L Girko proposed (see [25], Chapter 5) the statistics $\widehat{\eta}(t) = h(\theta)$, where θ is a root of the equation $t = \theta(1 - y + y\widehat{h}(\theta))$, and $y = n/N$. It is proved that this equation is solvable for $t > 0$ and defines a function that converges almost surely to $\eta(t)$ as $n \to \infty$ and $n/N \to \lambda$. Experimental investigation of this estimator is shown in Table 18. The empirical function $\widehat{\eta}(t)$ along with its mean square deviation was calculated from spectra of matrices C by averaging over $s = 100$ experiments. For comparison, we also show the theoretical value $\eta(t) = (1 + t)^{-1}$. In the last row, the empirical laws are shown for the deviation $\langle \eta(t) \rangle$ from $\eta(t)$.

This table demonstrates the convergence of G estimator to $\eta(t)$ with the inaccuracy $O(N^{-1})$.

Conclusions

The numerical experiments show first that spectral functions of sample covariance matrices of large dimension converge as $n \to \infty$ and $n/N \to \lambda$, where n is the dimension of observations and N is sample size. The empiric variance of the function $h_n(t) = n^{-1}\mathrm{tr}(I+tC)^{-1}$ decreases by the law near to $n^{-1}N^{-1}$ as predicted by the theoretical estimate [12]. The theoretical upper estimate of the function $h_n(t)$ variance by Theorem 1 shows only the decrease by the law N^{-1}, which seems to be related to taking no account of the independence of the observation vector components. The stronger theoretical estimation $n^{-1}N^{-1}$ of the asymptotic variance proves to be yet excessive by factor 5–10. In all experiments, we observed the convergence of $\widehat{h}(t)$ to the theoretical value $h(t)$ defined by the dispersion equation (7). The difference between $h(t)$ and averaged values of $\widehat{h}(t)$ decreases approximately as N^{-1}. The theoretical upper estimate of the inaccuracy of asymptotic formulas found in Chapter 3 proves to be substantially overstated (by hundreds times). A more precise upper estimate of this inaccuracy in [12] proves to be overstated by two to four times.

The maximum invariant fourth momentum (1) increases as n^2 as $n \to \infty$ for strongly coupled variables. The empirical dependence of the accuracy of the asymptotic equation (7) on coupling of variables (Tables 12–14) shows that the theoretical requirements to coupling bounds seem too stringent. The empirical law of dependence of inaccuracy on the correlation coefficient r does not agree with theoretical law for upper estimates of the remainder terms $\omega(t)$ and $\omega_1(t)$. Experiments show that, for normal populations, the equation (7) holds well even for $r \approx 0.8$ when $n = N = 50$. Thus, methods of estimation of the remainder terms developed in Chapter 3 provide weak upper estimates that require sharpening and revision. It remains not quite clear what minimal restrictions on the dependence of variables are yet necessary for the canonic equations (7) to be accurate.

Our experiments allow to make a general conclusion that the limit dispersion equations for large sample covariance matrices prove to be sufficiently accurate even for not large n and N: the average inaccuracy in different experiments decreases from 0.05

for $n = N = 10$ to 0.01 for $n = N = 50$. This result also keeps for nonstandard normal distribution and for other distributions including discrete distributions, and remains true for some cases of strongly coupled variables.

In Section 3.3 the Normal Evaluation Principle was offered stating that standard quality functionals of regularized multivariate statistical procedures are only weakly depending on moments of variables higher than the second and thus approximately are the same as for normal distributions. This property is observed in the experiment with binomial distribution (Tables 9 and 10) and is confirmed by a good fit between experimental and theoretical data.

Summarizing we may state that results of experiments substantiate the multiparametric approach and show that improved methods constructed in the above are expected to have the accuracy well acceptable for applications even for not large n and N (possibly, even for $n = N = 5 - 10$).

REFERENCES

1. D. J. Aigner and G. G. Judge. Application of pre-test and Stein estimation to economic data. Econometrica, 1977, pp. 1279–1288.

2. S. A. Aivazian, I. S. Yenyukov, and L. D. Meshalkin. Applied Statistics (in Russian), vol. 2: Investigation of Dependencies. Finansy i Statistika, Moscow, 1985.

3. S. A. Aivazian, V. M. Buchstaber, I. S. Yenyukov, and L. D. Meshalkin. Applied Statistics (in Russian). vol. 3: Classification and Reduction of Dimension. Finansy i Statistika, Moscow, 1989.

4. T. W. Anderson and R. R. Bahadur. Classification into multivariate normal distributions with different covariance matrices. Ann. Math. Stat. 1962, vol. 32, 2.

5. T. Anderson. An Introduction to Multivariate Statistical Analysis. J. Wiley, NY, 1958.

6. L. V. Arkharov, Yu. N. Blagoveshchenskii, and A. D. Deev. Limit theorems of multivariate statistics. Theory Probab. Appl. 1971, vol. 16, 3.

7. L. V. Arkharov. Limit theorems for characteristic roots of sample covariance matrices. Sov. Math. Dokl., 1971, vol. 12, pp. 1206–1209.

8. Z. D. Bai. Convergence rate of the expected spectral distribution of large random matrices. Ann. Probab. 1993, vol. 21, pp. 649–672.

9. Z. D. Bai and J. W. Silverstein. No eigenvalues outside the support of the limiting spectral distribution of large-dimensional sample covariance matrices. Ann. Probab. 1998, vol. 26, pp. 316–345.

10. A. J. Baranchik. A family minimax estimators of the mean of a multivariate normal distribution. Ann. Math. Stat., 1970, vol. 41, pp. 642–645.

11. T. A. Baranova. On asymptotic residual for a generalized ridge regression. In: Fifth International Conference on Probability and Statistics, Vilnius State University, Vilnius, 1989, vol. 3.

12. T. A. Baranova and I. F. Sentiureva. On the accuracy of asymptotic expressions for spectral functions and the resolvent of sample covariance matrices. In: Random Analysis, Moscow State University, Moscow, 1987, pp. 17–24.

13. A. C. Brandwein and W. E. Strawderman. Stein estimation: The spherically symmetric case. Stat. Sci., 1990, vol. 5, pp. 356–369.

14. M. L. Clevenson and J. V. Zidek. Simultaneous estimation of the means for independent Poisson law. J. Am. Stat. Assoc., 1975, vol. 70, pp. 698–705.

15. A. Cohen. Improved confidence intervals for the variance of a normal distribution. J. Am. Stat. Assoc., 1972, vol. 67, pp. 382–387.

16. M. J. Daniels and R. E. Kass. Shrinkage estimators for covariance matrices. Biometrics, 2001, vol. 57, pp. 1173–1184.

17. A. D. Deev. Representation of statistics of discriminant analysis, and asymptotic expansion when space dimensions are comparable with sample size. Sov. Math. Dokl., 1970, vol. 11, pp. 1547–1550.

18. A. D. Deev. A discriminant function constructed from independent blocks. Engrg. Cybern., 1974, vol. 12, pp. 153–156.

19. M. H. DeGroot. Optimal Statistical Decisions. McGraw-Hill, NY, 1970.

20. R. A. Fischer. The use of multiple measurements in taxonomic problems. Ann. Eugen. 7, 1936, pt. 2, pp. 179–188. Also in Contribution to Mathematical Statistics. J. Wiley, NY, 1950.

21. V. L. Girko. An introduction to general statistical analysis. Theory Probab. Appl. 1987, vol. 32, pp. 229–242.

22. V. L. Girko. Spectral Theory of Random Matrices (in Russian). Nauka, Moscow, 1988.

23. V. L. Girko. General equation for the eigenvalues of empirical covariance matrices. Random Operators and Stochastic Equations, 1994, vol. 2, pt. 1: pp. 13–24; pt. 2: pp. 175–188.

24. V. L. Girko. Theory of Random Determinants. Kluwer Academic Publishers, Dordrecht, 1990.

25. V. L. Girko. Statistical Analysis of Observations of Increasing Dimension. Kluwer Academic Publishers, Dordrecht, 1995.

26. V. L. Girko. Theory of Stochastic Equations. Vols. 1 and 2. Kluwer Academic Publishers, Dordrecht/Boston/London, 2001.

27. C. Goutis and G. Casella. Improved invariant confidence intervals for a normal variance. Ann. Stat., 1991, vol. 19, pp. 2015–2031.

28. E. Green and W. E. Strawderman. A James–Stein type estimator for combining unbiased and possibly unbiased estimators. J. Am. Stat. Assoc., 1991, vol. 86, pp. 1001–1006.

29. M. H. J. Gruber. Improving Efficiency by Shrinkage. Marcel Dekker Inc., NY, 1998.

30. Das Gupta and B. K. Sinha. A new general interpretation of the Stein estimators and how it adapts: Application. J. Stat. Plan. Inference, 1999, vol. 75, pp. 247–268.

31. Y. Y. Guo and N. Pal. A sequence of improvements over the James–Stein estimator. J. Multivariate. Anal., 1992, vol. 42, pp. 302–312.

32. K. Hoffmann. Improved estimation of distribution parameters: Stein-type estimators. Teubner, Leipzig 1992 (Teubner-Texte zur Mathematik, Bd 128).

33. K. Hoffmann. Stein estimation—a review. Stat. Papers, 2000, vol. 41, pp. 127–158.

34. P. Hebel, R. Faivre, B. Goffinat, and D. Wallach. Shrinkage estimators applied to prediction of French winter wheat yields. Biometrics, 1993, vol. 40, pp. 281–293.

35. I. A. Ibragimov and R. Z. Khas'minski. Statistical Estimation. Asymptotical Theory. Springer, Berlin, 1981.

36. W. James, C. Stein, Estimation with the quadratic loss. Proc. Fourth Berkley Symposium on Mathematical Statistics and Probability, 1960, vol. 2, pp. 361–379.

37. G. Judge and M. E. Bock. The Statistical Implication of Pretest and Stein Rule Estimators in Econometrics. North Holland, Amsterdam, 1978.

38. T. Kubokawa. An approach to improving the James–Stein estimator. J. Multivariate Anal., 1991, vol. 36, pp. 121–126.

39. T. Kubokawa. Shrinkage and modification techniques in estimation. Commun. Stat. Theory Appl., 1999, vol. 28, pp. 613–650.

40. T. Kubokawa and M. S. Srivastava. Estimating the covariance matrix: A new approach. J. Multivariate Anal., 2003, vol. 86, pp. 28–47.

41. A. N. Kolmogorov. Problems of the Probability Theory (in Russian). Probab. Theory Appl., 1993, vol. 38, 2.

42. J. M. Maata and G. Casella. Development in decision-theoretical variance estimation. Stat. Sci., 1990, vol. 5, pp. 90–120.

43. V. A. Marchenko and L. A. Pastur. Distribution of eigenvalues in some sets of random matrices. Math. USSR Sbornik, 1967, vol. 1, pp. 457–483.

44. A. S. Mechenov. Pseudosoltions of Linear Functional Equations. Springer Verlag, NY LLC, 2005.

45. L. D. Meshalkin and V. I. Serdobolskii. Errors in the classification of multivariate observations. Theory Probab. Appl., 1978, vol. 23, pp. 741–750.

46. L. D. Meshalkin. Assignement of numeric values to nominal variables (in Russian). Statistical Problems of Control, (Vilnius), 1976, vol. 14, pp. 49–55.

47. L. D. Meshalkin. The increasing dimension asymptotics in multivariate analysis. In: The First World Congress of the Bernoulli Society, vol. 1, Nauka, Moscow, 1986, pp. 197–199.

48. N. Pal and B. K. Sinha. Estimation of a common mean of a several multivariate normal populations: a review. Far East J. Math. Sci. 1996, Special Volume, pt. 1, pp. 97–110.

49. N. Pal and A. Eflesi. Improved estimation of a multivariate normal mean vector and the dispersion matrix: how one effects the other Sankhya Series A, 1996, vol. 57, pp. 267–286.

50. L. A. Pastur. Spectra of random self-adjoint operators. Russ. Math. Surv., 1969, vol. 28, 1–67.

51. Sh. Raudys. Results in statistical discriminant analysis: a review of the former Soviet Union literature. J. Multivariate Anal., 2004, vol. 89, 1, pp. 1–35.

52. G. G. Roussas. Contiguity of Probability Measures. Cambridge University Press, Cambridge, 1972.

53. S. K. Sarkar. Stein-type improvements of confidence intervals for the generalized variance. Ann. Inst. Stat. Math., 1991, vol. 43, pp. 369–375.

54. A. V. Serdobolskii. The asymptotically optimal Bayes solution to systems of a large number of linear algebraic equations. Second World Congress of the Bernoulli Society, Uppsala, Sweden, 1990, CP-387, pp. 177–178.

55. A. V. Serdobolskii. Solution of empirical SLAE unimprovable in the mean (in Russian). Rev. Appl. Ind. Math., 2001, vol. 8, 1, pp. 321–326.

56. A. V. Serdobolskii. Unimprovable solution to systems of empirical linear algebraic equations. Stat. Probab. Lett., 2002, vol. 60, 1–6, pp. 1–6, article full text PDF (95.9 KB).

57. A. V. Serdobolskii and V. I. Serdobolskii. Estimation of the solution of high-order random systems of linear algebraic equations. J. Math. Sci., 2004, vol. 119, 3, pp. 315–320.

58. A. V. Serdobolskii and V. I. Serdobolskii. Estimation of solutions of random sets of simultaneous linear algebraic equations of high order. J. Math. Sci., 2005, vol. 126, 1, pp. 961–967.

59. V. I. Serdobolskii. On classification errors induced by sampling. Theory Probab. Appl., 1979, vol. 24, pp. 130–144.

60. V. I. Serdobolskii. Discriminant Analysis of Observations of High Dimension (in Russian). Published by Scientific Council on the Joint Problem "Cybernetics", USSR Academy of Sciences, Moscow, 1979, pp. 1–57.

61. V. I. Serdobolskii. Discriminant analysis for a large number of variables. Sov. Math. Dokl., 1980, vol. 22, pp. 314–319.

62. V. I. Serdobolskii. The effect of weighting of variables in the discriminant function. In: Third International Conference on Probability Theory and Mathematical Statistics, Vilnius State University, Vilnius, 1981, vol. 2, pp. 147–148.

63. V. I. Serdobolskii. On minimum error probability in discriminant analysis. Sov. Math. Dokl., 1983, vol. 27, pp. 720–725.

64. V. I. Serdobolskii. On estimation of the expectation values under increasing dimension. Theory Probab. Appl., 1984, vol. 29, pp. 170–171.

65. V. I. Serdobolskii. The resolvent and spectral functions of sample covariance matrices of increasing dimension. Russ. Math. Surv., 1985, 40:2, pp. 232–233.

66. V. I. Serdobolskii and A. V. Serdobolskii. Asymptotically optimal regularization for the solution of systems of linear algebraic equations with random coefficients, Moscow State University Vestnik, "Computational Mathematics and Cybernetics," Moscow, 1991, vol. 15, 2, pp. 31–36.

67. V. I. Serdobolskii. Spectral properties of sample covariance matrices. Theory Probab. Appl., 1995, vol. 40, p. 777.

68. V. I. Serdobolskii. Main part of the quadratic risk for a class of essentially multivariate regressions. In: International Conference on Asymptotic Methods in Probability Theory and Mathematical Statistics, St. Peterburg University, St. Peterburg, 1998, pp. 247–250.

69. V. I. Serdobolskii. Theory of essentially multivariate statistical analysis. Russ. Math. Surv., 1999, vol. 54, 2, pp. 351–379.

70. V. I. Serdobolskii. Constructive foundations of randomness. In: Foundations of Probability and Physics, World Scientific, New Jersey London, 2000, pp. 335–349.

71. V. I. Serdobolskii. Multivariate Statistical Analysis. A High-Dimensional Approach. Kluwer Academic Publishers, Dordrecht, 2000.

72. V. I. Serdobolskii. Normal model for distribution free multivariate analysis. Stat. Probab. Lett., 2000, vol. 48, pp. 353–360. See also: V. I. Serdobolskii. Normalization in estimation of the multivariate procedure quality. Sov. Math. Dokl., 1995, v. 343.

73. V. I. Serdobolskii and A. I. Glusker. Spectra of large-dimensional sample covariance matrices, preprint 2002, 1–19 at: www.mathpreprints.com/math/Preprint/vadim/20020622/1

74. V. I. Serdobolskii. Estimators shrinkage to reduce the quadratic risk. Dokl. Math., ISSN 1064-5624, 2003, vol. 67, 2, pp. 196–202.

75. V. I. Serdobolskii. Matrix shrinkage of high-dimensional expectation vectors. J. Multivariate Anal., 2005, vol. 92, pp. 281–297.

76. A. N. Shiryaev. Probability. Springer Verlag, NY Berlin, 1984.

77. J. W. Silverstein and Z. D. Bai. On the empirical distributions of eigenvalues of a class of large-dimensional random matrices. J. Multivariate Anal. 1995, vol. 54, pp. 175–192.

78. J. W. Silverstein and S. I. Choi. Analysis of the limiting spectra of large dimensional random matrices. J. Multivariate Anal. 1995, vol. 54, pp. 295–309.

79. C. Stein. Inadmissibility of the usual estimator for the mean of a multivariate normal distribution, Proc. of the Third Berkeley Symposium, Math. Statist. Probab., 1956, vol. 1, pp. 197–206.

80. C. Stein. Inadmissibility of the usual estimator for the variance of a normal distribution with unknown mean. Ann. Inst. Stat. Math., 1964, vol. 16, pp. 155–160.

81. C. Stein. Estimation of the mean of a multivariate normal distribution. Ann. Stat., 1981, vol. 9, pp. 1135–1151.

82. V. S. Stepanov. Some Properties of High Dimension Linear Discriminant Analysis (in Russian), Author's Summary. Moscow States University, Moscow, 1987.

83. A. N. Tikhonov and V. Ya. Arsenin. Solution of Ill-Posed Problem. J. Wiley, NY, 1977.

84. H. D. Vinod. Improved Stein-rule estimator for regression problems. J. Econom., 1980, vol. 12, pp. 143–150.

85. A. Wald. On a statistical problem arising in the classification of an individual in one of two groups. Ann. Math. Stat., 1944, vol. 15, pp. 147–163.

86. A. Wald. Estimation of a parameter when the number of unknown parameters increases indefinitely with the number of observations. Ann. Math. Stat., 1948, vol. 19, pp. 220–227.

87. A. Wald. Statistical Decision Functions. J. Wiley, NY, 1970.

88. E. P. Wigner. On the distribution of roots of certain symmetric matrices. Ann. Math., 1958, vol. 67, pp. 325–327.

89. E. P. Wigner. Random matrices in physics. Ann. Inst. Math. Stat., 1967, vol. 9, pp. 1–23.

90. Q. Yin and P. R. Krishnaiah. On limit of the largest eigenvalue of the large-dimensional sample covariance matrix. Probab. Theory Relat. Fields, 1988, vol. 78, pp. 509–521.

91. S. Zachs. The Theory of Statistical Inference. J. Wiley, NY/London/Sydney/Toronto, 1971.

INDEX

admissible estimators 3, 4, 21

analytical properties of spectral functions 75, 89, 95, 107, 111, 154, 156, 161, 206, 247, 266, 272, 279

asymptotically unimprovable estimators, procedures 12, 20, 42, 105, 115, 130, 133, 138, 147, 159, 166, 192, 195, 209, 221, 230, 234–235, 240, 246–247, 261, 264, 266

biased estimators in multiparametric problems 7, 9, 33, 35, 38, 42, 53, 61, 83, 200–201, 221, 246, 257

Yu. N. Blagoveshchenskii 10

best-in-the-limit procedures 127, 193, 231

boundaries of limit spectra 9, 14 41, **88**, 112, 212, 254, 257

boundedly dependent variables 29, 30, 124

Burkholder 77, 249

canonical stochastic equations 14

classification errors 10, 195, 220, 233, 236

classes of normal evaluable functionals 21, 33, 114, 118, 181

component-wise estimation 9, **56**, 57, 59, 147

consistent estimation and procedures 3, 11, 17, 22, 132, 139, 239

contigual distributions 10, 196

counting function 12

contribution of variables to discrimination 11, 193, 200, **209**, 218–219, 230, 269, 281, 283

convergence of spectral functions 12–13, 45, 54, 71–74, 91, 106–107, 115, 131–136, 142–146, 149–152, 159, 165, 251, 255, 257, 272, 279

correlation coefficient 55, 82, 98, 168, 190

A. D. Deev 10, 11, 199

degenerate estimators 3, 17–18, 46, 129–130, 132, 167–168, 191, 220

Printed and bound by CPI Group (UK) Ltd, Croydon, CR0 4YY

03/10/2024

01040428-0008